Digital Bipolar Integrated Circuits

Digital Bipolar Integrated Circuits

MOHAMED I. ELMASRY
Department of Electrical Engineering
University of Waterloo, Ontario

A Wiley-Interscience Publication

JOHN WILEY & SONS

New York • Chichester • Brisbane • Toronto • Singapore

Library of Congress Cataloging in Publication Data:

Elmasry, Mohamed I., 1943–
Digital bipolar integrated circuits.

"A Wiley-Interscience publication."
Includes bibliographical references and index.
1. Digital integrated circuits. 2. Bipolar transistors. I. Title.

TK7874.E5 1983 621.381′73 82-21868
ISBN 0-471-05571-9

Printed in the United States of America

10 9 8 7 6 5 4 3 2 1

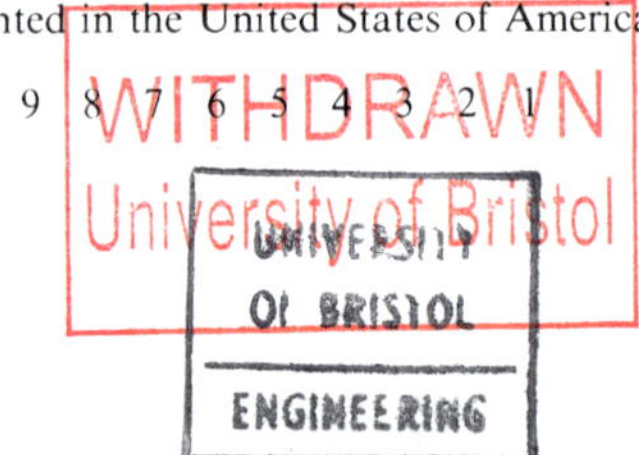

To my wife Elizabeth
and our children Carmen, Samir, Nadia, and Hassan

Preface

One of the most exciting and challenging engineering careers in recent years has been integrated circuit (IC) engineering. Many reasons contribute to this fact. In integrated circuits an engineer is given a rare opportunity to be innovative within the constraints imposed by the rules of physics. This has led to an exponential growth in many areas, for example, circuit density and performance. This in turn has widened the area of applications of ICs, and reduced the manufacturing cost. Much electronic equipment can now be built economically for the masses and the word IC is a familiar word to the public. This has no doubt enhanced IC engineers' sense of achievement and attracted many talented people to the field.

This book addresses one important aspect of IC engineering, that is, circuit design and analysis of ICs—more specifically, *circuit design and analysis of digital bipolar ICs.*

It is assumed that the reader is familiar with the basics of semiconductor physics, particularly *pn* junctions. Thus the book can be used as a text or a reference for senior undergraduate or graduate courses.

The emphasis of this book is circuit analysis and design, and this is covered in Chapters 4, 5, and 6. Because circuit designers have to know the process they design with, the model of the device they are using, and the system constraints they design within, these topics are discussed in Chapters 2, 3, and 7, respectively. In the introductory first chapter, the link between the different aspects of IC design is explained. Appendixes A, B, C, and D deal with processing steps, ion implantation, design projects, and symbolic-design layout rules for bipolar circuits.

Chapter 1 discusses the trade-offs involved in the design of bipolar digital integrated circuits among such factors as power dissipation, speed, and area. A design example is given which demonstrates that different bipolar logic families can be used on the same chip to optimize the design. It is suggested that the reader go through this example twice, once at the beginning of the course and a second time at the end, after becoming familiar with the two families under consideration, I^2L and EFL.

Chapter 2 discusses some basic bipolar processing steps and technologies. This is to serve as background information to circuit designers. Thus, no details of processing steps are given. Both junction- and dielectric-isolation bipolar technologies are discussed. Integration of transistors, resistors, and Schottky diodes is included.

Chapter 3 deals with the characterization and modeling of the bipolar transistor. The approach is aimed toward computer-aided design (CAD) and new material is included. Examples are given. The course instructor can generate similar problems using available circuit-analysis computer programs.

Chapter 4 first discusses some basic general definitions for digital circuit design. This is followed by the analysis of the transistor–transistor logic (T^2L) family. Different circuit configurations are considered. Schottky T^2L (ST^2L) is also studied. A circuit application is given which uses PLA (programmable logic arrays).

Chapter 5 deals with the analysis of integrated injection logic (I^2L). The upward mode of operating a bipolar transistor is studied and the application of I^2L in logic and memory realizations is discussed. Typical layouts are given and I^2L-related families, for example, ISL, are also discussed.

Chapter 6 first discusses emitter-coupled logic (ECL). This is followed by the analysis of emitter-function logic (EFL). The applications of EFL in logic and storage element design and typical layouts are given.

Chapter 7 deals with the system/subsystem aspects of IC design. A design example is discussed in some detail.

Appendix A discusses in general the different steps involved in the design of an IC. Appendix B discusses in some detail the use of ion implantation in bipolar ICs. Appendix C includes a list of 30 design projects for the student reader. Finally, Appendix D offers universal stick diagram/layout design rules that can be used in association with the projects of Appendix C.

It is my hope that this book offers the reader an in-depth study of digital bipolar integrated circuits and conveys some of the know-how regarding the design of these circuits.

MOHAMED I. ELMASRY

Waterloo, Ontario
April 1983

Acknowledgments

The author would like to thank Professor R. W. Dutton of Stanford University for his critical review of the manuscript and for his many valuable comments. The contributions of H. H. Muller of Motorola to Chapter 2, Professor R. W. Dutton to Chapter 3, and Dr. K. A. Pickar of Signetics to Appendix B are greatly appreciated. The comments of Stan Zkohan of Hewlett-Packard on Chapter 6 and Jon Levi of Burroughs on Appendix D were extremely valuable. The comments of Professor D. J. Roulston of the University of Waterloo on an early version of the manuscript were extremely useful. The interest, encouragement, and the feedback comments of many colleagues and students at Waterloo, Carleton, Manitoba, Columbia, Illinois, Berkeley, MIT, and Stanford are greatly appreciated.

M.I.E.

Contents

Chapter 3. Bipolar Device Models, 43

R. W. DUTTON

Chapter 4. Logic Families for MSI / LSI: T^2L and ST^2L, 99

Chapter 7. Digital LSI / VLSI Subsystems, 224

1.3 IC SUBBLOCK DEFINITION AND SPECIFICATION

In a digital-system design environment registers are essential building blocks. Throughout the subsequent chapters a variety of bipolar circuit technology realizations of registers will be used as examples. By way of introduction and viewing the problem top-down, Fig. 1.1*a* shows a four-bit register implemented with JK flip-flops (FF). Figure 1.1*b* shows the gate-level representation of each JK/FF using NOR and AND gates. The IC designer views the problem at even a more primitive level of abstraction as discussed later. At the device level one can postulate that each input and output reflects the need for one transistor. For MSI realizations using conventional transistor–transistor logic (T^2L) gates the number may be twice the nominal estimate, whereas, for circuit technologies such as integrated injection logic (I^2L) gates where a wired-AND function is used extensively, the actual transistor count may in fact be much lower than the nominal estimate. However, using the device count of one per input–output (I/O) as a reference, each JK/FF as shown in Fig. 1.1*b* contains nominally 26 transistors and the 4-bit register requires 104 devices. To gain some insight as to what these numbers imply from an area perspective and to correctly interpret the power–speed balance several back-of-the-envelope calculations are appropriate.

There exists a tight loop of reasoning as one seeks to resolve the questions of area–speed–power. Stated briefly the arguments go as follows. Power is proportional to the product of current times power supply voltage:

$$P = V_{CC} I \tag{1.1}$$

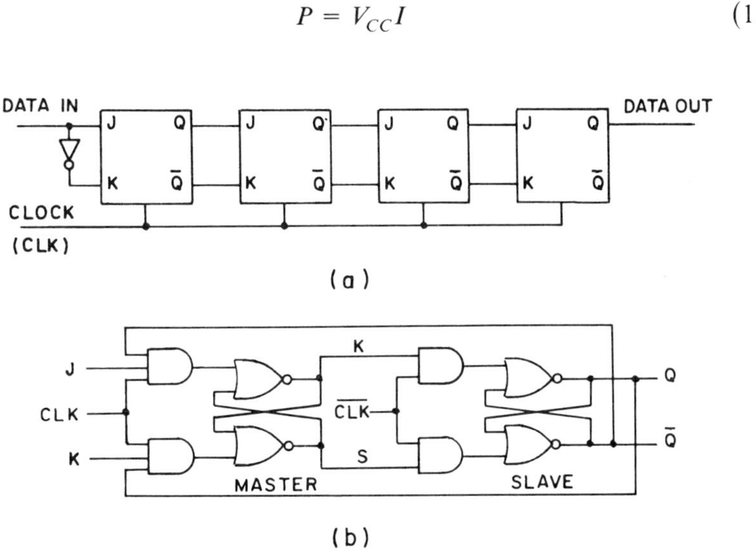

Figure 1.1 (*a*) Four-bit shift register implemented using JK flip-flops; (*b*) JK master–slave flip-flop at gate level.

The signal delay can be represented as the time needed to store or remove a packet of charge by means of a source current:

$$\tau_D \propto \frac{Q}{I} \tag{1.2a}$$

and if the charge represents only that stored on a parallel plate capacitance:

$$\tau_D \propto \frac{CV_l}{I} \tag{1.2b}$$

where V_l is the logical swing of voltage across C. Combining the above equations the power delay product, which represents energy required per switching event, is given as

$$P\tau_D \propto CV_{CC}V_l \tag{1.3}$$

As yet we have not introduced the area constraint. IC chip area is bounded by two factors: packaging and fabrication yield. The primary constraint in packaging is the amount of power a package can dissipate. Current state of the art is 1–10 W/package. Hence, given the number of devices per chip, each dissipating a given amount of power, one can determine the chip area that bounds the capacity of package. The fabrication yield problem imposes a different constraint on area. Random as well as technology-specific defects occur during IC fabrication and are measured in terms of number/cm^2. As the number of defects and die per square centimeter become comparable, the yield goes to zero. Thus one always seeks to minimize defects and keep die size within an acceptable range to produce a reasonable yield. Figure 1.2 shows a typical plot of expected yield versus chip size for a bipolar process circa 1976. Now applying each of these area factors to the power-delay equation, the generic extremes of maximum speed and maximum functional density can be defined.

Consider a 1-W package and two different IC chips—one designed to maximize speed and the other designed to maximize transistor count. For the

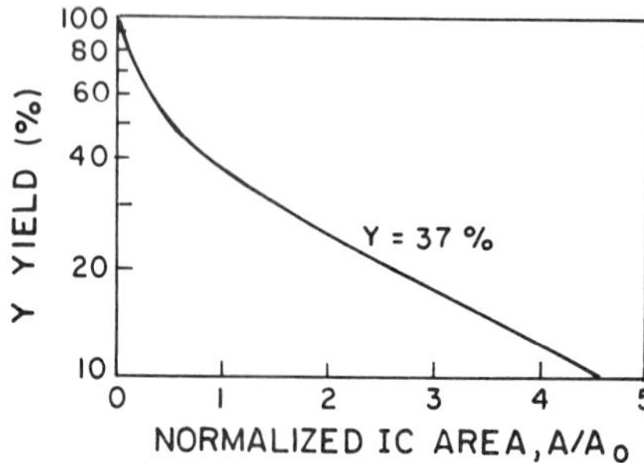

Figure 1.2 Yield logarithm versus normalized area for a bipolar process circa 1976.

case of maximum speed the following assumptions and calculations represent state-of-the-art:

$$\tau_D(\text{min}) \approx 20 \text{ ps} = 20 \times 10^{-12} \text{ s}$$

$$\text{assume } V_l \approx 0.5 \text{ V and } V_{CC} = 3 \text{ V*}$$

there is now a trade-off between P and C. Nominally C is set by circuit fan-out and wire capacitance to other gates. For state-of-the-art minimum geometry devices and typical wiring, C ranges from 10 to 100 fF (fF $= 10^{-15}$ farads). Using Eq. (1.3) the average power per gate becomes 0.75–7.5 mW so that given the package constraint on total power, the high-speed chip can contain on the order of 130–1300 gates. These numbers have the obvious flaw involved in assuming that *all* gates are active and the minimum delay is required everywhere on the chip. Nonetheless, the small gate counts indeed reflect the worst-case cost of high speed.

Turning to the other extreme case of maximum functional complexity, Eq. (1.3) can be used in reverse. Namely, assume

$$P = 1\ \mu\text{W per gate}$$

$$V_l = 0.5 \text{ V}, \quad V_{CC} = 3 \text{ V}$$

so that the total gate count now becomes 1×10^6 gates/chip. Again using the capacitance bounds of 10–100 fF, the delays range between 15 and 150 ns. Clearly for the extreme of micropower to maximize component count the gate delays have been compromised substantially. It is clear that choices between 1000 gates with subnanosecond delays and 10^6 gates with hundreds of nanoseconds delays are the rule rather than the exception. Moreover, for most systems only a portion of the circuitry must run at maximum speed. Hence, it is possible to have designs with both high-speed sections as well as micropower sections. Such an example is presented shortly. First, however, it is useful to define the range of BIDIC technologies that presently span the performance space between nanosecond delays and very high density.

1.4 BIPOLAR LOGIC FAMILIES

In Chapter 2 a review of technology specifics for bipolar circuitry is presented. It is assumed that the reader has a basic familiarity with the planar process and the general meaning of dielectric isolation and Schottky-clamped circuitry. The purpose of this section is to introduce the overall framework of bipolar circuit technologies so that some of the system level trade-offs concerning speed and

*Whereas these choices are somewhat arbitrary, it is likely that for bipolar devices the limits of $V_l \approx$ 4–5 kT/q, i.e., 100–125 mV and $V_{CC} \approx$ 5–10V_l i.e., $\simeq$ 0.5–1.5 V will be close to fundamental.

power can be viewed generally. The subsequent chapters will explore the various circuit families in much greater depth.

Figure 1.3 shows a plot of gate delay τ_D versus power per gate. The solid diagonal lines represent constant energy per switching event which is in units of picojoules. In essence the lines represent the fact that the right-hand side of Eq. (1.3) can remain constant at a given energy level and the P and τ_D terms on the left-hand side can be traded off. Also shown on the figure are experimental points for various bipolar circuit technologies and levels of circuit integration ranging from small scale and medium scale integration (SSI and MSI) to large scale integration (LSI) of circuit functions. Although Fig. 1.3 shows values of τ_D and P corresponding to a given state-of-the-art (1976), the general features of the circuitry are discussed briefly.

Transistor–transistor logic (T^2L) is the first bipolar digital integrated circuitry to achieve MSI functional complexity, primarily because it became government standardized in the early 1960s. Figure 1.4*a* shows the basic MSI T^2L NAND gate with the multiple-emitter (Q_1) input transistor and the so-called totem-pole output driver pair (Q_3-Q_4). The name reflects the fact that both input and output devices are transistors as compared with earlier circuit families such as RTL (resistor–transistor) or DTL (diode–transistor) logic circuits. The delays for T^2L are typically less than 10 ns and the power delay product ranges from 10 to 100 pJ. The corresponding power levels reflect the fact that functional densities greater than 100 gates are possible within 1-W package. From the basic circuit one can see that four active devices as well as several passive components are needed. The use of several devices increases power consumption but gives no added functional advantage other than

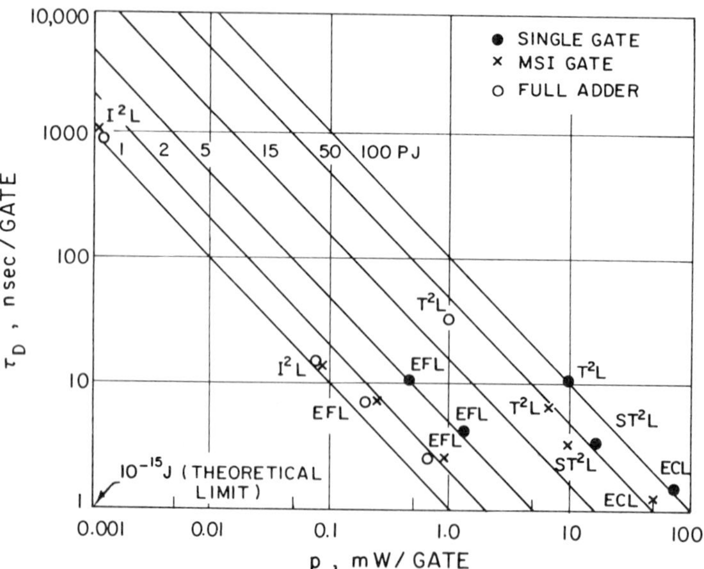

Figure 1.3 Gate delay τ_D versus power per gate for bipolar logic families.

off-chip current driving capability. Subsequent developments of digital IC families in the 1970s have both reduced component count per gate and increased speed.

One set of circuit changes for T^2L came with the development of Schottky clamping and the use of smaller on-chip voltage supplies. For example, the gate component count can be cut in half and a 3-V supply used with a logic swing of 1.5 V. These changes result in more than a tenfold increase in components per chip. Figure 1.4*b* shows one such gate configuration where the conventional Schottky notation is used. Chapters 3 and 4 present further details concerning the device and circuit impact of Schottky clamping. The primary point to be emphasized here is the fact that reduced voltage levels and capacitance both, as reflected through Eq. (1.3), result in a reduced power-delay product.

Integrated injection logic (I^2L) is the successor to T^2L with a substantial potential for LSI applications, both digital and as an analog-compatible circuitry. In the early 1970s the technology was reported with special emphasis on micropower. The measured results shown in Fig. 1.3 clearly reflect this micropower emphasis with power delay products of 1 pJ or less. The power levels of 0.001–0.1 mW reflect the potential for LSI and even VLSI since gate counts could reach 10^6 for a 1-W package. The obvious deficiency is the gate delays, which typically are greater than 5–10 ns. Details of the I^2L circuitry

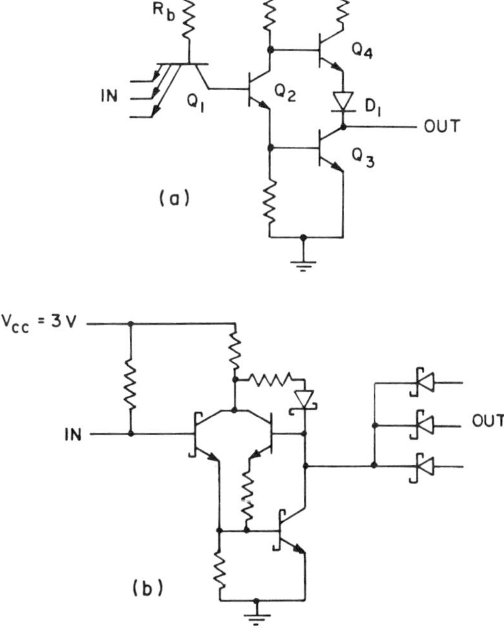

Figure 1.4 (*a*) Basic T^2L NAND gate with totem-pole output stage. (*b*) A Schottky T^2L gate.

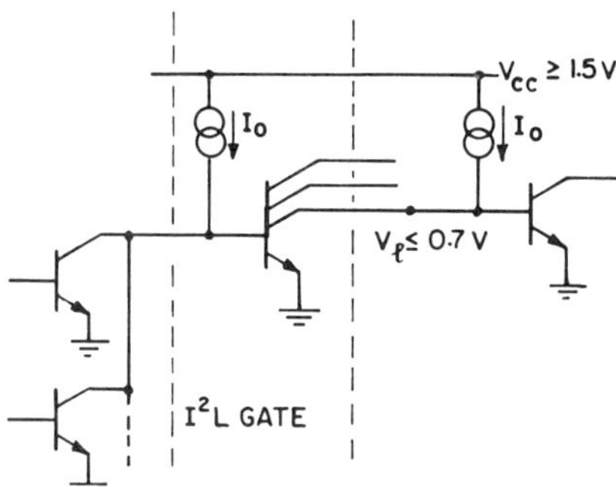

Figure 1.5 Basic I^{2}L gate with driving gates and a loading gate.

are discussed in Chapter 5. Figure 1.5 shows an overly simplified gate representation, which illustrates two key points:

1. The gate is based on a single transistor-current source pair with multiple inverting outputs and a wired-AND input node.
2. The scaled-down V_{CC} and V_l values make the micropower operation possible, although the ultimate speed limitations as predicted by Eq. (1.3) hinge on the node capacitance values.

Comparing the I^{2}L and T^{2}L gates two dominant differences are apparent: a major reduction in both subcircuit components and voltage levels. The I^{2}L technology has set the stage for further LSI/VLSI innovations in digital bipolar circuits that emphasize both decreased gate size and power consumption.

Emitter-coupled logic (ECL) is the highest-speed bipolar technology that has been developed to date. Compared with the T^{2}L and I^{2}L bipolar families, ECL avoids saturated-device effects by limiting the logic swing and steering a fixed current through the load resistors. Figure 1.6 shows the basic NOR/OR gate. The sources I_{EE} and V_{EE} set the power level; resistors R_{C1} and R_{C2} set the logic swing. The circuit shown in Fig. 1.6 is only the skeleton—a schematic—since the voltage reference, current source, and output emitter-followers have been omitted. These extra components, like for T^{2}L, consume area and power but add only MSI off-chip advantages. As will be shown in Chapter 6, there are on-chip ECL-type circuits such as emitter-function logic (EFL) that offer more logic functions per unit area and per unit source dissipation. Nonetheless, neither ECL/EFL nor T^{2}L bipolar logic families can match the density advantages of I^{2}L. Especially for ECL/EFL, the need for both a current source and voltage reference are major drawbacks in terms of density. However, since the objective with ECL/EFL is clearly its speed advantages, the sacrifice in density is most often masked by the dominance of package power limitations.

As a final comment for this section, combinations of the above three families on a single chip are common and have major functional advantages. In

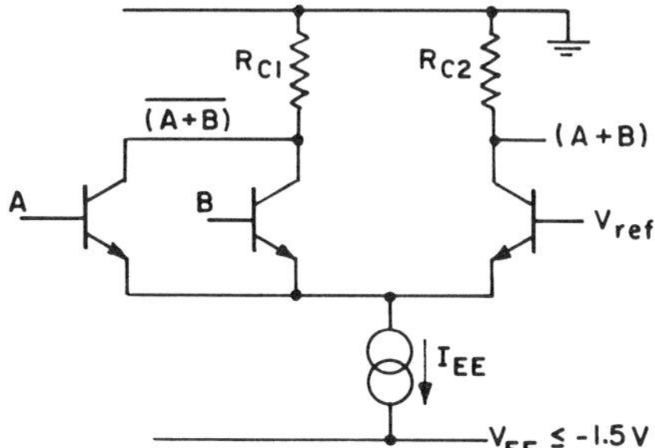

Figure 1.6 Basic ECL gate.

terms of the MSI world, conversions from ECL to T^2L and the converse are necessary functions. To achieve T^2L-compatible LSI functions with a modest speed sacrifice it is reasonable to use reduced on-chip logic levels with I^2L and have input/output T^2L buffering. Another example is the combination of I^2L and EFL. In this case one may clearly want ECL/EFL performance, but for noncritical control functions the I^2L gates provide an excellent power and area savings. The next section discusses a specific chip example, which uses EFL and I^2L. The chip is a 100-MHz multiple-register universal counter.

1.5 AN EXAMPLE: A MULTIPLE-REGISTER-COUNTER (MRC) CHIP*

Microprocessors have broadened performance capability in many new counter products. Intricate computing abilities are now readily obtainable in commercially available microprocessor chip sets. However, microprocessors alone cannot carry out complex counting functions. In addition to the processor unit, various circuit elements, both digital and analog, are required to accept, store, and regenerate data for different phases and different types of measuring functions.

As a group, these circuit elements cover a wide range of performance requirements. To optimize their performance, it is highly desirable if not totally essential to implement them in a special IC chip. For this reason, a custom bipolar LSI chip known as the multiple-register counter (MRC) has been developed for the new generation of microprocessor-controlled counter products. Except for the controller (microprocessor), the display circuitry, the signal input amplifiers, and the instrument power supply, all of the circuits required to perform all necessary counting functions are included in the MRC chip.

The decision to implement the MRC chip in the bipolar LSI process is mainly based on three factors. First, the process can accommodate mixed logic families with divergent performance characteristics, namely emitter function logic (EFL) (see Chapter 6), and integrated injection logic (I^2L) (see Chapter

*This section, with permission, is based on a paper by B. W. Wang and W. Jackson, A High Performance Bipolar LSI Counter Chip Using EFL and I^2L Circuits, *Hewlett-Packard Journal* **10**, 12–17 (1979).

5). Second, the process can accommodate miscellaneous circuit configurations, both digital and analog. Third, the dual-layer metallization feature of the process provides an efficient means of solving the complex high-density logic interconnection problem.

1.5.1 A Data-Aquisition Chip

The MRC chip has been specifically designed for universal counter use. The chip performs all the data-acquisition functions of a universal counter or timer when controlled by a microprocessor. For instrument performance flexibility and chip design simplicity, data computation features have not been included in the MRC. Instead, the processor performs all necessary calculations, with measurement data furnished by the MRC chip.

The MRC is designed not for a single counter but rather for a number of different counter products with a variety of performance requirements. Thus, its performance capability is broad and flexible. Depending on the type of processor and/or accessory circuits used, the MRC can provide a wide range of measurement capability, including different modes of frequency, period, time interval, ratio, totalizing, and other measurements.

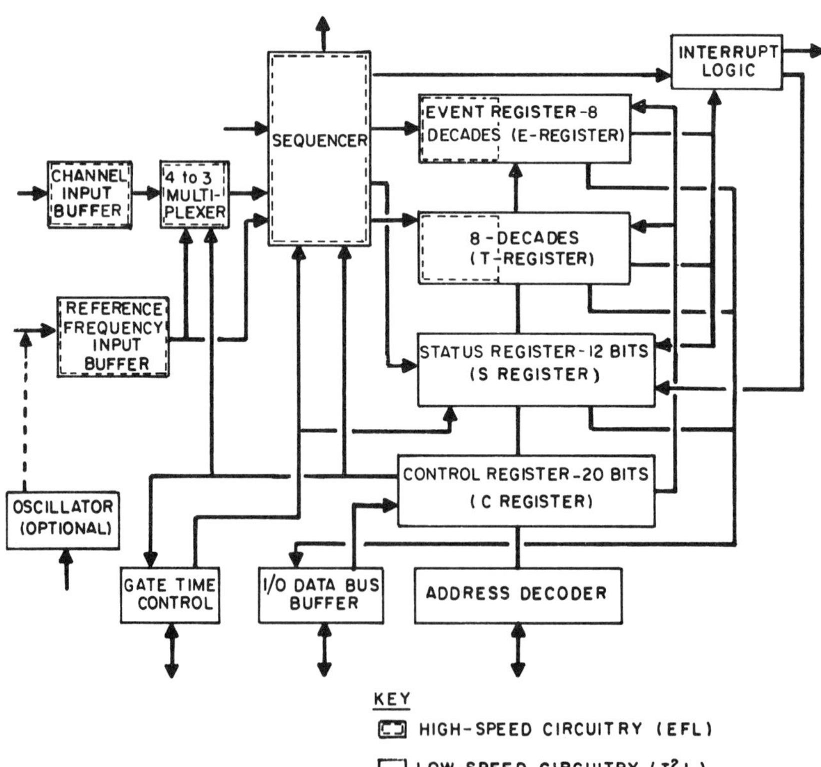

Figure 1.7 Multiple-register counter (MRC) chip function block diagram.

1.5.2 Multiple-Register Structure

The name "multiple-register counter" is derived from the chip's logic organization, a parallel multiregister structure. Figure 1.7 is the basic functional block diagram of the MRC. The four registers—the *E* (event), *T* (timing reference), *C* (control), and *S* (status)—operate in parallel for I/O access.

The *E* and *T* registers are chains of eight decades each, with the front-end (first) decades capable of counting signals over 100 MHz. Because of the division factor of 10^8 resulting from the eight-decade serial arrangement, the signal frequency coming out of the tail-end (last) decades is only 1 Hz for a 100-MHz input frequency.

To optimize speed, power, and device density, it is obvious that different types of circuits should be used for different segments of the chains. Accordingly, the first two decades of the *E* and *T* registers are implemented as EFL circuits to meet the speed requirements, while the remaining six decades are I^2L circuits for high device packing density and low power consumption.

Signal counting takes place in these two registers when gated by the sequencer block. An overflow flip-flop is provided for each of the two registers to extend the effective length of the counting chains.

The 20-bit control register serves as a storage buffer for program instructions received from the processor through the I/O block. As the name implies, it controls the general operation of the MRC chip. Specifically, the jobs of the *C* register are to define the particular mode of the measuring functions, to set up proper synchronization and timing of the sequencer block, and to carry out the reset routine before a measurement is initiated.

Operational status of the MRC, including measurement cycle arming and gating, overflow of the two counting (*E* and *T*) registers, measurement interrupt and time-out, and so on, is recorded in the 12-bit status register and is available to the processor when properly addressed.

To minimize package pin count, the four-bit I/O bus is bidirectional (see Fig. 1.7). The bus is shared by the four registers. Through the address decoder block, any of the four registers can be selected. The *C* register uses the bus for the data input function (i.e., the write function), and the *E*, *T*, and *S* registers use the bus for the data output function (i.e., the read function). Once a read or write function has been requested and a particular register has been selected individual decades (in the case of the *E* and *T* registers) or individual words (in the case of the *C* and *S* registers), are accessed sequentially by scanning the 3-to-8 nibble-select logic within the address decoder block.

1.5.3 Gating Functions

Of all of the functional blocks in the MRC, the sequencer logic block is most critical to the complex counting capability of the chip. In the sequencer block, two key elements that enable the MRC to excel in performance are a gating synchronization technique and the realization of this technique with EFL and other circuits in a high-speed bipolar LSI process.

The gating technique involves three stages of successive synchronization: arming synchronization, synchronization for E-register gating, and synchronization for T-register gating. This scheme provides maximum measurement resolution in the sense that measurement error is limited to ± 1 count of either the input frequency or the reference oscillator frequency, whichever is higher.

The bipolar LSI process features dual-layer metallization, which reduces on-chip parasitics, and an f_T of 1 GHz for fast device switching.

Complementing the sequencer logic block are the high-speed three-channel input buffer and the high-speed 4-to-3 multiplexer. Signals to be measured are received through the channel inputs. They are then directed into the E and/or T counting registers via the multiplexer and the sequencer. The multiplexer block is responsible for proper signal routing based on the selected measuring function, while the sequencer logic block is responsible for gating synchronization.

1.5.4 Circuit- and Mask-Design Requirements

The diverse circuit performance requirements, including wide deviations of device speed, variable packing density, mixing of logic families, on-chip interplay among various types of digital and analog circuits, and narrow noise margins (because of the low supply voltage used), have made implementation of the MRC more than a trivial task. Aside from coming up with an adequate assortment of digital and analog circuit elements to realize various circuit functions, considerable engineering effort has also been given to planning and design of the mask topology.

Special care has been taken to minimize cross coupling between adjacent high-speed input signals, to provide adequate ground connections (or taps) to minimize IR drop for the ground-potential-sensitive I^2L circuit blocks to alleviate heat-gradient problems resulting from uneven power consumption of different types of circuits in different locations of the chip, to ensure adequate noise margin by generating on-chip temperature-compensated reference voltages for the EFL circuits, and to provide additional test probe pads to reduce the normally lengthy time required for testing multidecade chains.

Figure 1.8 is a photomicrograph of the MRC chip. To show the difference in device density, the cell layouts of an EFL decade and an I^2L decade are marked.

Aside from conventional logic elements, such as multiinput logic gates, flip-flops, decades, and so on, which are implemented extensively in EFL or I^2L throughout the chip, a variety of other special-function circuits, both digital and analog, are used in the MRC. The digital circuits include miscellaneous types of I/O buffers, high-speed and low-speed multiplexers, different types of delay circuits for timing use, special purpose complementary clock drivers, matrix decoders, and EFL to I^2L level converters and vice versa. The analog circuits include precision voltage regulators, Schmitt triggers, oscillators, differential amplifiers, and voltage-level detectors.

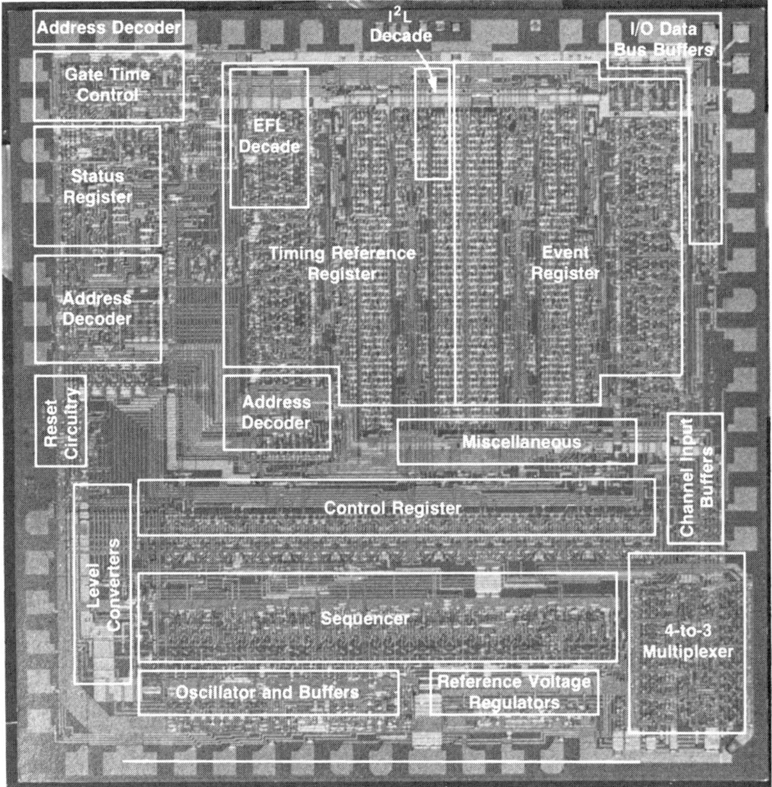

Figure 1.8 MRC chip photomicrograph. EFL circuits are used where high speed is needed. I^2L is used for low power consumption and high density. (Courtesy *Hewlett-Packard Journal*)

A few of the MRC circuits are unusual enough to warrant further discussion. These include the timing delay circuit, matrix decoder, and level converters.

1.5.5 I^2L Delay Circuit

The timing delay circuit is implemented in low-current I^2L gates. In the past, IC designers have had difficulty implementing well-controlled long-delay circuit elements with conventional bipolar circuit techniques. The difficulty is mainly due to two factors, namely, the relatively fast transit time of conventional bipolar transistors and the impracticality of making large-value on-chip capacitors and resistors. It takes a moderately complex circuit and tight process control to generate a signal delay of up to a few microseconds. Longer delays usually require off-chip RC trimming.

By taking advantage of the intrinsically slower propagation delay of I^2L transistors and by appropriately controlling the magnitude of the injector

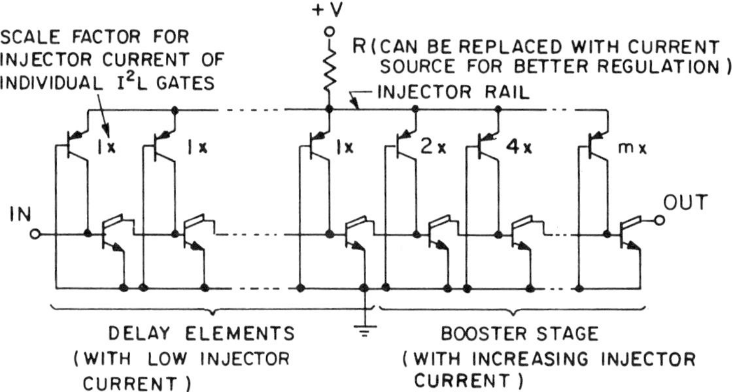

Figure 1.9 I^2L delay circuit provides delays in the millesecond range.

current in the I^2L gates, the MRC delay circuit, implemented in special serially connected I^2L gates, provides delay times not only of microseconds but of milliseconds. Figure 1.9 shows the relatively simple circuit. The delay is strictly a function of the number of series I^2L gates used and the magnitude of the injector current into each individual gate.

For long delay requirements, the injector current should be held to a minimum for the bulk of the delay string. However, for adequate drive to other circuit blocks, the final section of the delay string should have increased injector current, as shown in Fig. 1.9. If the delay circuit is properly designed, the delay time will be within $\pm 20\%$ of target value. Further accuracy can be achieved by replacing the resistor R with a well-regulated current source.

1.5.6 Matrix Decoder

Figure 1.10 shows the 3-to-8 address decoder, another circuit conveniently implemented in I^2L. Because of the array layout, the multicollector inverse *npn* transistors are longer than those of regular I^2L gates. To ensure proper current drive for the worst-positioned collectors, which are the ones farthest from the injector rail, a double-injector scheme has been used, with injector rails flanking both ends of the long inverse *npn* transistors.

The mask layout sketch and cross section for the matrix decoder, Fig. 1.10*b* and 1.10*c*, also reveal two mask-topology features, the dense packing layout of the I^2L gates, and the convenience of logic hook-up using dual-layer metallization.

1.5.7 EFL-to-I^2L Converter Circuits

Since I^2L gates operate at moderately slow speeds, the interface between EFL and I^2L gates is reasonably simple. Figure 1.11 shows two different voltage conversion schemes, one from EFL levels to I^2L and the other from I^2L to EFL.

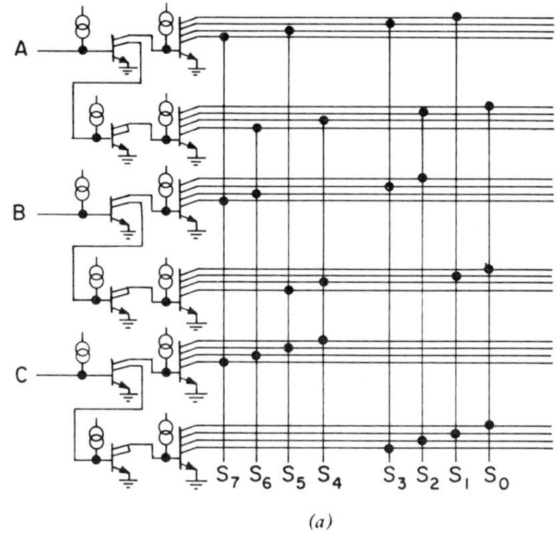

(a)

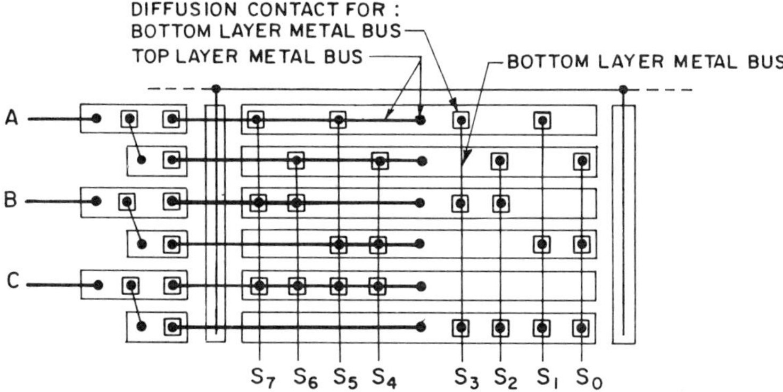

(b)

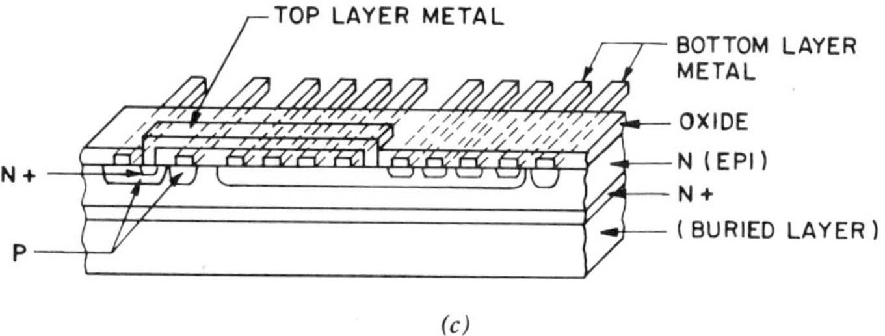

(c)

Figure 1.10 The 3-to-8 address decoder uses a double-injector scheme (see Chapter 5). (*a*) Circuit schematic, (*b*) mask layout diagram, and (*c*) cross section of mask layout voltage-level converters.

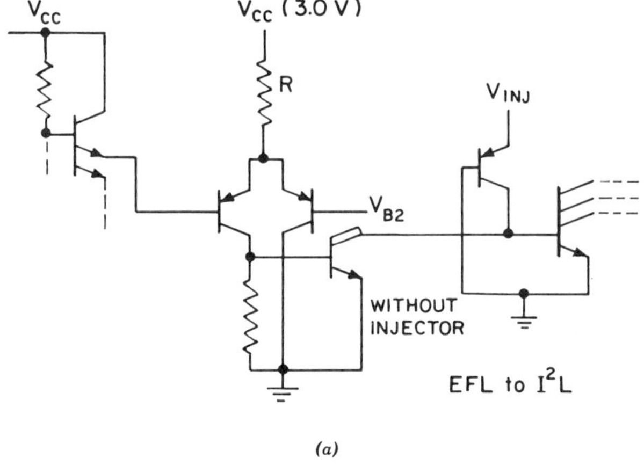

(a)

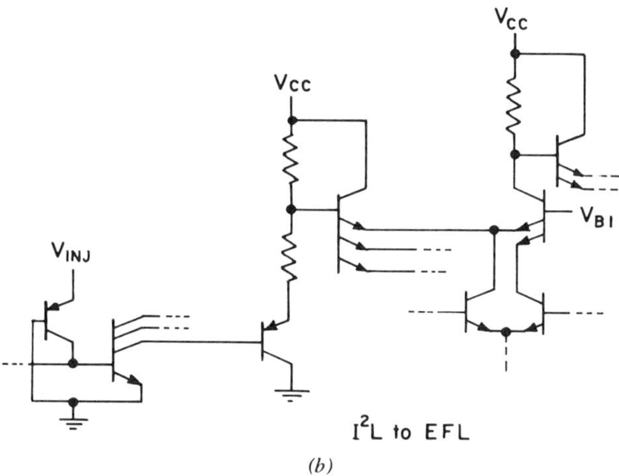

(b)

Figure 1.11 (*a*) EFL to I^2L and (*b*) I^2L to EFL voltage conversion schemes.

In the EFL to I^2L converter, the output transistor is a multicollector inverse *npn* transistor, that is, a typical I^2L gate without an injector. Since switching speed is not a prime requirement, lateral *pnp* transistors have been used conveniently in both converters.

1.5.8 System Interface

Since the MRC is designed for a number of different counter products with a wide range of performance specifications, interface flexibility has been emphasized in the architectural design. The performance capability of the MRC depends largely on the type of controller. A prescale option has been provided for high front-end counting frequencies and extended resolution. All input–output pins, except for the analog areas, are either TTL or ECL compatible. Also, a chip enable–inhibit control input is available for convenient system multiplexing.

1.6 CONCLUSIONS

The MRC example demonstrates an important point: It is the understanding of what each bipolar digital family can offer that the design of a given subsystem can be optimized for LSI. Chapters 2–6 of this volume offer the necessary material for such an understanding and, in conclusion, we shall go back in Chapter 7 to reemphasize some system/subsystem design aspects of LSI.

Chapter 2

Bipolar Process Technologies and Structures*

2.1 INTRODUCTION

This chapter will limit its coverage to the use of bipolar silicon integrated circuit technology as it applies to high-speed, high-density, and low-delay-power product digital circuits. By definition, therefore, we will focus on low-voltage usage of structures that are built with shallow junctions on thin epitaxial layers. Only a cursory review of processing will be given, and it is assumed that the reader is, in principle, familiar with the basic techniques of artwork generation; photomask processing; crystal structure and growth; impurity diffusion and implantation; and metallization techniques. Several books covering the appropriate material are recommended [1–4]. Two critical aspects of IC technology are covered in the appendices—Appendix A presents briefly the factors involving lithography and Appendix B gives further details on ion implantation.

Gold doping as a means of controlling minority-carrier lifetime will not be considered, and the discussion will assume that either nonsaturated circuits are to be implemented or that saturation is controlled by clamping techniques.

Given this rather limited view of bipolar technology, noise characteristics of active or passive devices are omitted and the reader is referred to reference 5. Lateral *pnp*s are not considered; the reader is referred to references 6–8 and to Chapters 3 and 5.

Bipolar technologies tailored especially toward a mixture of digital and linear circuits, e.g., super β transistors or laser trimming of resistors, or specialized technologies required to build a single class of devices, for example, fusible link technologies for PROM, are omitted as they lie beyond the scope of this book.

2.2 BIPOLAR IC TECHNOLOGY WITH JUNCTION ISOLATION

Over the past decade, bipolar processing technology of integrated circuits has continuously been refined to meet the technological requirements of higher

*This chapter is in part based with permission on a review article by H. H. Muller.

speed, lower delay-power product, and higher levels of integration. As a result, rather elaborate process flow charts have developed, and a typical process is depicted in Fig. 2-1, which will be reviewed briefly. The chart, however, is by no means comprehensive as in-process electrical test, cleaning cycles, etc., have been omitted to concentrate on the principle. The resulting bipolar structure is characterized by:

1. p-type substrate $\langle 100 \rangle$ or $\langle 111 \rangle$, approximately 10 $\Omega \cdot$ cm.
2. n-type epitaxial layer, 1.5–3 μm thick, approximately 0.5 $\Omega \cdot$ cm.
3. p-type junction isolation.
4. p-type base, 0.5–1.5 μm deep, 100–500 $\Omega/\square$ (unpinched, i.e., not under the emitter region) enhanced with 50–100 $\Omega/\square$ p-type contact regions.
5. n-type emitter less than 0.4 μm deep.
6. Dual-level metallization.
7. Glass passivation.

This process will result in the following typical *npn* transistor performance:

1. A cutoff frequency f_T of 1–3 GHz.
2. β_F of 50–200.
3. Breakdown voltages that allow circuit operation at $\leqslant$ 8 V.

The process begins with the p-type starting material on which an initial oxide is grown to be used as the actual mask against the subsequent diffusion. During oxide growth, Si is converted into SiO_2 on the surface. Approximately 45% of the oxide thickness results from consumption of silicon; therefore, each time the wafer is subjected to an oxidizing atmosphere after is has been patterned, steps will result in the silicon. The details of these steps are not shown in Fig. 2.1 as they do not contribute to the principle at hand.

Using a photolithographic process the pattern image is transferred from the first mask onto the silicon by selective etching of the exposed SiO_2 areas, while the photoresist covering the remainder of the wafer serves as a mask against the etch. This photolithographic technique will subsequently be used to pattern each layer in preparation for the etching process.

Subsequent to the growth of an initial masking oxide layer a window is patterned for the deposition and drive-in of an n^+-type layer. This region will subsequently be buried by an epitaxial layer, and thus is termed a "buried layer." The n^+ dopant can be introduced by solid phase or gaseous deposition and in some cases by ion implantation. Arsenic is generally chosen as a dopant due to its small diffusion coefficient. This minimizes structural changes of this buried layer during subsequent heat cycles. The function of the buried layer is primarily to form low-resistance paths beneath the final surface and to prevent parasitic vertical *pnp* action since higher n-type doping (the base of the parasitic *pnp*) reduces the current gain of the *pnp* devices.

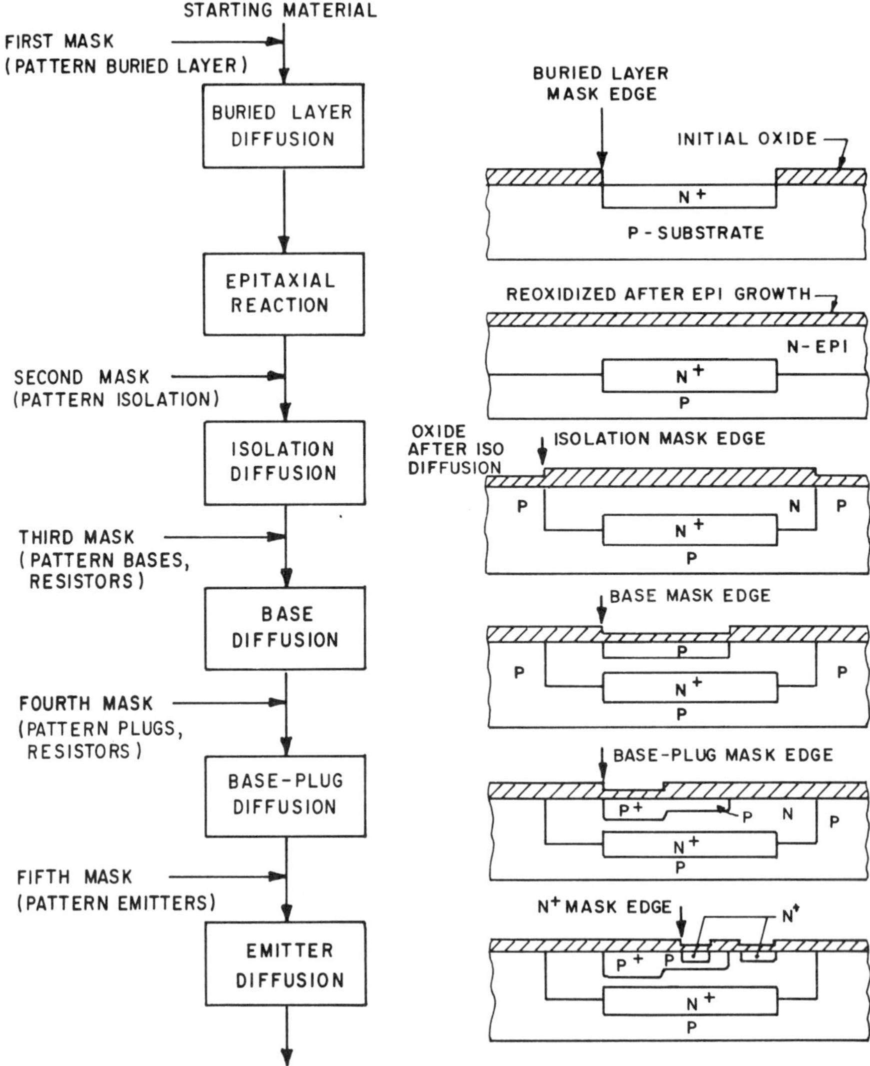

Figure 2.1 Simplified process flow chart for a bipolar IC technology with junction isolation.

After formation of the buried layer, all oxide is removed and a thin *n*-type layer is grown epitaxially. The epitaxial layer will serve as the collector region of *npn* transistors and will form one side of the isolation junction—the *p*-type substrate and other connecting *p* regions being the other side. The surface is reoxidized, and, using the second mask, the isolation pattern is transferred. A diffusion of *p*-type impurities, driven deep enough to touch the *p*-type substrate, generates isolated islands of *n*-type material surrounded on all sides but the surface by *p*-doped silicon. If these *p*-type areas are electrically connected

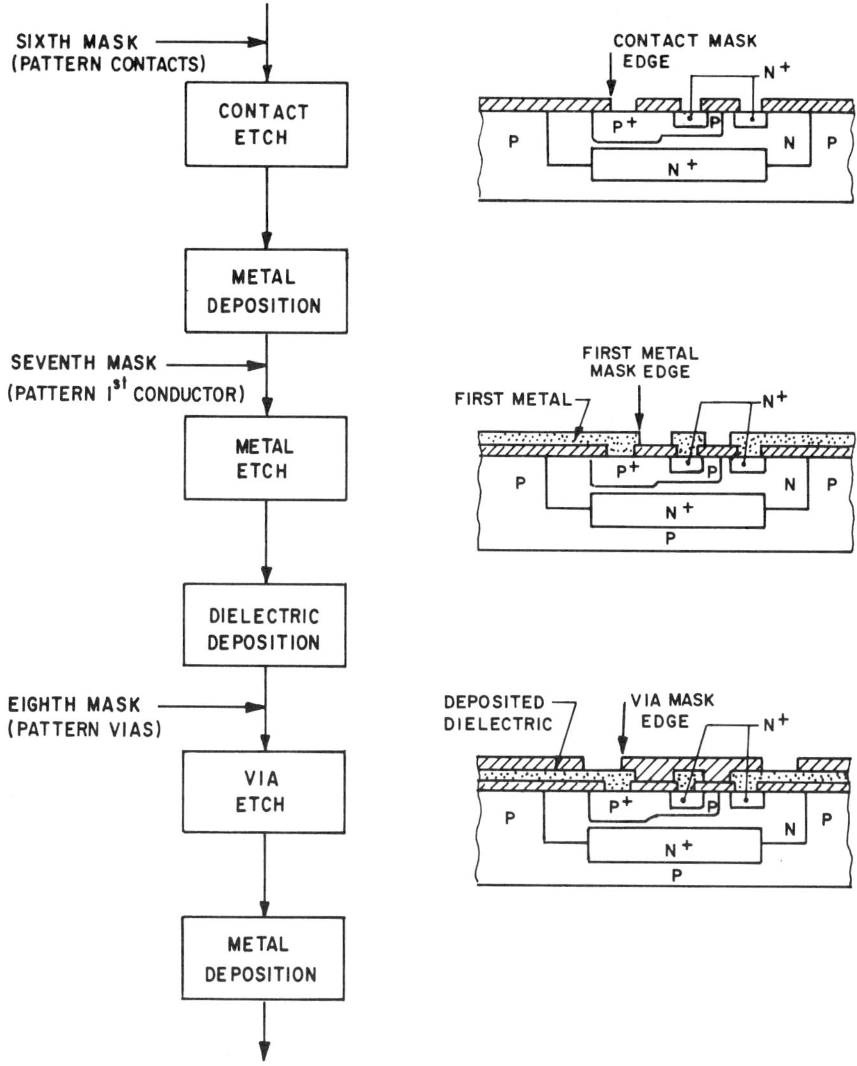

Figure 2.1 (*Continued*)

to the most negative voltage utilized in operation, any two *n*-type islands will be isolated via back-to-back diodes. During the isolation diffusion, oxide is regrown on the surface, and the wafer is ready for patterning by the third mask.

After patterning by the third mask, *p*-type impurities are diffused to form *p*-type resistors and the base regions of *npn* transistors. In general, a limited-source boron diffusion is employed while higher-speed and higher-accuracy requirements have recently given rise to ion-implanted base-emitter cycles (see

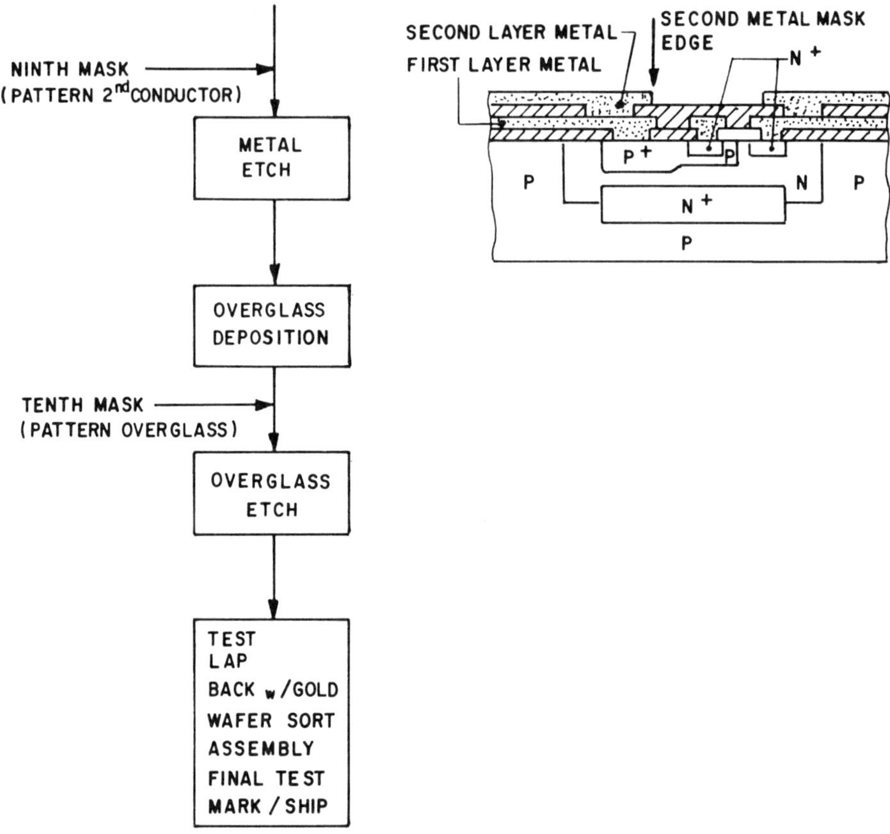

Figure 2.1 *(Continued)*

Appendix B). In order to improve switching performance of the transistor, a high-doped base-plug diffusion is added after patterning by the fourth mask. This region reduces extrinsic base resistance and can supply a lower p-type sheet resistivity for lower value of diffused resistors as required by circuit needs.

The fifth mask creates the emitter pattern of *npn* transistors when a subsequent diffusion or implantation of either arsenic or phosphorus takes place. It is essential that the combined base emitter cycles be well controlled since the resulting junction depths, base width, and sheet resistivities determine the major dc and ac characteristics of the transistor.*

*For high-speed applications, a "washed" emitter cycle is used and denotes that essentially no thermal reoxidization is allowed in the n^+ region and a short "wash" rather than a contact window etch is used to prepare the emitter for metallization. The purpose of these steps is to avoid the need to realize the metal contact mask—rather to use the existing opening as the metal contact opening—and thereby reduce device area, which also improves speed since capacitance is decreased. The subsequent process steps will *not* assume a washed-emitter process since other changes in masking are required and special precautions are needed in the metallization itself.

A sixth mask is used to pattern the contacts to the first level of metal interconnections. After the contacts are etched, the first conductor level is evaporated, patterned with the seventh mask, and etched. A layer of glass is vapor deposited to form a dielectric and patterned with the eight mask to form the conductor pattern between the first and second level of metallization. The second layer of metal is deposited and patterned with the ninth mask. The contact windows between the first and second levels of metal are called " vias." These patterned openings are a critical factor in both reliability and density. The deposited glass used for metal-layer isolation can result in nonuniformities and etching problems. Since lateral etching of these windows is difficult to control, greater tolerances must be allowed in the patterning of vias. On the other hand, the availability of two-layer metal can substantially improve overall chip layout, and any penalties paid for the via design rules are frequently more than offset by improved overall chip density.

A final layer of vapor-deposited glass is patterned with the tenth mask to open the bonding areas on the pads, the wafer is subjected to sintering temperature to ensure ohmic contact between metal and semiconductor, certain basic electrical device characteristics are tested, the wafer is mechanically lapped on its backside to reduce its thickness and thus facilitate scribing and heat transfer, and gold is deposited on the backside to simplify eutectic die attach.

After this review of conventional bipolar planar IC technology, the following areas of specific interest within the scope of this chapter will be reviewed.

2.2.1 Epitaxy

The technology of epitaxial growth gave rise to the economical manufacture of monolithic ICs as it allows a high-quality thin film of opposite conductivity to be grown on top of a substrate, with the same crystal orientation. Junction-isolation techniques became possible due to the thin high-quality epitaxial layers on a *p*-type substrate.

Epitaxy is a gas-phase deposition of silicon and impurity atoms on top of a crystalline surface. The epitaxially deposited layer with its impurities does not compensate the substrate impurities, although doping levels during the growth phase may be altered so that nonuniform epitaxial doping can be used to improve device performance (e.g., I^2L current gain).

Epitaxial layers for digital applications must be chosen to allow maximum circuit density with the least number of mask operations and to achieve high-speed devices with low values of intrinsic collector resistance while affording good values of breakdown voltage. In addition, the layers must meet the following general requirements:

1. Accuracy in resistivity and thickness should be within $\pm 5\%$ each.
2. Uniformity of resistivity and thickness across one wafer and from lot to lot should be less than 3% each.

3. Defect density introduced by epitaxial growth due to contamination, mechanical damage, nonheterogenous reaction, etc., should be less than $2\ \text{cm}^{-2}$.
4. Negligible pattern shift—this refers to distortion of buried layer patterns such that they appear shifted at the surface of the epitaxial layer. As alignment must be made with reference to the image appearing at the surface of the epitaxial layer, any shift reduces the distance between buried layer and isolation junction thus causing higher collector-to-substrate capacitance and reduced breakdown voltage.

Different steps in the fabrication process affect the shape of the doping profile. A rather sharp transition between the substrate and the growth epitaxial layer is desired; however, the temperature at which the reaction occurs will cause impurity diffusion—both buried layer and boron in the substrate. When a lightly doped film is grown on a heavily doped substrate (assumed to be of infinite extent), the resulting profile will resemble the complementary error function applicable to unlimited source diffusion:

$$C(x, t) = \frac{1}{2} C_0 \operatorname{erfc}\left(\frac{x}{2\sqrt{Dt}} \right) \tag{2.1}$$

where C_0 is the surface concentration of the substrate or buried layer doping, D is the diffusion coefficient, and x and t are depth into silicon and time, respectively.

Although the hydrogen reduction of silicon tetrachloride

$$SiCl_4 + 2H_2 \overset{1200°C}{\rightleftharpoons} Si + 4HCl \tag{2.2}$$

has been used for many years as a source for epitaxial film growth, modern circuit requirements cannot tolerate the "stretched" epi–substrate junction caused by the high reaction temperature of 1200°C. At higher temperatures the diffusion of substrate impurities into the growing epitaxial layer cause the n^+ buried layer as well as the boron in the substrate to outdiffuse toward the surface. This outdiffusion frequently requires thicker epitaxial layer to be grown over the n^+ region.*

Owing to the high temperature and autodoping drawbacks of $SiCl_4$, the silane reaction is used predominantly for epitaxial silicon growth:

$$SiH_4 \overset{1000°C}{\rightleftharpoons} Si + 2H_2 \tag{2.3}$$

*In addition to outdiffusion there are surface processes that transfer dopant into the growing layer directly. The term "autodoping" is commonly used to describe this phenomena. This autodoping, both vertical and lateral, is more severe when silicon tetrachloride is used owing to the HCl produced in the reaction.

This reaction achieves the sharp doping transition required for thin epitaxial layers—by a factor of 100 sharper compared to $SiCl_4$ owing to the lower reaction temperature—and avoids pattern shift due to the absence of HCl as a by-product.

2.2.2 Diffusion and Implantation

In contrast to the epitaxial reaction discussed above, diffusion creates an impurity gradient from the surface downward into the crystal. The one-dimensional diffusion may be described by Fick's second law:

$$\frac{\partial C(x,t)}{\partial t} = D\frac{\partial^2 C(x,t)}{\partial x^2} \tag{2.4}$$

where C is the concentration at each vertical point x as a function of time t and D is the impurity diffusion coefficient. Figure 2.2 gives the values of D for common dopant impurities in silicon.

Solving Eq. (2.4) for the "limited-source" boundary condition yields a Gaussian profile [3, 10]:

$$C(x,t) = \frac{Q}{\sqrt{\pi D t}}\exp\left(\frac{x^2}{4Dt}\right) \tag{2.5}$$

where Q is the total amount of impurities (limited source) as provided by a previous processing step such as ion implantation.

For the "unlimited-source" boundary condition, the surface concentration C_0 remains constant and the solution to Eq. (2.4) becomes a complementary error function [3, 10]:

$$C(x,t) = C_0 \operatorname{erfc}\left(\frac{x}{2\sqrt{Dt}}\right) \tag{2.6}$$

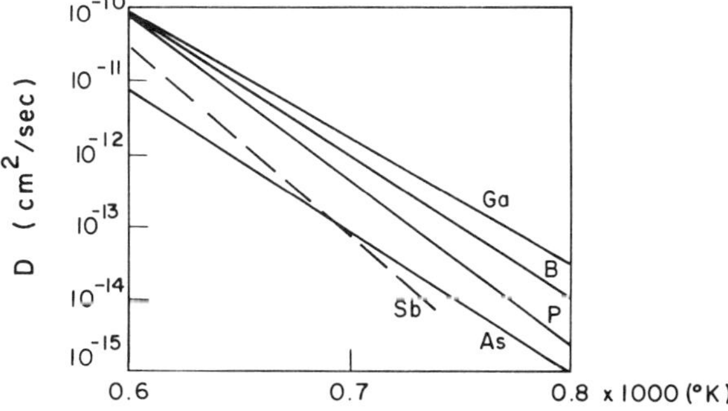

Figure 2.2 Impurity diffusion coefficients as a function of temperature in silicon.

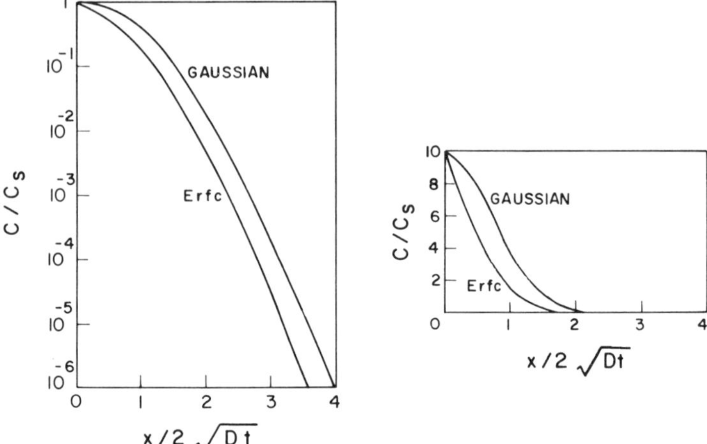

Figure 2.3 Comparison of Gaussian and erfc distribution on semilog and linear scales. Both concentration and distance are normalized.

Figure 2.3 compares the two normalized concentration profiles on both semi-log and linear scales [10]. It is clear that for a given peak concentration and "Dt" the complementary error function profile is steeper than the Gaussian profile. The results of Eqs. (2.5) and (2.6) can be approximated by exponential or linearly graded profiles over a *limited* range of doping. While these approximations are common for device modeling, they do not necessarily have a technological counterpart based on the solution of Eq. (2.4).

A junction is formed whenever impurity compensation occurs, that is,

$$C(x_j)_{\text{donor}} = C(x_j)_{\text{acceptor}} \tag{2.7}$$

Compensation is accompanied by degradation of such material properties as carrier mobility and lifetime. For present technologies it has been determined that the practical limit for the maximum number of compensations in silicon is approximately three. Figure 2.4 shows a typical concentration diagram for a shallow diffused vertical *npn* structure as used in high-speed bipolar integrated circuits. Several features are apparent from the figure:

1. The n^+ buried layer has compensated the p substrate.
2. The n^+ buried layer outdiffused profile into the n epi is rather abrupt.
3. The n epi has been compensated by the boron base diffusion.
4. The base has, in turn, been compensated (a second time, often referred to as double-diffused) to make it n type again—recall the epi layer is n type.

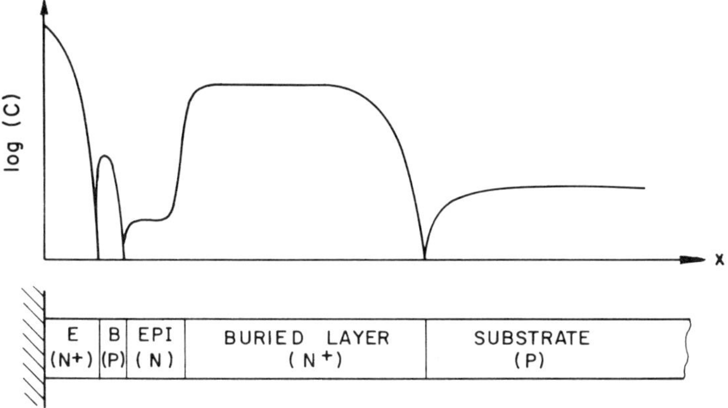

Figure 2.4 Typical doping diagram of an *npn* bipolar transistor.

Ion implantation is another means to introduce impurities into silicon, especially for low and moderate concentration levels. The major advantages of ion implantation are listed below.

Since the dopant impurities are introduced as ions:

1. It is possible to control the amount of charge introduced. For high doses, reproducibility is within 2%.
2. Variability of implant energy allows flexibility in tailoring doping profile to specific device requirements.
3. Directionality of beam reduces lateral spread.

The nonthermal nature of ion implantation allows:

1. Better depth uniformity and control.
2. Profiles to be placed deeper than would be possible using diffusion from the surface. Moreover, subsurface concentrations greater than solid solubility are possible under certain conditions.
3. Interchangeability of doping and processing steps such as implanting the emitters *prior* to implantation of the base.
4. Masking by dense materials not limited to high-temperature-tolerant materials such as SiO_2.

Under ideal conditions ion implantation generates a Gaussian distribution of impurities [compare Eq. (2.5)], its peak does not lie at the surface but rather below the surface by R_p (projected range) as shown in Fig. 2.5. Data for implantation profiles for all common dopant impurities are tabulated [11]. Appendix B gives further details concerning ion implantation. A typical set of results of doping profiles for boron in silicon are given in Fig. 2.6 as a function

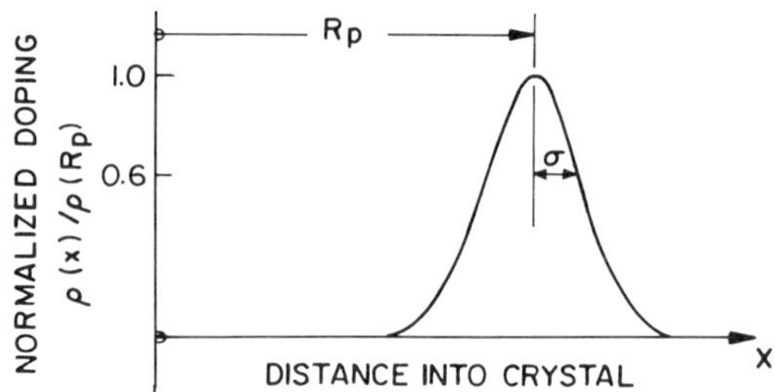

Figure 2.5 Typical ion-implanted profile.

of implant energy [12]. These profiles might well represent different base profiles for an *npn* transistor.

2.2.3 Metal Technology and Schottky Barrier Diodes (SBD)

Conductive layers in silicon technology have several purposes as well as constraints. For the first level of metallization there is need for both ohmic and rectifying (or barrier) contacts to the doped layers. The primary constraints are control of proper conductivity while avoiding excessive penetration of the metal into the bulk. For the second level of metal the dominant objective is to maximize conductivity and fine-line patterns and at the same time provide compatibility with underlying layers.

In many technologies aluminum is used for both layers of metallization. Thus all the considerations mentioned must be addressed for the aluminum–silicon system. The constituents commonly added to aluminum are silicon and copper. Silicon concentrations of less than 2% are added to provide an in-metal source of silicon for sintering and thus protect the underlying shallow emitter junction from shorts—the aluminum consumes silicon during

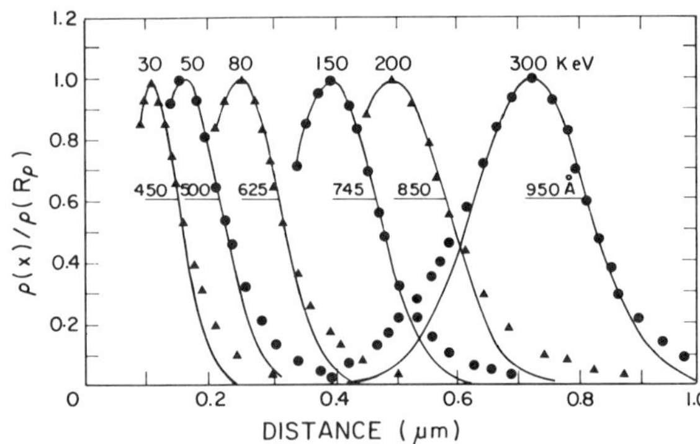

Figure 2.6 Normalized doping profiles of boron in silicon for several energy levels. The solid lines show the Gaussian approximations.

sintering. Copper is added in small amounts to improve electromigration characteristics. The current density limit of the described structure is well above 200 kA/cm^2, the sheet resistivity for the commonly used thickness of 10 kÅ is 30–50 mΩ/□, and the films are readily compatible with standard thermocompression bonding with aluminum wires.

While the above discussion provides information about the mechanical and electrical properties of aluminum in intimate contact with silicon, the barrier properties must also be discussed. Aluminum forms both ohmic and rectifying barriers to doped silicon, depending on the doping type and level. Ohmic contact results for aluminum or heavily doped p-type material, and a 0-V breakdown contact between aluminum and n^+ layers exhibits ohmic behavior. For lightly doped n- and p-type layers Schottky barriers are formed. The barrier commonly used for Schottky T^{2}L technology discussed in Chapter 4 is to contact the aluminum directly to the n-epi layer. On the other hand, to make ohmic contact to the epi layer, an n^+ emitter diffusion must be introduced.

Many metals form rectifying contact to silicon. For purposes of discussion here we will not discuss the barrier or conduction mechanisms and the reader is referred to reference 31. The rectifying contact to n-type materials, excluding heavily doped regions, is the primary application considered here. In Chapters 4 and 5 a variety of barrier diode circuit configurations are exploited. In Chapter 3 the modeling aspects of these rectifying contacts are presented. The nature of barrier contacts or Schottky barrier diodes (SBD) results from a potential barrier ϕ_B between the metal and semiconductor. The barrier height is a function of the metal used and of the semiconductor doping to a lesser extent [32, 33]. Figure 2.7a shows barrier height versus the metal or silicide used for the contact. Aluminum has a barrier height of 0.7 eV, whereas metals such as paladium (Pd) and gold (Au) have barrier heights greater than 0.8 eV. The effect of the barrier height on current–voltage characteristics is described based on thermionic emission of carriers across the barrier, and the equation that describes the current is

$$I_F = RAT^2 \exp\left(-\frac{q\phi_B}{kT}\right)\left[\exp\left(\frac{qV_F}{nkT}\right) - 1\right] \tag{2.8}$$

where A is the diode area, R is the Richardson constant, T is the absolute temperature, ϕ_B is the barrier height, k is Boltzmann's constant, q is the electronic charge, V_F is the forward voltage, and n is the ideality factor, which ranges between 1 and 1.4 [33]. It is useful to observe that small differences in barrier height result in exponential changes in current. For example, given two diodes of equal area and forward voltage but different barrier heights ϕ_{B1} and ϕ_{B2}, the ratio of currents is

$$\frac{I_{F1}}{I_{F2}} = \frac{e^{-q\phi_{B1}/kT}}{e^{-q\phi_{B2}/kT}} = e^{q(\phi_{B2}-\phi_{B1})/kT}$$

Thus for $\phi_{B2} - \phi_{B1} = (2\text{–}3)kT/q = 0.077$ V, the ratio I_{F1}/I_{F2} is *10*. On a

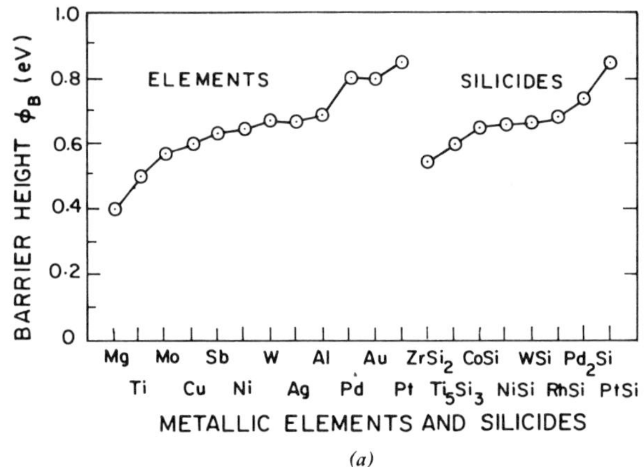

(*a*)

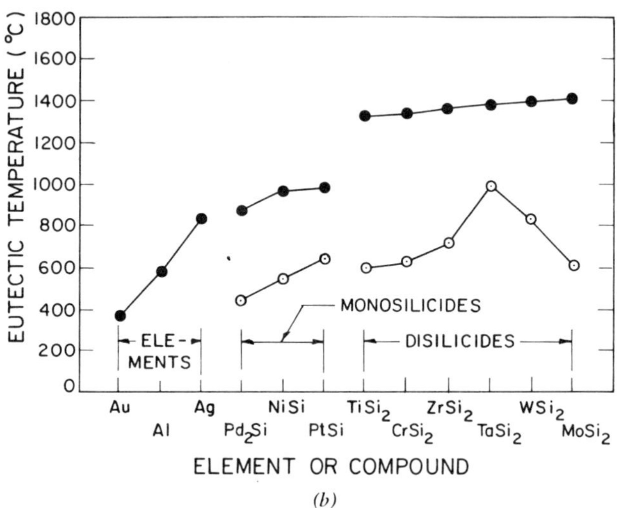

(*b*)

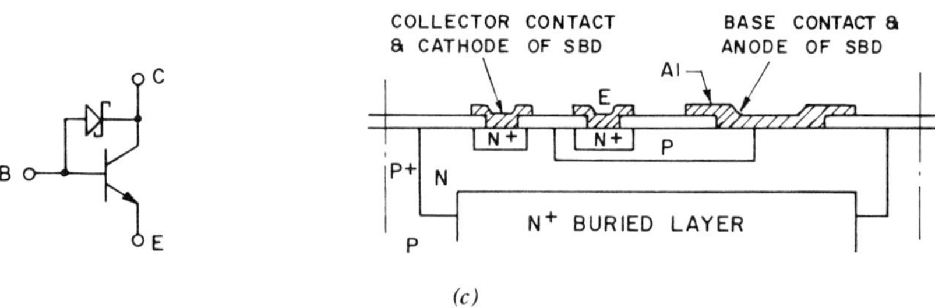

(*c*)

Figure 2.7 (*a*) The barrier heights for metals and for metals silicides on *n*-type silicon (after Andrews [32]). (*b*) Eutectic melting temperatures for various metalization materials: ● melting temperature, ○ typical sintering temperature (after Andrews [32]). (*c*) STTL composite device.

semilog plot of I_F vs V_F the two curves are vertically separated by an order of magnitude for all bias levels. It should be emphasized that ϕ_B is not the "on" voltage associated with conduction; however, the on voltage for the SBD is less than that for the *pn* junction diode. The comparison of the SBD with a bulk *pn* junction is discussed in Chapter 3.

Figure 2.7*a* shows barrier heights for both metal contacts and silicides. Silicides are formed when the silicon and metal reach the eutectic melting temperature as shown in Fig. 2.7*b* or by means of depositing proper mixtures of the constituents and sintering the material. There are several advantages of using silicides for barrier contacts. First, the problem of metal spiking* as observed with aluminum is avoided. Second, as can be seen from Fig. 2.7*b*, the silicides can withstand higher temperatures before melting occurs. As a result it is possible to oxidize the silicides to form native silicon oxide passivation layers rather than having to deposit dielectrics for second-layer metallization.

SBD are used in numerous applications taking advantage of the smaller forward voltage compared to junction diodes and of the absence of minority-carrier storage. The clamping of collector–base junctions in Schottky TTL, on-chip diode (AND) decoding, and cell clamping in bipolar memories are probably the most widespread in the manufacture of integrated circuits. Figure 2.7*c* depicts the outline, the cross section, and the electrical equivalent of a typical SBD clamped *npn* transistor as it is used in TTL stages. Note that the metal (assume aluminum) contacts both the *p*-type base and *n*-type collector regions directly. The aluminum contact to the *p*-type region becomes the ohmic base contact, whereas the aluminum contact to the *n*-type region creates the SBD barrier as shown by the parallel-connected diode that shunts the collector–base diode. In the saturation region the collector junction is now slightly forward biased. The SBD forward voltage drop is much lower than the on voltage of the base–collector junction and most of the excess base current flows through the diode. Saturation time of the above composite structure is reduced to about 10% of that of the original transistor. A detailed derivation of speed improvements due to SBD clamping may be found in reference 16.

2.3 BIPOLAR IC STRUCTURES USING DIELECTRIC ISOLATION

The technologies that will be described in the following are compatible with the processing techniques outlined in Section 2.2. They utilize the same starting material, burial layer diffusion, and epitaxial growth. The differences lie in the method of isolating adjacent expatial regions. Processing beyond the isolation step again follows the pattern outlined in Fig. 2.1.

In Fig. 2.8 the active area of a transistor is shown to be only as wide as its emitter and as long as dictated by the edges of base and collector. The

*Spiking refers to the preferential diffusion and reaction of aluminum along silicon crystal faults. For shallow junctions this spiking causes junction shorts to occur.

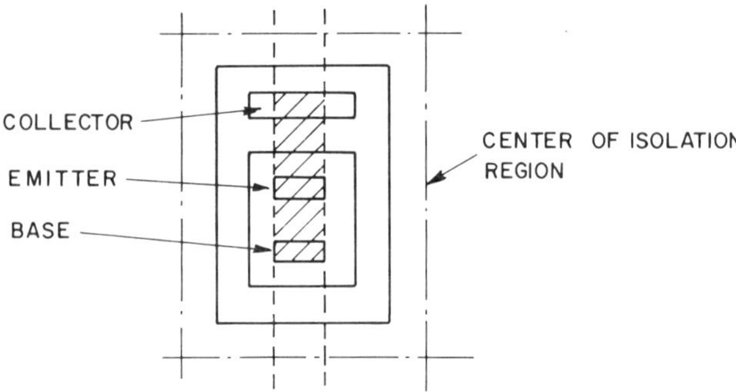

Figure 2.8 Comparison of the active area of an *npn* transistor with its surface area requirement.

cross-hatched rectangle depicts the active transistor, that is, the region which determines the dc characteristics of the structure. Areas extending beyond the dashed boundaries may be considered "parasitic"; they are required to safeguard against misalignment and outdiffusion and are used to maintain the appropriate distance between junctions. The larger area of the conventional planar transistor translates into larger values of parasitic components, thus reducing the ac performance of the structure. But it also means that packing density of LSI circuits may be increased if the isolation technology is improved, without reducing photolithographic tolerances and without changing circuit design techniques.

An improvement in isolation technology has been achieved in many ways, of which the most commonly used ones are described. They all have one thing in common: a dielectric is substituted for the two back-to-back diodes, which perform the isolation function in conventional planar technology. The following features may be listed:

1. Reduction in area by a factor 4–6.
2. Reduction in photolithographic and alignment requirements.
3. Reduction in parasitic capacitance by a factor of 3–8.
4. Improvement in cut-off frequency by a factor of 2.5–6.

The seemingly controversial first two items—reduction in area and a relaxation of photolithographic requirements—will become clear as soon as the self-aligning nature of dielectric isolation is explained.

2.3.1 Air Isolation

The process described is best known under its trade name, V-ATE (*v*ertical *a*nisotropic *e*tch, Raytheon) [17]. Utilizing *p*-type ⟨100⟩ substrate, standard buried layer, and silane epitaxy, the process begins to differ from the conven-

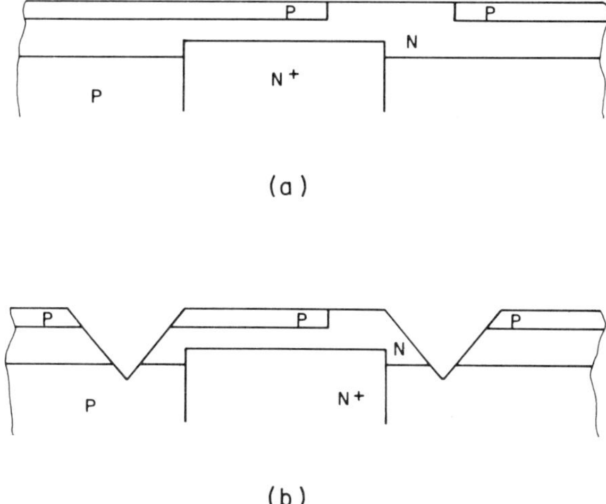

Figure 2.9 Air isolation (V-ATE).

tional one described in Section 2.2 after epitaxial growth. Now the base and *p*-type resistor regions are diffused as shown in Fig. 2.9*a*.

Next, the isolation pattern is transmitted and etched into the protective oxide layer. At this point, the vertical anisotropic (nonuniform) etch creates the air isolation channels. The etch moves fastest along the ⟨100⟩ crystal face, that is, downward, and forms very precise vee grooves as shown in Fig. 2.9*b*. What follows may be described as standard processing of a high-performance base–emitter cycle. The metallurgical system described in reference 17 is titanium–platinum–gold, but anodized aluminum is known to have been used successfully.

Figure 2.10 illustrates the process of anisotropic etching of a channel and the feature of self-stopping [18]. Owing to the selectivity of the etchant, another interesting property is the fact that sidewall edges are smoother than the borders of the oxide mask. This is important as it reduces metal-coverage problems.

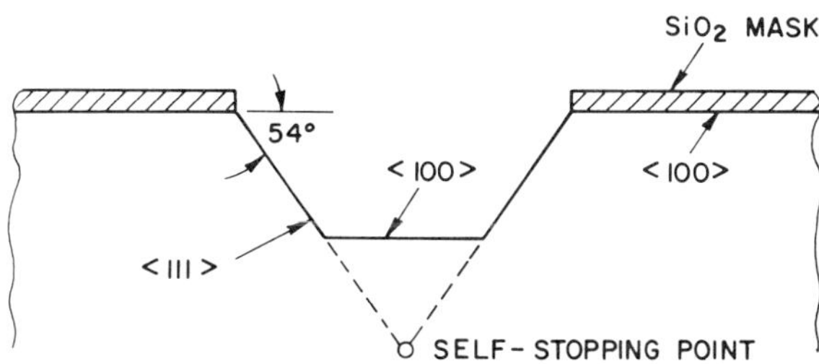

Figure 2.10 Cross section of anisotropically etched channel.

2.3.2 Embedded Oxide Isolation

Versions of the embedded oxide isolation process are best known under the trade names Isoplanar (Fairchild) or Oxim (Bell Laboratories). Here again standard processing is used until the epitaxial film is deposited [20]. Then a nitride oxide layer is deposited as shown in Fig. 2.11*a*. After patterning with the isolation mask, grooves are etched isotropically to two-thirds of the thickness of the epitaxial layer, and oxidation of the exposed silicon fills up the grooves and generates SiO_2 isolation between adjacent islands and a planar surface as shown in Figs. 2.11*b* and *c*.

Following these steps is again a high-speed compatible but standard process flow as described in Section 2.2.

It was mentioned above that the mask tolerance requirements are relaxed. This is caused by the fact that all geometries are defined by one mask, namely, the mask that defines the isolation pattern and, secondly, by the fact that patterns may be generated that are smaller than their photolithographic images. These features are illustrated in Fig. 2.12. The cross-hatched areas in Fig. 2.12*a* correspond to the *n*-epi regions. In Fig. 2.12*b* the base mask covers one entire area with no alignment tolerances other than to cover the opening and avoid the neighboring collector regions. Similarly, the emitter mask shown in Fig. 2.12*c* completely covers the collector opening while covering enough of the base region to provide the needed tolerance for emitter metallization.

A large percentage of the chip surface will be converted into a thick layer of SiO_2 and will thus render the integrated circuit much less subject to defects caused by pinholes or other photoresist- and mask-related problems. In addi-

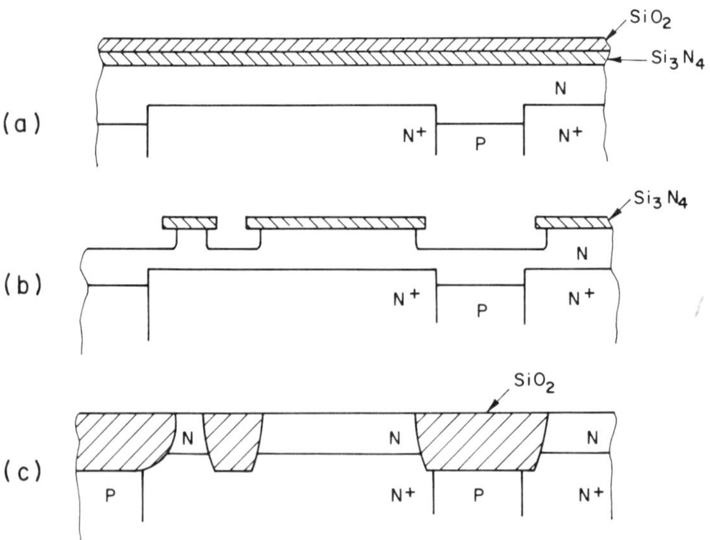

Figure 2.11 Embedded oxide isolation.

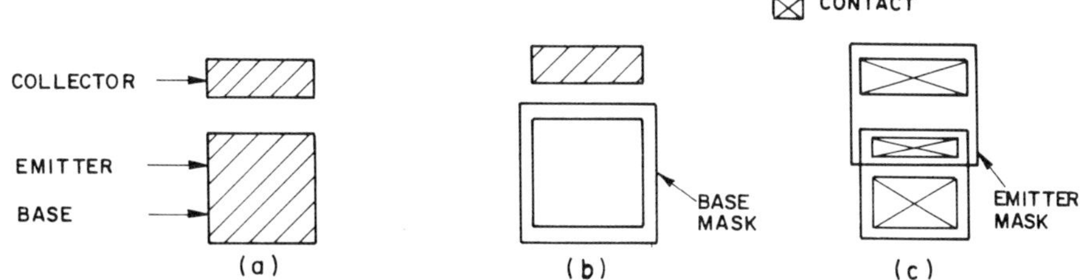

Figure 2.12 Self-alignment feature and generation of small patterns in dielectric isolated bipolar technology. (*a*) Isolation mask applied. Cross-hatched areas are *n*-type epitaxial Si islands surrounded by thick-field oxide. (*b*) Base mask applied. The actual base area is smaller than the mask. Alignment becomes less critical. (*c*) Emitter mask applied. The emitter area is substantially smaller than mask pattern size.

tion, the reduced emitter area of these structures improves the yield, which is limited by the total area of all emitters in a circuit. Thus, for comparable die size, yields are increased over junction-isolated structures.

Owing to the isotropic characteristic of the etch, undercutting occurs, which means that the resulting epi island is smaller than the mask image. This oxide encroachment has to be taken into account when critical areas, for example, resistor values, are defined. But not only does the oxide encroach sideways into the epitaxial regions, it also pushes upward at the oxide–episurface interface as shown in Fig. 2.13. This indicates that the final surface will not be "planar" and that it is possible to find a compromise between the step height—the amount of upward protrusion—and the steepness of the oxide slope. An empirically developed [21] relationship is shown in Fig. 2.14, which displays height and step as a function of epitaxial and oxide thickness.

Figure 2.14 in connection with the choice of etchant to achieve the desired slope of the oxide enables process designers to achieve surface conditions that are compatible with metallization.

Whenever a silicon–silicon dioxide interface is generated, positive charge resides in the oxide near the interface [3]. If, as in our case, oxide is grown in

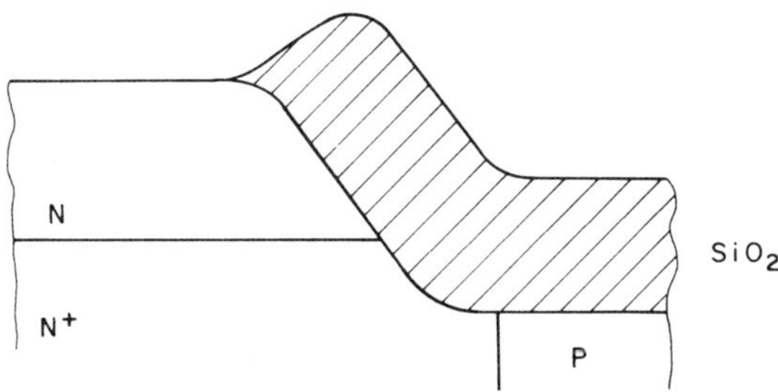

Figure 2.13 Cross section of embedded oxide structure.

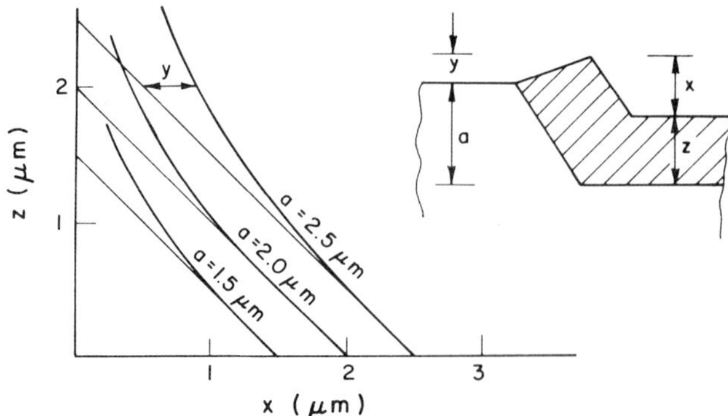

Figure 2.14 Optimization of embedded oxide structure.

lightly doped *p*-type silicon substrate, this positive charge may cause the substrate surface to invert, resulting in an *n*-type channel between two adjacent buried layers. The channel will manifest itself as leakage between otherwise isolated collector regions. A second such case is conceivable at the oxide–base sidewall. If the emitter diffusion is set in from the base–oxide interface, the channel does not influence transistor performance. If, however, full use is made of the self-aligning features of embedded oxide processing and if the emitter is "walled" with the oxide up to three sides as shown in Fig. 2.12, an inversion along the oxide sidewall may introduce collector–emitter leakage.

In order to prevent inversion, the *p* doping at the silicon–oxide interface must be enhanced to beyond 10^{18} cm^{-3} [3]. It is not possible to increase base and substrate doping to this level as parasitic collector–base and collector–substrate capacitance would be increased. Several methods of "channel stopping" have therefore been introduced.

In one method, p^+ rings are diffused surrounding and separating buried layers [22]. This method requires an extra masking step and does not suppress inversion along the sidewall, thus preventing the use of walled emitter structures. Figure 2.15 depicts the resulting cross section.

More recently, ion implantation has been used nonselectively, that is, without requiring the masking step, after etching of the isolation area and before oxide growth. The resulting structure is shown in Fig. 2.16.

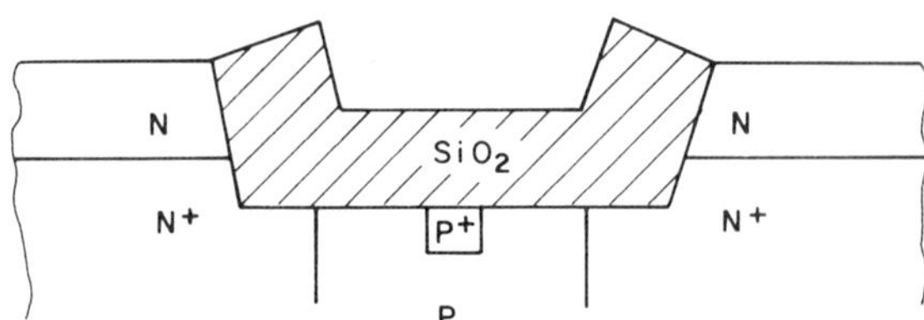

Figure 2.15 Patterned diffused channel stopper.

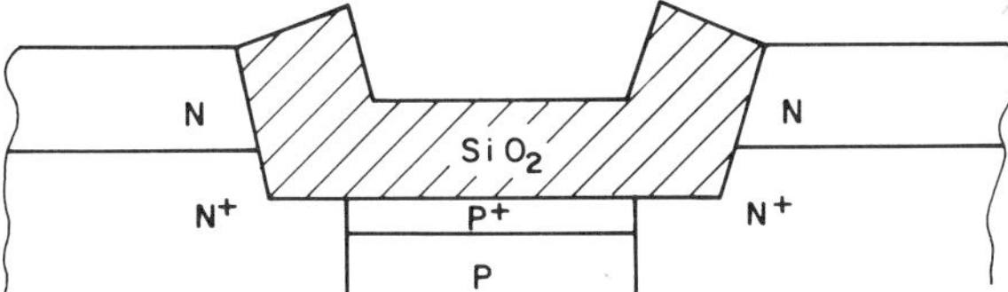

Figure 2.16 Nonselective ion-implanted channel stopper.

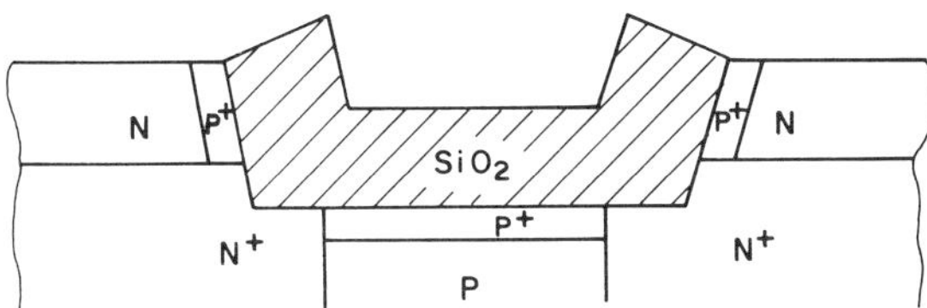

Figure 2.17 Nonselective diffused channel stopper.

In another technique, a nonselective boron diffusion is made just before oxide growth, that is, after etching of the isolation area. During oxidation, the boron diffuses into the silicon at the bottom and on the sidewalls of the isolation moat and thus enhances the doping concentration of all *p* layers. The technique increases parasitic capacitance but levels are smaller than those for uniform heavily doped base and substrate. Good substrate contacts are now possible from the top surface and the *p*-type sidewalls may be used as a guard ring for SBD. No extra mask is required.

Table 2.1 Comparison of Parameters of an *npn* Transistor

	Technology			
			Embedded Oxide Isolation	
Parameter	Junction Isolation (washed emitter)	Air Isolation	Nonwalled Emitter	Walled Emitter
Area (μm^2)	3100	710	1550	770
Emitter (μm)	5 × 20	5 × 20	5 × 20	3 × 8
C_{CB} (pf)	0.27	0.1	0.15	0.08
C_{BE} (pf)	0.15	0.05	0.15	0.03
C_{CS} (pf)	0.47	0.07	0.4	0.14
f_T (MHz)	1200	1500	1200	5000
Beta (@1 mA)	120	70	70	70
LVCEO (V @1 mA)	9	7	7	4
BVCBO (V @10 μA)	22	20	22	16

As a refinement, gallium may be applied *after* oxide growth in a nonselective diffusion [23]. The resulting layer of enhanced *p* is relatively shallow and lighter doped than in the boron method, thus reducing the effect on junction capacitance and breakdown voltage. Figure 2.17 displays the result of either the boron or the gallium technique.

Table 2.1 summarizes and compares the size and the electrical characteristics of *npn* transistors processed on the various technologies described previously.

Dielectrically isolated integrated circuits have been used to build high-speed, high-density bipolar random access memories (RAM) with static storage capacities of 1024 bits [17, 20] and dynamic high-speed storage capacity of 4096 bits [24]. Ultra-high-speed gating circuits with subnanosecond delays [25] and sequential circuits with count frequencies beyond 1 GHz [26] have been built utilizing the improved performance characteristics of the bipolar structures. Both trends will undoubtedly continue.

2.4 INTEGRATED RESISTORS

2.4.1 Diffused or Implanted Resistors

A diffused or implanted resistor is formed by a diffusion or implantation of *p*-type material into an *n*-type area (or vice versa). Figure 2.18 shows the cross section of a *p*-type diffused resistor.

Metal contacts are made to the end of the resistor strip whose resistance is determined by the geometry and the average conductivity of the structure:

$$R = \frac{l}{\bar{\sigma} x_j w} \tag{2.9}$$

where l is the distance between contacts, x_j is the junction depth, w is the width of the resistor body, and $\bar{\sigma}$ is the average conductivity. As it is more convenient to express the resistivity of a thin layer in terms of sheet resistivity ρ_s ($\Omega/\square$), a simpler relationship results:

$$R = \rho_s \frac{l}{w} \tag{2.10}$$

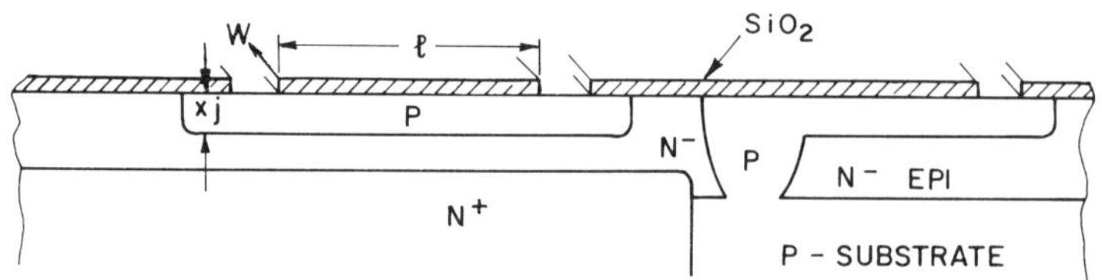

Figure 2.18 Cross section of *p*-type diffused transistor.

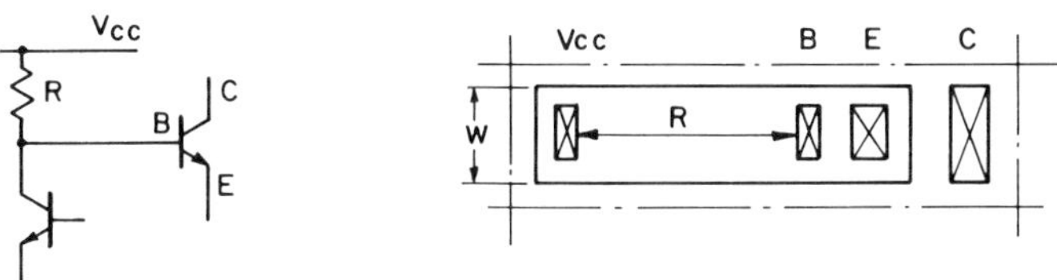

Figure 2.19 Merging of diffused resistor into base of driven device.

Diffused resistors generally have a manufacturing tolerance of about $\pm 15\%$ while ratios of resistors can be held to $\pm 2\%$ if the matching structures are carefully designed and preferably are of identical geometry. Ratios other than 1 may be obtained by series and parallel combinations of identical resistors.

The area requirement for resistors may be reduced by merging. A load resistor is merged with the base of a driven device as shown in Fig. 2.19. This technique is used in several circuit examples in subsequent chapters. Table 2.2

Table 2.2 Diffused Resistors

Range (Ω)	Diffusion	Sheet Resistance (Ω/□)	Temperature Coefficient (ppm/°C)
200–20K	Base	120–500	2500
20–2K	Base plug	50–100	1500
5–500	Emitter	10	500

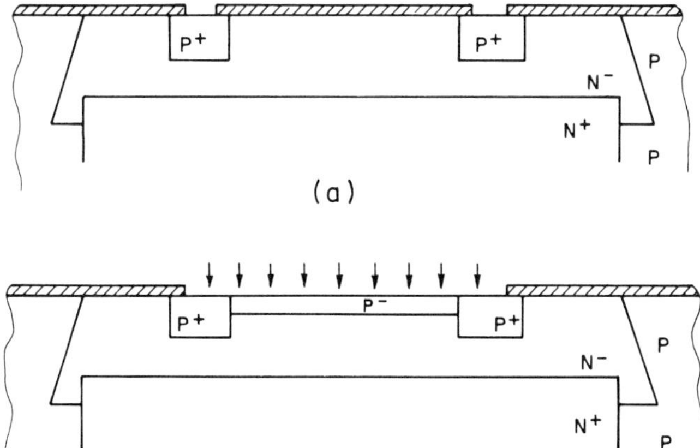

Figure 2.20 Implanted resistor.

gives a comparison of resistor ranges and temperature coefficients for different diffusions.

Owing to the improved control of junction depth and sheet resistivity, ion implantation is being used for higher-valued resistors, and control of resistance values may be improved by at least a factor of 2. An implanted resistor structure is shown in Fig. 2.20. Contact plugs are diffused into the ends of the resistor (Fig. 2.20*a*), and a subsequent implantation connects the contact areas (Fig. 2.20*b*).

Note that the buried layer is found under the resistor structure and that the resistor body is surrounded by a reverse-biased junction. The buried layer reduces the current gain of the vertical parasitic *pnp* transistor, and the parasitic junction capacitance gives rise to a distributed *RC* equivalent network. In the case of air or embedded oxide isolation, the capacitive parasitic effect can be greatly reduced, as seen in Table 2.1.

2.4.2 Double-Diffused (Pinched) Resistors

In order to reach very high values of resistance, a *p*-type resistor manufactured with base diffusion may be pinched with an *n*-type (emitter) diffusion as shown in Fig. 2.21. Effective sheet resistivities of 10–20 kΩ/□ can be achieved at the cost of tolerance and temperature coefficient. The resistance value of the above structure will vary over the process and the commercial temperature range by −50–+100%. Owing to the JFET [3] characteristic of the structure, the resistance will be a strong function of applied voltage, and the device may prove to be light sensitive, which may have to be eliminated by a metal cover.

2.4.3 Epitaxial Resistors

Another common method to generate high resistor values is to utilize the epitaxial layer as material for the resistor and to define the resistor body by isolation. In case of diffused isolation, Fig. 2.22 shows the resulting cross section perpendicular to the long dimension of the resistor.

It will be noted that the isolation reduces the effective width of the resistor below the mask width *w*. This encroachment has to be taken into account during design and it will add to the tolerance of resistance. In addition, the resistor value will depend on the thickness and resistivity of the epitaxial layer. The temperature coefficient is as high as 5000 ppm/°C, and over the range of

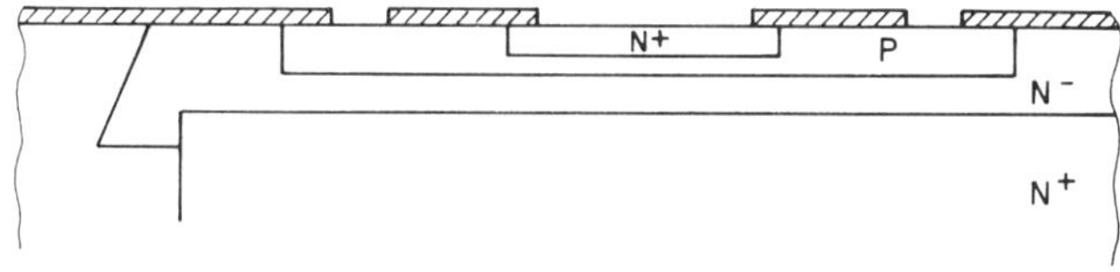

Figure 2.21 Double-diffused (pinched) resistor.

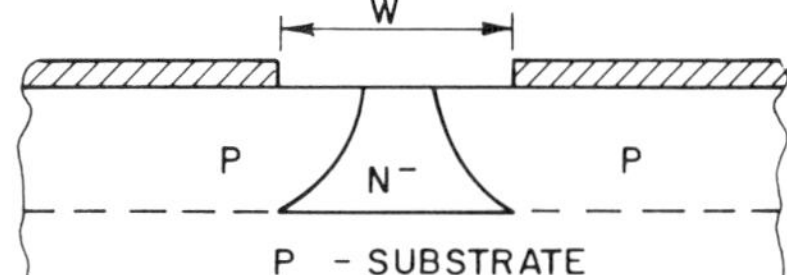

Figure 2.22 Cross section of epitaxial resistor.

temperature and manufacturing tolerances, resistor variations of −30% to +80% will be encountered.

REFERENCES

1. Staff, Motorola Inc., *Integrated Circuits, Design Principles and Fabrication*, McGraw-Hill, New York, 1965.
2. J. Lindmayer and C. Y. Wrigley, *Fundamentals of Semiconductor Devices*, Van Nostrand, Princeton, 1966.
3. A. S. Grove, *Physics and Technology of Semiconductor Devices*, Wiley, New York, 1967.
4. S. M. Sze, *Physics of Semiconductor Devices*, 2nd ed., Wiley, New York, 1981.
5. R. D. Thornton, D. DeWitt, E. R. Chenette, and P. E. Gray, *Characteristics and Limitations of Transistors*, Wiley, New York, 1967.
6. H. H. Berger and S. K. Wiedman, Merged Transistor Logic, a Low-Cost Bipolar Logic Concept, *IEEE J. Solid-State Circuits* vol **SC-7**, 340–346 (1972).
7. R. Mueller, Current Hogging Injection Logic: A New Logic with High Functional Density, *IEEE J. Solid-State Circuits* **SC-10**, 348–352 (1975).
8. K. Hart and A. Slob: Integrated Injection Logic: A New Approach to LSI, *IEEE J. Solid-State Circuits* **SC-7**, 346–351 (1972)
9. P. M. Burger and R. P. Donovan, eds.: *Fundamentals of Silicon Integrated Device Technology*, Vol. 1, Prentice Hall, Englewood Cliffs, NJ, 1967.
10. H. S. Carslaw and J. C. Jaeger, *Conduction of Heat in Solids*, 2nd ed., Oxford University Press, 1959.
11. R. G. Wilson and G. R. Brewer, *Ion Beams*, Wiley, New York, 1973.
12. T. E. Seidel, Proceedings of the Second International Conference on Ion Implantation, (Garmisch-Partenkirchen, Germany) Springer-Verlag, 1971.
13. K. Chino, *Solid State Electronics* **16**, 119 (1973).
14. A. Y. C. Yu and E. H. Snow, *J. Appl. Phys.* **39**, 3008 (1968).
15. H. C. Card, Digest of the International Electronics Device Meeting, *IEEE J. Solid-State Circuits*, 288–290, 1975.
16. Y. Tarui, Y. Hayshi, H. Teshima, and T. Sekigawa, *IEEE J. Solid-State Circuits* **SC-4**, 3–12 (1969).
17. J. Mudge and K. Taft, V-ATE Memory Scores a New High in Combining Speed and Bit Density, *Electronics* **45**, 65ff. (1972).
18. D. B. Lee, Anisotropic Etching of Silicon,*J. Appl. Phys.*, **40**, 4596–4574 (1969).
19. M. J. DeClerca, L. Gerzberg, and J. D. Meindl, Optimization of the Hydrazine-Wafer Solution of Anisotropic Etching of Silicon in Integrated Circuit Technology, Stanford Electronics Laboratory, Stanford University, CA.
20. D. Peltzar and W. Herndon, Isolation Method Shrinks Bipolar Cells for Base, Dense Memory, *Electronics* **44**, 53–55 (1971).

21. U. Schwabe, R. Rathbone, and H. Murrmann, Oxide Isolation Technology for High Performance ICs with *N*-Epitaxy, Abstract No. 181, Spring Meeting of the Electrochemical Society, May 1975.

22. J. A. Appels, W. H. Cornelis, and G. Verkuijlen, U.S. Patent No. 3,900,350, Aug. 19, 1975.

23. A. S. Grove, O. Leisliko, and C. T. Sah, Diffusion of Gallium through a Silicon Dioxide Layer, *J. Phys. Chem. Solid* **25**, 985 (1964).

24. W. B. Sander, J. E. Early, and T. A. Longo, A 4096 × 1 Bipolar Dynamic RAM, *Digest of International Solid-State Circuits Conference*, Feb. 1976, Lewis–Winner, New York, p. 182.

25. V. A. Dhaka, J. Muschinske, and W. K. Owens, Subnanosecond Emitter-Coupled Logic Gate Circuit Using Isoplanar II, *IEEE J. Solid-State Circuits* **SC-8**, 368–372 (1973).

26. H. H. Muller and J. W. Wu, A 1 GHz IC Counter using Isoplanar II, *Digest of International Solid-State Circuit Conference* 1974, Lewis–Winner, New York, Vol. 7 pp. 20–21.

27. W. D. Roehr and D. Thorpe, Ed., *Switching Transistor Handbook*, Motorola Semiconductor Products Inc., Phoenix, 1966.

28. J. L. Moll, Large-Signal Transient Response of Junction Transistors, *Proc. IRE* **42**, 1971 (1954).

29. Staff, Motorola Inc., *Analysis and Design of Integrated Circuits*, McGraw-Hill, New York, 1967.

30. J. J. Muller, W. K. Owens, and P. W. J. Verhofstadt, Fully-Compensated Emitter-Coupled Logic, *IEEE J. Solid-State Circuits* (1973), **SC-8**, 362–367 (1973).

31. R. S. Muller and T. I. Kamins, *Device Electronics for Integrated Circuits*, Wiley, New York, 1977.

32. J. M. Andrews, The Role of the Metal-Semiconductor Interface in Silicon Integrated Circuit Technology, *J. Vac. Sci. Technol.* **11**, 972–984 (1974).

33. V. L. Rideout, A Review of the Theory, Technology and Applications of Metal-Semiconductor Rectifiers, *Thin Solid Films* **48**, 261–291 (1978).

Chapter 3

Bipolar Device Models

R. W. DUTTON

Stanford Electronics Laboratories
Stanford University, Stanford, California

3.1 INTRODUCTION

In the preceding chapter the IC technology steps used to fabricate a broad range of bipolar structures have been discussed. Key technology features used to increase device performance and packing density include:

1. Ion implantation.
2. Dielectric isolation.
3. Special contacts—both ohmic and Schottky.

In addition, Chapter 2 provides a menu of passive component structures and examples pertinent to subsequent chapters. This chapter will draw the focus much more tightly on the electrical effects of integrated transistor structures. However, in the process of focusing the discussion we begin with a somewhat global discussion of the impact of the technology on bipolar device considerations. Three factors can readily be identified that go hand-in-hand with technology choices:

1. *Intrinsic* effects controlled by vertical impurity distributions for *npn* devices and lateral profiles for *pnp*s.
2. *Isolation* effects and associated parasitics.
3. *Distributed* device effects including parasitics.

Figure 3.1 shows three bipolar structures that will be discussed later in this chapter, the circuit applications will be considered in subsequent chapters. The three devices are (*a*) a junction-isolated T^2L *npn* with Schottky clamp, (*b*) an oxide-isolated ECL transistor, and (*c*) an I^2L transistor with n^+ collar and Schottky collector contact. The large number of components that can be extracted from the section views—both desired *intrinsic* as well as parasitic—is a direct consequence of technology options discussed in the previous chapter. Let us briefly summarize key technological points with the intention of drawing primary focus toward the intrinsic *npn* and *pnp* devices. Figure 3.1*a* is a junction-isolated technology with the primary added device feature of a Schottky clamping diode created by laterally overlapping the metal on both *p* and n^- regions. The intrinsic *npn* device is identified with the cross-sectional arrows as is the Schottky barrier diode (SBD). The impact of isolation and distributed device effects are to add components to the terminal schematic shown in Fig. 3.1*d*. The bulk resistances are shown in the section view of Fig. 3.1*a*, and the substrate diode represents the *parasitic* contribution of the epi-substrate junction. Turning to the ECL structure in Fig. 3.1*b*, the intrinsic *npn* device and series resistances are virtually identical to those in Fig. 3.1*a*. The oxide isolation, however, has clipped-off substantial sidewall epi-substrate parasitics as well as provided useful self-alignment features to shrink device area. As shown by the dashed component in Fig. 3.1*e* a new parasitic path now can exist between emitter and collector due to the oxide sidewall. It is clear that to have a viable bipolar circuit technology, this effect must be made negligible. Turning to the final example shown in Fig. 3.1*c*, a substantially different set of intrinsic and parasitic effects are observed. The two intrinsic devices are shown with sectional arrows. The *vertical npn* uses a lightly doped surface collector and the subsurface buried n^+ emitter. In addition, a Schottky diode contact appears in series with the collector. The *lateral pnp* device is the second intrinsic element shown, which in fact shares the *p* diffused region with the *npn* device. The *npn* base region resistor is similar in nature to the other examples. The isolation considerations for this technology, however, are substantially different. In particular, the emitter-base region is now defined by the

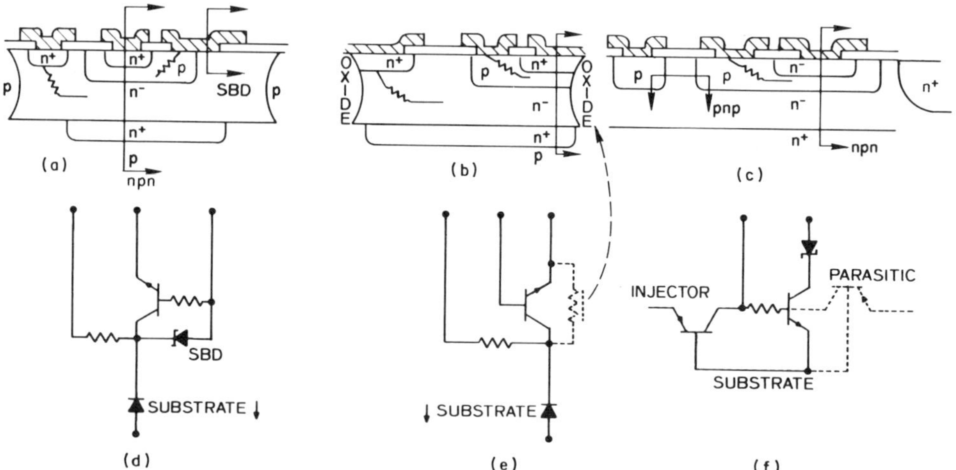

Figure 3.1 Section views of three *npn* bipolar device technologies: (*a*) Schottky clamped diffusion isolated, (*b*) oxide isolated, (*c*) I^2L type with a Schottky diode in series at the collector. (*d*)–(*f*) The equivalent circuit models for the structures shown in (*a*)–(*c*). Note in particular the parasitic substrate diodes (*d* and *e*), the oxide surface conductance (*e*), and parasitic lateral *pnp* (*f*).

lower most *pn* junction in the *npn* section. Under forward bias, holes are injected laterally into the n^- epitaxial region giving rise to parasitic *pnp*s due to neighboring *p* regions *other* than the *intrinsic pnp* shown. Shown in Fig. 3.1*c* is a so-called n^+ collar used to repel injected holes. An alternative is to use oxide isolation in essentially the same manner as for the ECL technology shown in Fig. 3.1*b*. A common factor in all three examples shown in Fig.

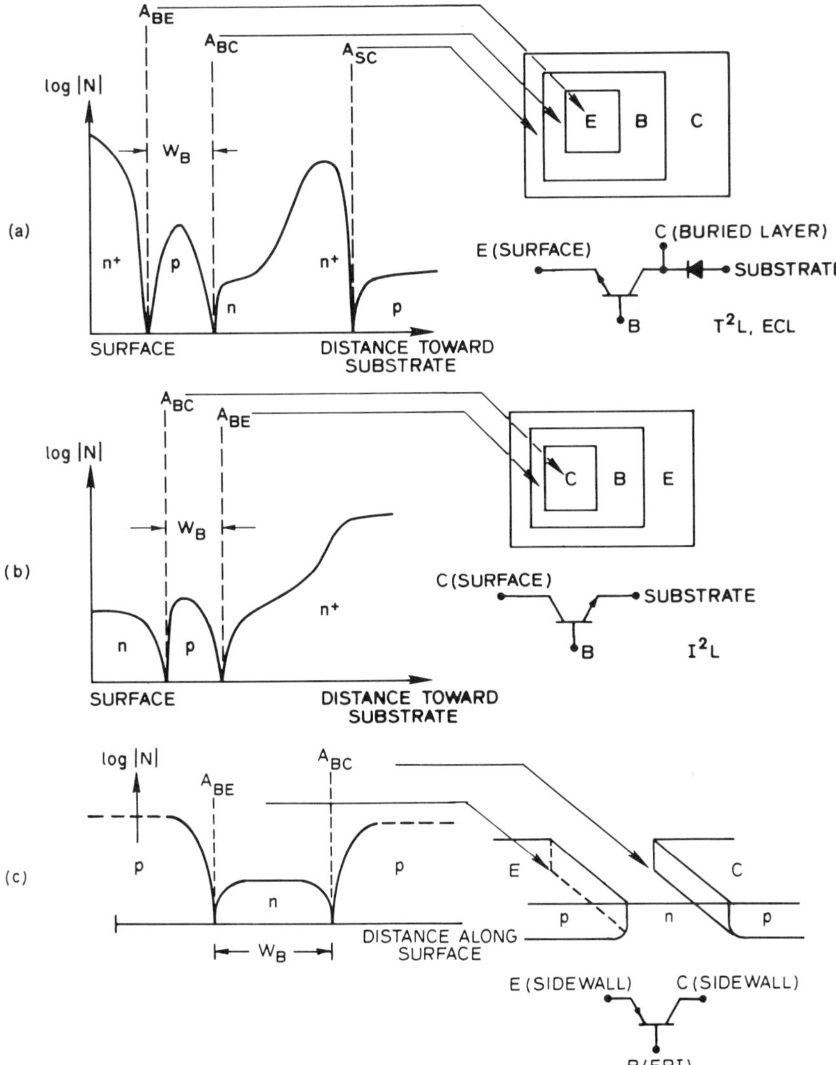

Figure 3.2 Impurity profile sections for two technologies and typical surface geometries, (*a*) normal double diffused, epitaxial structure with buried layer. Typical surface geometries for T^2L and ECL give $A_{BE} < A_{BC}$. (*b*) An I^2L-type process with a lightly doped surface *n*-type region. For this case $A_{BE} > A_{BC}$. (*c*) The lateral *pnp* structure for an I^2L process $A_{BE} = A_{BC}$.

3.1*d–f* is the intimate connection of the intrinsic effects and the parasitic device contributions due either to isolation techniques or to the inherent geometrical factors. At the device level one wishes to optimize the intrinsic device performance while minimizing the parasitics. As will become evident in later chapters, auxiliary constraints imposed by circuit layout can affect choices of device level performance trade-offs.

The discussion of model considerations, which cover the range of technologies shown in Fig. 3.1, must begin with consideration of the impurity profiles shown in Figure 3.2—both vertical and lateral. Figure 3.2*a* represents the profile for both T^2L- and ECL-type devices. The emitter is at the surface and the area ratios are such that $A_{SC} > A_{CB} > A_{BE}$, where S, C, B, and E represent the $pn^+n - p - n^+$ regions of the substrate, collector, base, and emitter. Figure 3.2*b* shows an I^2L configuration where the collector is at the surface—a so-called inverted structure—and $A_{CB} < A_{BE}$. One must always determine explicitly the area factors for the junctions as well as capacitance per unit area and the integral of the base doping. A subsequent section will show how these parameters can be determined directly from knowledge of the process. Figure 3.2*c* shows the lateral surface profile corresponding to the *pnp* device in Fig. 3.1*f*. In contrast with the vertical *npn* devices, the emitter and collector are defined by the subsurface junction sidewall areas shown. Since normally these p regions are created in the same fabrication sequence, $A_{BE} \approx A_{BC}$. However, one major difference to note between the vertical *npn*s and the *pnp* is the lateral extensions of the emitter and collector regions. Although this does not affect the transport injection, it does definitely contribute to junction capacitance and diode leakage.

3.2 BIPOLAR TRANSISTOR MODEL

Consider the double-diffused *npn* device shown in Fig. 3.3*a*. To understand the transistor as a circuit component, the vertical parameters that limit its performance must be defined. Figure 3.3*b* shows a simplified representation of the transistor given in Fig. 3.3*a*. The notation of emitter (E) and collector (C) are added with subscripts to denote direction of electron injection relative to the surface—upward toward or downward from the surface. The *BE* junction is assumed to be forward biased and the *BC* junction is reverse biased. Electrons are injected downward across the shaded region and collected at the region indicated by x's. The model for both dc and transient effects of the device excluding the saturation mode of operation is shown in Fig. 3.3*c*. The diode and capacitance represent two effects at the *BE* junction. The diode models base recombination current as well as reverse injection into the emitter regions. The capacitance represents charge storage due to both majority-carrier space charge and neutral region minority charge effects. The capacitance and dc properties of the diode are both proportional to the *BE* junction area; that is, the capacitance scales linearly with junction area as does base current.

To understand the current relationship associated with the BE junction we begin by defining the transport saturation current I_S. The value of I_S is related to the minimum BE or BC area. Current which passes between the two junctions is constricted by the minimum of the two areas $A_{\min} = \min\{A_{BE}, A_{BC}\}$. For the structure of Fig. 3.2*a*, $A_{\min} = A_{BE}$, whereas for Fig. 3.2*b*, $A_{\min} = A_{BC}$. Based on the respective value of $A_{\min}$, I_S for an *npn* device is given by

$$I_S = \frac{qA_{\min}n_i^2}{\int_0^{W_B}[N_A(x)/D_n]\,dx} \tag{3.1}$$

where the integral is taken over all acceptors $N_A(x)$ within the neutral base of width W_B, q is the electron charge, n_i is the intrinsic carrier concentration, and D_n is the electron diffusion coefficient. This result is the well-known Moll–Ross relationship [1]. As will be seen shortly, for a *pnp* device—and, in particular,

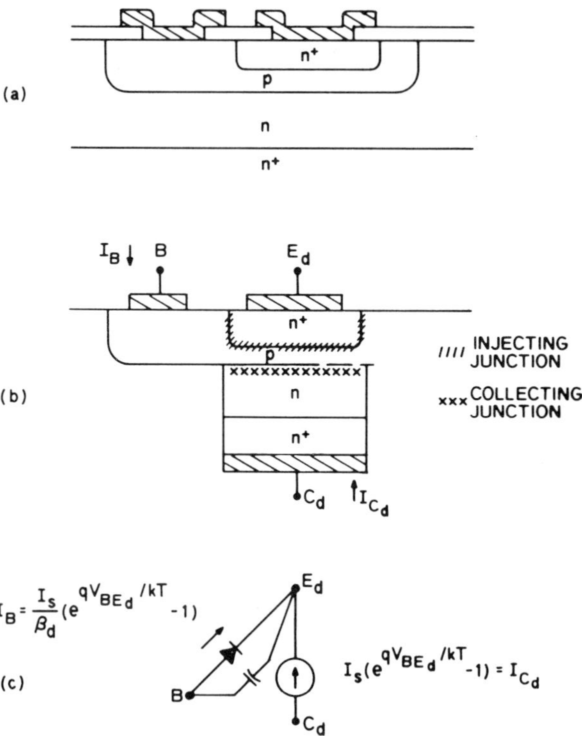

Figure 3.3 The section view and device model for bipolar injection from the surface *pn* junction in the "downward" direction. (*a*) Technology view of the device, (*b*) modeling view with regions of injection and collection region, (*c*) the device model considering only the injection due to applied bias V_{BE_d}.

lateral transistors—Eq. (3.1) also applies; however, the denominator is the integral of $N_D(x)/D_p$ over the *donors* in the neutral base. The importance for this discussion can be stated briefly. The linear dependence of I_S on $A_{\min}$ is apparent. Also note the dependence of the denominator on integrated majority charge in the base, that is, Q_B is proportional to the product of base width and doping density $[Q_B = q\int_0^{W_B} N_A(x)\,dx]$. Now in terms of this transport saturation current we can define the two dc elements shown in Fig. 3.3*c*. Given a bias of V_{BE} across the base–emitter and assuming $V_{BC} = 0$, the resulting collector current is given by

$$I_{C_d} = I_S(e^{qV_{BE}/kT} - 1) \tag{3.2}$$

The base current is some fraction of the collector current and can be defined in terms of β_d as follows:

$$I_B = \frac{I_{C_d}}{\beta_d} = \frac{I_S}{\beta_d}(e^{qV_{BE}/kT} - 1) \tag{3.3}$$

The quantity β_d depends on recombination processes, which contribute to base current. For the given driving force, base current flows to provide hole injection from the base into the n^+ emitter as well as recombination components in neutral and space-charge regions. The bias conditions shown would correspond to the integrated device structure operated for either T^2L or ECL applications.

The discussion to this point has elucidated several physical effects. The device, operating as described in Fig. 3.3, is controlled by two physical quantities. First, the transport current (I_S) is a measure of area and integrated base doping. Given the base–emitter bias, a current source exists between emitter and collector, and I_S is the proportionality constant for the resulting exponential voltage-dependent current flow. Second, the base-current and charge-storage effects are proportional to the respective junction area. As we consider the operational mode of the device under different bias conditions, these principles will be found to also apply.

Figure 3.4*a* shows the same structure with a forward bias now applied to the *np* junction (again labeled *BE*) and a reverse bias applied to the n^+p junction. Now the minority-carrier transport is upward. The shaded region again represents injection and the *x*'s indicate collection. The resulting device model is shown in Fig. 3.4*b*. Similar to the result above, the *BE* charge-storage effects and I_S/β_u are proportional to the *BE* shaded junction area. Using the minimum area criteria and the fact that Figs. 3.3 and 3.4 are assumed to have the same distribution of majority carriers in the base region, the I_S value shown in Fig. 3.4*b* is also given by Eq. (3.1). Given a bias of V_{BE} across the n^+np base–emitter junction and assuming $V_{BC} = 0$, the resulting collector current is given by

$$I_{C_u} = I_S(e^{qV_{BE}/kT} - 1) \tag{3.4}$$

The base current is a fraction of this value as determined by β_u as follows:

$$I_B = \frac{I_{C_u}}{\beta_u} = \frac{I_S}{\beta_u}\left(e^{qV_{BE}/kT} - 1\right) \tag{3.5}$$

The term β_u again characterizes recombination and hole injection properties associated with the BE; this time it is assumed to be n^+np junction. The bias conditions depicted in Fig. 3.4 are characteristic of integrated devices as used for $\mathrm{I^2L}$ applications. At this point it should again be emphasized that both Figs. 3.3 and 3.4 correspond to the same physical and geometrically defined device. The definition of the n^+p base–emitter used in Fig. 3.3 is characteristic of $\mathrm{T^2L}$ and ECL applications, whereas Fig. 3.4 assumes the n^+np region is normally forward biased as is the case for $\mathrm{I^2L}$. Hence, in the normal "forward" active mode of operation $\beta_F(\mathrm{T^2L}) = \beta_F(\mathrm{ECL}) = \beta_d$, whereas $\beta_F(\mathrm{I^2L}) = \beta_u$. The subscript F, used to denote forward active mode, is the one used normally for the Ebers–Moll device equations.

At this point it is appropriate to merge the results from Figs. 3.3 and 3.4 and define the classical bipolar transport model, which applies for any arbitrary bias across the junctions. Figure 3.5 again defines the device cross section; however, this time only junctions are labeled. The junction voltage polarities are such that a positive voltage results in forward bias. The notation

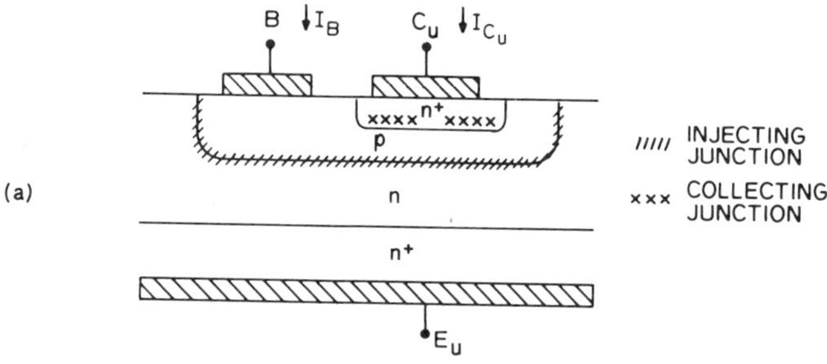

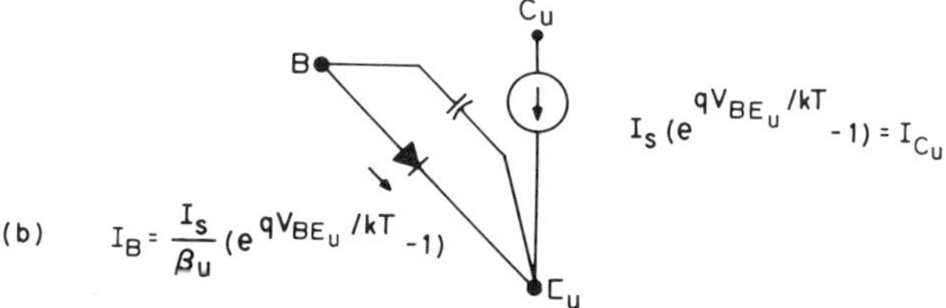

Figure 3.4 A section view and model equivalent for the case of bipolar injection from the subsurface pn junction: (a) the indication of injecting and collection junctions; (b) the equivalent model assuming injection controlled only by V_{BE_u}.

(a)

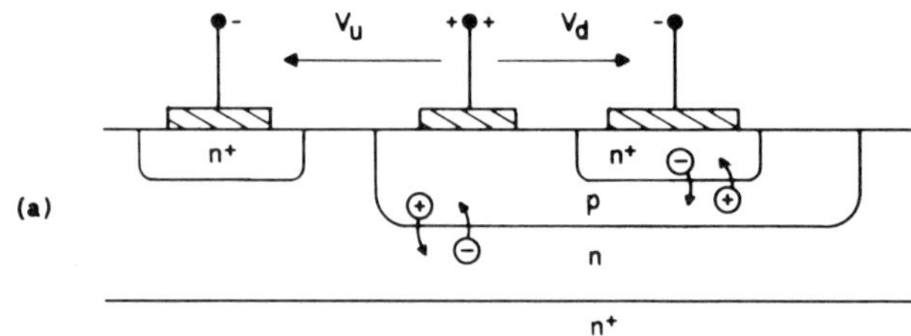

(b)

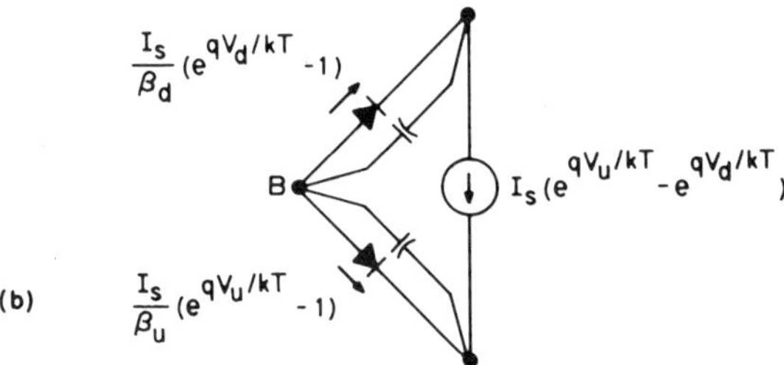

Figure 3.5 The complete bipolar device section and model for applied voltages V_u and V_d; (*a*) the section view with carrier injection shown; (*b*) the model assuming both junctions injecting.

of up and down components of current are again used. It is important to note two things concerning the current source. First, I_S has the same value as given in Eq. (3.1). This makes sense since both components cause injection of carriers into the same volume defined by W_B and A_{min}. Current that flows into other regions not intersected by A_{min} becomes diode current rather than transport current, which is related to either β_d or β_u. Second, reversing the magnitudes of V_u and V_d causes the sign of the current to change, but exactly the same current magnitude will be obtained for reciprocal voltage values. One will note that base currents can be quite different for the two cases owing to the difference in β_u and β_d. The key point is that the transport current as defined by I_S is exactly reciprocal with bias conditions. The model configuration shown in Fig. 3.5 will be used subsequently to define the bipolar model for computer-aided design (CAD)*. One can realize that for the saturated mode of device operation both β_u and β_d enter into bias considerations. Before considering the conditions involved with saturated device operation, we will now discuss junction charge-storage effects.

3.2.1 Space-Charge Effects

Having defined the complete set of dc transport relationships for the double-diffused bipolar transistor, we can now consider the charge-storage effects of

*The term CAD is generic for circuit analysis using the computer to solve the nonlinear and time-dependent equations describing node voltages and terminal currents in a circuit.

the device. The importance of these effects is paramount. If there were no charge storage, then the devices would be infinitely fast. That is, there would be no charge inertia and currents could be changed in zero time. The two forms of charge storage are space charge and minority-carrier injection. Space charge at each *pn* junction results from the depletion regions between materials of opposite carrier type [2]. This same effect can occur at a metal–semiconductor contact which is nonohmic, a Schottky barrier, for example [3]. The general form for the stored charge due to depleted carriers is

$$q_{sc}(V) = qA\left[\overline{N}_D \cdot x_n(V)\right] \tag{3.6a}$$

where $\overline{N}_D$ is the average donor density over the n side of the junction and x_n is the voltage-dependent depletion depth on that side of the junction. For a *pn* junction there is an equal but opposite charge due to $\overline{N}_A$ on the p side with depletion depth given by x_p. The Schottky diode has only the spatial depletion in the semiconductor. To obtain the space-charge capacitance it is necessary to differentiate the space charge with respect to voltage:

$$C_{sc}(V) = \frac{dq_{sc}}{dV} = \frac{C_{j0}}{(1 - V/\phi)^m} \tag{3.6b}$$

$$q_{sc}(V) = \frac{\phi C_{j0}}{1 - m}\left(1 - \frac{V}{\phi}\right)^{1-m} \tag{3.6c}$$

where C_{j0} is the zero-bias value, ϕ and m are process-dependent coefficients, and V is the forward bias across the *pn* junction (V is negative for reverse bias). For typical bipolar devices m is bound between 0.2 and 0.6 and ϕ the built-in voltage ranges from 0.5 V to 0.8 V. The coefficient m results from the spatial dependence of $N_D(x)$ in solving Poisson's equation across the space charge. For example, an abrupt *pn* junction with constant N_A and N_D gives $m = \frac{1}{2}$, whereas a linearly graded doping profile going from p to n gives $m = \frac{1}{3}$. The range for ϕ is set by N_A and N_D at the space-charge edges based on the equation

$$\phi = \frac{kT}{q}\ln\left[\frac{N_D N_A}{n_i^2}\right] \tag{3.6d}$$

where n_i is the intrinsic concentration.

It is useful to define a circuit equivalent, different from Eqs. (3.6), which properly represents the charge-storage effects. A suitable choice is a constant-valued capacitor, which transfers the correct amount of charge over the switching event. To write this more precisely:

$$\overline{C}_{sc} = \frac{\int_{V_1}^{V_2} C_{sc}(v)\, dv}{V_2 - V_1} \tag{3.7}$$

For most circuits one can consider the maximum forward-bias limit to be the built-in voltage ϕ and the maximum reverse-bias limit to be $-R\phi$, where the factor R is chosen such that $R\phi$ equals the maximum reverse bias. Hence using Eq. (3.6) in Eq. (3.7), the result is:

$$\overline{C}_{sc} \approx \frac{C_{j0}}{(1-m)[1+R]^m} \tag{3.8}$$

For a typical example of $m = \frac{1}{2}$ and $R = 1$, $\overline{C}_{sc} = 2C_{j0}/\sqrt{2}$. One can note that the average value will not vary drastically from the C_{j0} value.

Now focusing our attention on the associated charge storage and transfer during switching, the change in charge will be

$$\Delta Q_{sc} = \overline{C}_{sc}\Delta V \tag{3.9}$$

By design we have guaranteed that this change will be identical to that resulting from the integral given in Eq. (3.7). A key point can be stated now. For switching events in terms of changes in space charge, storage equation (3.9) can be used as the controlling relationship. We will consider this point in greater detail concerning approximations for device switching.

3.2.2 Saturation and Charge Storage

Charge storage results from excess minority carriers *injected* (subscript I) across the junctions in forward bias. This charge is proportional to the total current injected across a junction and hence is exponential in voltage. The proportionality constant between the current and charge is a time constant that represents the minimum time required to either store or remove the charge. Hence

$$q_I = \tau_I I(V) \tag{3.10}$$

where $I(V)$ is an exponential function of V. Figure 3.6 shows a plot of both q_{sc} and q_I as a function of V. Two important observations can be made based on Fig. 3.6. For reverse bias and small forward bias, q_{sc} is the dominant charge storage. For moderate forward bias and beyond, the injected charge dominates and hence q_{sc} effects become negligible. This second observation is extremely important, since as V approaches ϕ the differential space-charge capacitance becomes large (but negligible charge compared to q_I). The plot of q_{sc} in Fig. 3.6 may at first seem puzzling. Equation (3.6a) was written in differential form as a capacitance, however, the plot refers to charge as given by Eq. (3.6c). The key point to be made is that the space charge is *finite* and its value approaches zero as V approaches ϕ as shown in Eq. (3.6c). The derivative, on the other hand, goes to an infinite value at $V = \phi$ as shown in Eq. (3.6b).

While space charge is a two-terminal effect, minority-charge injection and storage are not. Equation (3.10) describes a diodelike relationship in a two-

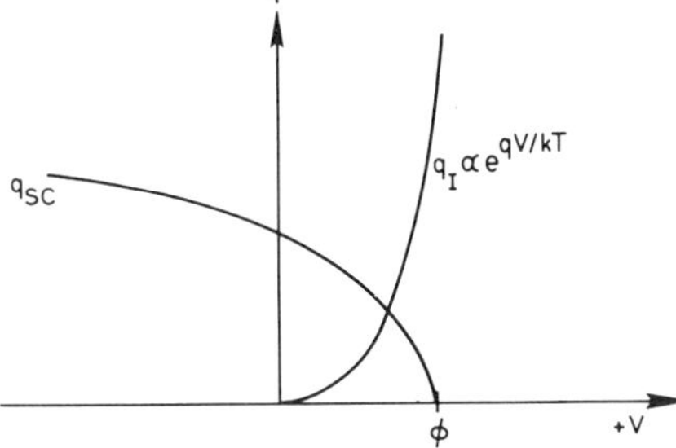

Figure 3.6 Plot of space-charge (sc) and minority-carrier-injection (I) contributions to stored charge versus forward applied junction voltage.

terminal sense. The problem becomes more complex when the injected charge can result from several injecting terminals. For the bipolar device both *pn* junctions can be forward biased and hence contribute to minority storage in the base. Consider the case when the injection of two junctions leads to a saturation condition. Figure 3.7*a* defines the current flows and junction notation of a typical switching device. The sense of I_C (and I_E) is such that the junction with subscript F can be considered the emitter. For nonsaturated conditions ($V_R < 0$), $I_B = I_C/\beta_F$, whereas in saturation $I_B \gg I_C/\beta_F$. It is convenient to define the *excess base current* as

$$I_{BS} = I_B - \frac{I_C}{\beta_F} \tag{3.11}$$

The meaning of I_{BS} can be interpreted as follows: It is the base current in excess of the minimum needed to saturate the device with a collector current I_C. Just at the saturation point the injected charge is stored only due to the forward bias of V_F and is given by

$$q_F = \tau_F I_C \tag{3.12}$$

As the device goes into saturation ($I_{BS} > 0$) the junction with subscript R

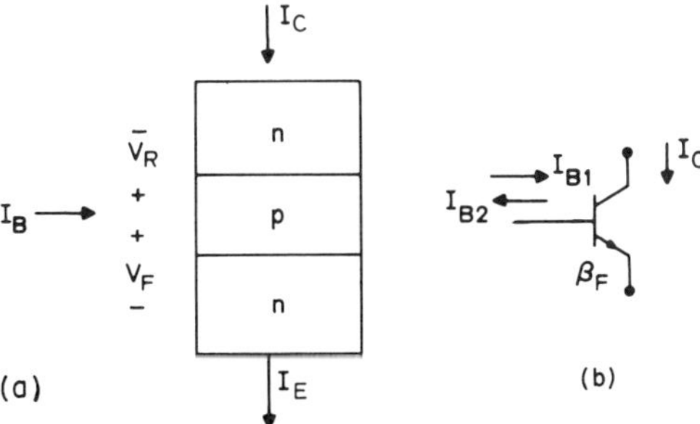

Figure 3.7 Terminal notation for voltages and currents used to model transistor-saturated switching.

becomes forward biased and the voltage on the junction labeled F also increases. The result is an additional charge storage that cannot be represented simply in terms of a single-junction effect. The detailed analysis is treated elsewhere [2, 3], but the results can be stated simply and physically.

The excess stored base charge is given by

$$q_s = \tau_s I_{BS} \tag{3.13}$$

where I_{BS} is given in Eq. (3.11) and

$$\tau_s = \frac{\beta_F \tau_F (1 + \beta_R) + \beta_R \tau_R \beta_F}{1 + \beta_F + \beta_R} \tag{3.14}$$

One can see that τ_s depends on the injection properties of both junctions. However, the key point is that in saturation the charge can be described in terms of the excess base current and this τ_s. The subscript notation of F and R used above has a specific technology translation as we consider different logic families. For I^2L the correspondence of the up and downward components gives $\beta_F = \beta_u$ and $\beta_R = \beta_d$. The point now to emphasize is that u and d imply technology constraints, whereas F and R are the electrical circuit choices associated with bias and connection options of circuit design. Turning to the T^2L and ECL technologies the correspondence of notations gave $\beta_F = \beta_d$ and $\beta_R = \beta_u$.

Having defined the expressions for stored charge just prior to saturation [Eq. (3.12)] and for deep saturation [Eq. (3.13)], the changes in charge that are important in switching applications can now be defined. Figure 3.7 shows the notation to be used. The notation assumes a given emitter, and junction potentials are omitted. The base currents I_{B1} and I_{B2} represent forced currents from "sources" that will turn on (saturate) and turn off the device, respectively, and I_C represents the saturated collector current. The on time to obtain I_C flowing (neglecting $\overline{C}_{sc}$ effects) is simply

$$t_{\text{on}} = \frac{q_F}{I_{B1}} = \tau_F \frac{I_C}{I_{B1}} \tag{3.15}$$

Any effects of storage build-up going into saturation occur after t_{on} but make no difference to the output current I_C. For the off-time calculation the problem is somewhat more complicated. Considering first the saturation delay to remove q_s, the following equation describes the dynamic performance:

$$I_B - \frac{I_C}{\beta_F} = \frac{q_s}{\tau_s} + \frac{dq_s}{dt} \tag{3.16}$$

The left-hand side was defined as I_{BS} in steady state [Eq. (3.11)] and if $dq_s/dt = 0$, one can also obtain Eq. (3.13) by solving for q_s. In the transient case:

$$I_B = I_{B1}, \qquad t < 0$$
$$I_B = -I_{B2} \qquad t \geqslant 0 \tag{3.17}$$

Solving Eq. (3.16) for t_{sd} such that $q_s = 0$ subject to the conditions given in Eq. (3.17), it follows that

$$q_s(t) = \left[(I_{B1} + I_{B2}) e^{-t/\tau_s} + \left(-I_{B2} - \frac{I_C}{\beta_F} \right) \right] \tau_s \tag{3.18}$$

and $q_s = 0$ when

$$t_{sd} = \tau_s \ln\left(\frac{I_{B1} + I_{B2}}{I_{B2} + I_C/\beta_F} \right) \tag{3.19}$$

A useful set of approximations, which may seem arbitrary at this point, are as follows. If $I_{B2} \gg I_{B1} \gg I_C/\beta_F$, then:

$$t_{sd} \approx \tau_s \ln\left[1 + \frac{I_{B1}}{I_{B2}} \right] \approx \frac{\tau_s I_{B1}}{I_{B2}} \tag{3.20}$$

This can be interpreted as follows. If $q_s \approx \tau_s I_{B1}$, then the saturation delay can be approximated as q_s/I_{B2}. Once the saturated charge q_s is removed, the q_F contribution must in turn be removed. Hence

$$t_{\text{off}} = \frac{q_F}{I_{B2}} = \tau_F \frac{I_C}{I_{B2}} \tag{3.21}$$

Again the correspondence of technology subscripts for the I^2L device gives $F \to u$ and $R \to d$, whereas for T^2L and EFL technologies the junction roles are reversed so that $F \to d$ and $R \to u$.

Although the switching equations (3.15), (3.19), and (3.21) give an estimate of time delays, they have two serious drawbacks. First, they ignore such effects as space-charge storage. Second, there are many changes occurring simultaneously in a digital circuit and one single event may not necessarily go to completion before another can begin. This is especially true when nonlinear devices appear as loads. For diode loads, a change of 60 mV can be all that is required to switch from an on to an off state. The next section shows some appropriate approximations for certain regions of digital-device operation.

3.2.3 Digital Switching

The above sections concerning transport and charge storage give the necessary detail to estimate device dc and switching behavior. However, to apply these results properly in the context of practical circuit examples it is useful to show schematically the appropriate model simplifications used for digital-device-modeling examples. Figure 3.8*a* illustrates a typical configuration with a grounded-emitter transistor. The output voltage might exhibit the waveform shown in Fig. 3.8*b* during a typical switching event. The regions labeled I–IV indicate different operating conditions for the transistor. In region I the device is typically off and only junction capacitance with its associated q_{sc} needs to be considered. As the device turns on and enters region II the base–emitter diode, current transport between emitter–collector, and minority-charge storage q_F must be considered as well as the space charges q_{sc}. Finally, after the device has saturated—as is the case for I^2L and standard T^2L—the saturation charge q_s and the base–collector diode now are added to the model as can be seen in III. During this phase of switching operation (see Fig. 3.8*b*) the saturated charge must be removed before the output voltage can begin to rise. As the device comes out of saturation, the model considered for region II now again applies for operation as shown in region IV.

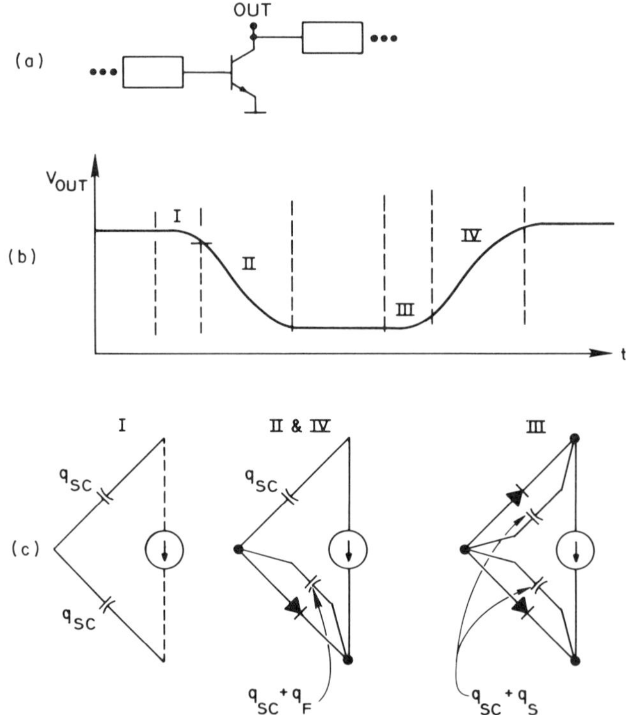

Figure 3.8 A bipolar device embedded in a switching circuit. The connections are shown in (*a*). Shown in (*b*) is a typical output node waveform and in (*c*) the corresponding model simplifications over regions of the switching.

The details of how these simplified models can be used to estimate switching performance of digital circuits will be discussed in Section 3.5. However, to make the most effective use of all device data under conditions where the regions I–IV are not as neatly defined or characterized, one must use computer circuit simulation—often called CAD (computer-aided design). By CAD the following is implied:

1. A complete circuit topology is specified (Rs, Cs, supplies, pulses, and transistors).
2. The transistor-model information pertinent to Fig. 3.5 is used.
3. The complete set of nonlinear equations for the given circuit is solved by iterative matrix-inversion techniques [4] for:

   (a) dc conditions.

   (b) transient conditions at user-specified time intervals.

For CAD efficiency, *macromodels* may be used to represent gates, which have a large number of internal devices, but not all of these devices require the model complexity shown in Fig. 3.5 from a terminal performance point of view. Hence, the internal gate model is reduced to a minimum while still reproducing

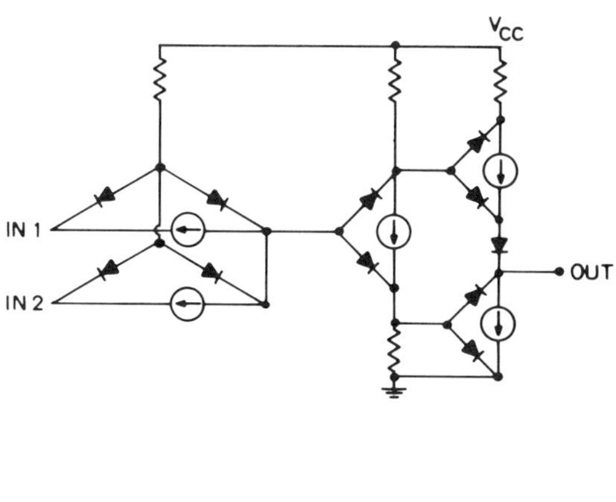

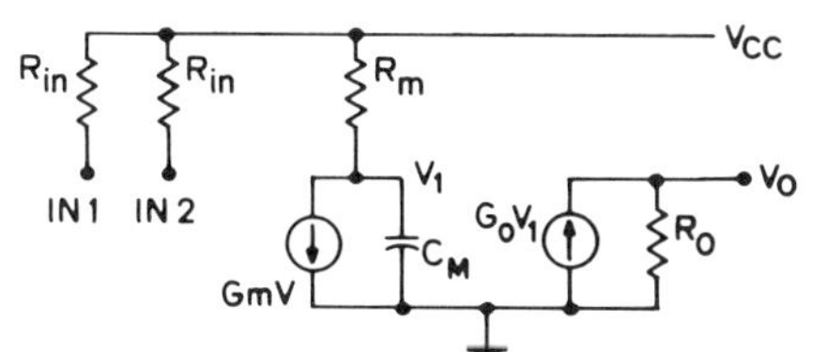

Figure 3.9 A device-level equivalent circuit for a T^2L gate and a simplified macromodel [4], which can give first-order equivalent terminal performance.

terminal nonlinearities and transient performance features. Figure 3.9 shows a comparison of the detailed (transistor-level) and macromodel representations of a T^2L NAND gate used for CAD [5]. In Section 3.4 the general features of a bipolar transistor model used in the CAD program SPICE [6] as well as input parameters used to pass the needed information to the program are outlined. The model will evolve from the basic topology shown in Fig. 3.5 (and associated equations) and be expanded to include effects pertinent to low- and high-current performance.

3.3 DIODE MODEL

At this point it is useful to consider diode characteristics for both the *pn* junction and Schottky devices. Primarily we will consider devices other than those included in the intrinsic *npn* transistor structure discussed in Section 3.2. In particular, we begin by focusing on the diodes shown in Fig. 3.1*d*.

The Schottky device labeled SBD and the *pn* substrate junction diode both exhibit exponential current–voltage characteristics, although their dynamic response can be markedly different. In this section we will begin with dc model considerations and then discuss charge-storage effects. In the last parts of this section CAD models and applications will be presented.

3.3.1 Transport Equations

The fundamental dc transport equation for a diode is given by the general equation set:

$$I = I_{S1} e^{qV/nkT} + I_{S2} e^{qV/kT} \tag{3.22}$$

where I and V are the terminal parameters, I_{S1} and I_{S2} are injection coefficients, and n is the emission coefficient of the device. This equation is a superposition of transport phenomena controlled within junction space charge (term one) and those determined by events in the so-called neutral region(s) (term two). The first term in Eq. (3.22) applies for low currents in *pn* junctions and over the entire current range for Schottky devices. The emission coefficient n deviates from unity based on two physical phenomena:

1. Injected minority carriers recombine within the space-charge region for *pn* junction devices.
2. The space charge and barrier-lowering effects for Schottky devices alter the emission properties of the device.

For the *pn* junction case typically $n = 2$, whereas for the Schottky device $n \approx 1.2$ [7]. The preexponential coefficients depend directly on junction area.

For the *pn* junction case I_{S1} depends directly on the width of the space-charge region (W_{sc}). For the Schottky device the I_{S1} term depends primarily on barrier height (ϕ_B) control effects. The two limiting cases are as follows:

$$I_{S1} = A_{\text{SBD}} e^{-\phi_B/kT}[RT^2] \qquad \text{(Schottky)} \qquad (3.23a)$$

$$I_{S1} = A_{pn} W_{sc} \left[\frac{qn_i}{2\tau} \right] \qquad (pn \text{ junction}) \qquad (3.23b)$$

where in addition to the respective area and width factors R is Richardson's constant, T is the absolute temperature, and τ is the minority-carrier lifetime [2].

Typical results based on Eqs. (3.23a) and (3.23b) are superposed in Fig. 3.10 to show two effects. First, the slopes differ markedly in the low-current region. However, as the current increases the slopes become nearly equal. Second, for a fixed voltage bias level the current through the Schottky is typical more than two orders of magnitude greater than for the junction device. This difference in magnitude occurs primarily because of the effectively low barrier height for majority carriers in the Schottky device compared with the band-gap-controlled phenomena for the minority-carrier *pn* junction.

In terms of Eq. (3.22) and its relationship to Fig. 3.10 several points can be emphasized. For the *pn* junction device both terms in Eq. (3.22) are apparent. However, for the Schottky diode only the first component is observed. As suggested above in the context of discussing Eqs. (3.23), the physical mecha-

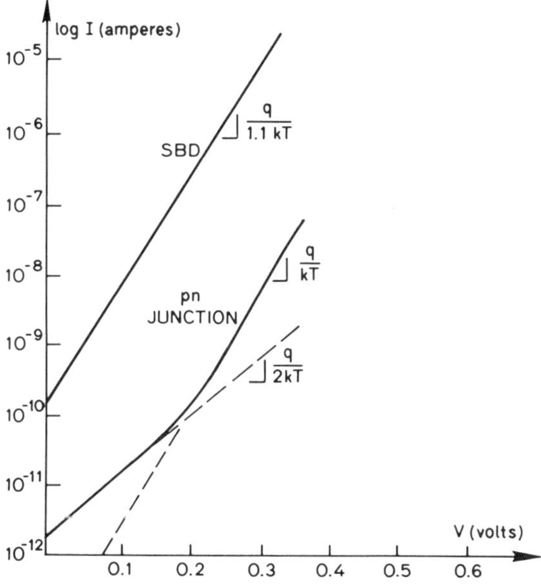

Figure 3.10 Log current versus voltage plots for SBD and diffused *pn* junction device.

nisms involved for Eqs. (3.23a) and (3.23b) are quite different. If one considers the second term in Eq. (3.22) to represent the neutral-region minority-carrier recombination current, then it is clear that this term can be neglected for normal Schottky devices. At least based on our first-order approximations this is true, and the assumption will be consistent with subsequent discussions of charge storage.

The basic transport equation associated with the second term in Eq. (3.22) is diffusion and for the *pn* junction is given by

$$I_{S2} = qA_{pn}\left[\frac{D_p p_{n0}}{W_{p\text{eff}}} + \frac{D_n n_{p0}}{W_{n\text{eff}}}\right] \tag{3.24}$$

where p_{n0} is the minority hole concentration on the n side (conversely for n_{p0}) and the designation "effective" for the lengths W indicate appropriate diffusion distances for the injected minority carriers before they recombine. Clearly the well-known dependence of heavily doped asymmetric junctions where $p_{n0} \gg n_{p0}$ or conversely where $p_{n0} \ll n_{p0}$ can easily be distinguished using Eq. (3.24).

The overall importance of Eqs. (3.22)–(3.24) can be summarized as follows. The observed characteristics of both *pn* and Schottky diodes depicted in Fig. 3.10 can be represented by Eq. (3.22). For the Schottky device the first term with coefficient I_{S1} represents adequately all relevant physics for digital-circuit applications. The barrier-controlled transport [Eq. (3.23a)] for majority carriers exhibits substantially greater current flow characteristics for a given bias than for the *pn* junction device. On the other hand, the *pn* junction device requires both terms expressed in Eq. (3.22). Minority-charge transport and recombination effects dominate, and, hence, the spatial dependencies of W_{sc} and W_{eff} control these recombination processes [Eqs. (3.23b) and (3.24)]. The superposition of the effects given in Eq. (3.22) and shown in Fig. 3.10 are typical of *pn* junction devices. However, for many applications one can ignore the composite effect and assume a single characteristic determined by the appropriate coefficient in Eq. (3.22) and one asymptotic form shown in Fig. 3.10.

3.3.2 Charge Storage

Having established the basic transport relationship given by Eq. (3.22) as well as identified the relevance of majority-carrier versus minority-carrier transport phenomena, the details of charge storage can be discussed in a straightforward way. Drawing upon the material discussed in Section 3.2.1 the following charge-balance equation can be used:

$$\Delta q_{\text{total}} = \Delta q_I + \Delta q_{sc} \tag{3.25}$$

where Δq_{total} is the composite incremental charge-storage change with bias and the two right-hand terms represent the injected minority-charge and space-charge contributions, respectively. As discussed above, for the Schottky device the injected term is negligible, and all barrier-associated space-charge effects are modeled by the second term. Drawing upon the previous section, Eq. (3.6c) as well as the approximations given by Eqs. (3.7)–(3.9) can be used freely. The space-charge effects for the *pn* junction can as well be modeled by these same equations.

For the *pn* junction the additional charge storage due to minority-carrier injection must be accounted for. It is well known that for the same device area ($A_{pn} = A_{\mathrm{SBD}}$) the *pn* junction device is slower owing to this additional charge storage. Again, drawing upon the previous modeling results for the *npn* device, Eq. (3.10) can be used directly. Essentially, given the current as determined by Eq. (3.22), there is an appropriate time constant τ_I that describes the decay of excess injected minority charge in the device. Since many IC *pn* junction diodes are fabricated by special interconnecting metallization of *npn* transistors, the τ_I becomes essentially either the τ_F or τ_R terms described in Section 3.2.2. Figure 3.11 shows two commonly used diode-connected transistors along with their associated dc and charge-storage parameters. The capacitor C_s for both devices represents the parasitic charge storage that is associated with the *n* well to substrate for junction-isolated devices. As shown in Figs. 3.1*d* and 3.1*e* this substrate junction is more properly modeled as a diode.

In fact, for the structure shown in Fig. 3.11*b* the possibility exists that hole injection from the *p* region into the *n* well can be collected by the substrate and form a parasitic transistor as shown in Fig. 3.12*a*. This *pnp* device has a saturation current given by Eq. (3.1) where the integral is taken over the *n* region $[N_D(x)]$. Since the first objective is to minimize this parasitic device, one can reduce I_S by increasing the integral $N_D(x)$ as shown in Fig. 3.12*b*. Here the added n^+ buried layer effectively reduces I_S to zero. But there is an added expense in that the C_s capacitance becomes larger. As shown in Fig. 3.12*b* the

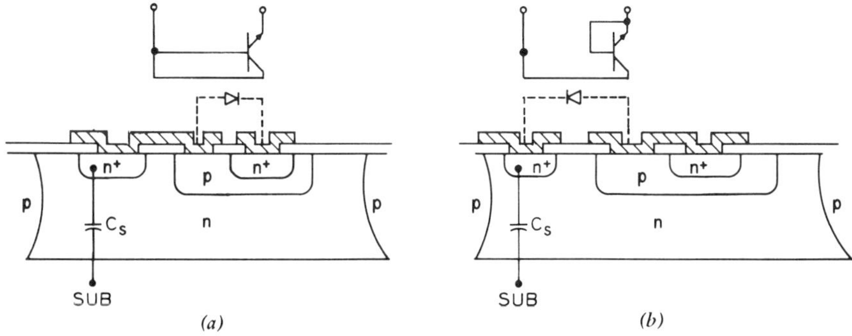

Figure 3.11 Section views of diode-connected devices (*a*) using the surface n^+p junction with relatively smaller effective active area and (*b*) using the epitaxial *np* region with larger junction area. Also shown are general trends in dc and ac parameter scaling.

substrate diode is present; however, because of the typical reverse bias and heavy n^+ doping, the contribution to charge transport is only a leakage effect, which might best be modeled as a small-value current source. Hence, for suitable isolation and use of buried layers the substrate diode and potential *pnp* device can be reduced to simply the C_s term shown originally in Figs. 3.11*a* and 3.11*b*.

3.3.3 Diode Modeling for Computer-Aided Design (CAD)

The two previous sections have presented both dc and charge-storage considerations for Schottky as well as *pn* junction devices. In addition the rudimentary considerations of parasitic junction effects have been outlined for one particular *pn* junction. The objective here is to formalize the notation into that used for computer-aided circuit analysis (CAD) programs such as SPICE [6].

The current approach in CAD device modeling is to use two types of specifications for each device type:

1. Connectivity and topology data.
2. Electrical model parameters.

Circuit simulators need to have uniquely numbered nodes throughout the circuit in order to form the admittance matrix used to solve for all node voltages, both dc and in the time domain. Figure 3.13*a* shows the polarity notation for a diode and Fig. 3.13*b* shows a typical CAD input for connectivity, where DNAME is an identifier and the three other parameters are node numbers and an area factor used to correctly scale the data given in model specification. The built-in device models for CAD require only that the user specify the coefficients for the equations such as given by Eqs. (3.6), (3.7), and (3.22). The equations are already present within the program and require only that values be given—if no specifications are made, then the program assumes its built-in defaults. The AREA term discussed above is used to scale the

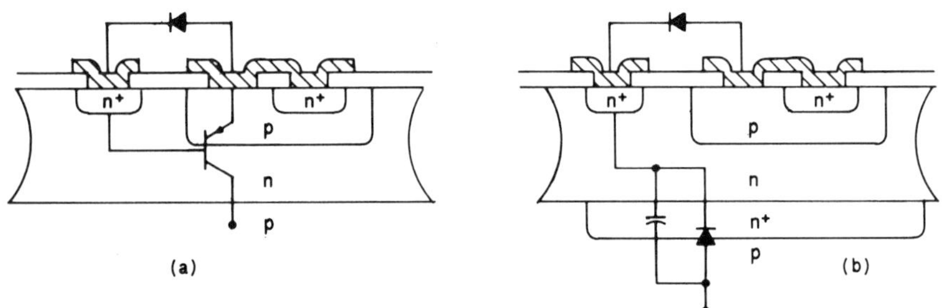

Figure 3.12 Section view of diode-connected device and parasitic structures: (*a*) the substrate vertical *pnp*; (*b*) an added n^+ layer to suppress *pnp* action leaving only the substrate diode.

(a) V CαAREA IαAREA N⁺ N⁻

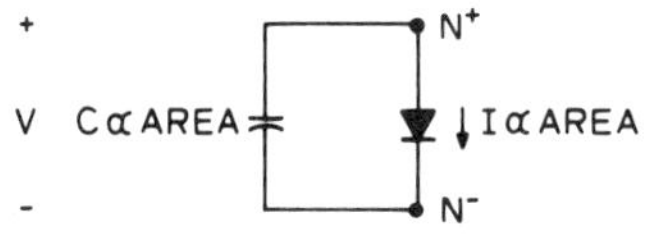

(b) DN N^+ N^- DNAME AREA

(c) .MODEL DNAME IS = 1E − 14 N = 1.3
+CJ0 = .1PF PHI = .7
+TT = .1NS

Figure 3.13 The CAD diode model and specifications: (*a*) terminal connections; (*b*) CAD specification including model reference; (*c*) details of model specification.

values given in the model specification. For example:

$$I_S = \text{AREA} \cdot I_S\,(\text{model})$$

$$C_{j0} = \text{AREA} \cdot C_{j0}\,(\text{model})$$

Figure 3.13*c* shows a typical input set for the diode model; the DNAME specification gives reference to the diode which is being modeled. The definition of mnemonics given in Fig. 3.13 relative to the equations above are listed in Table 3.1. For example, IS will be either I_{S1} or I_{S2}. As is typical of CAD,

Table 3.1 Diode Model

AREA	Name	PARAMETER	DEFAULT	EXAMPLE
*	IS	Saturation current	1.0E-14	1.0E-14
*	RS	Ohmic resistance	0	10
	N	Emission coefficient	1	1.0
	TT	Transit time	0	0.1NS
*	CJ0	Zero-bias junction capacitance	0	2PF
	PHI	Junction potential	1	0.6
	m	Grading coefficient	0.5	0.5
	EG	Activation energy	1.11	1.11 Si
				0.69 SBD
				0.67 Ge

The DC characteristics of the diode are determined by the parameters IS and N. An ohmic resistance, RS, is included. Charge-storage effects are modeled by a transit time, TT, and a nonlinear depletion layer capacitance, which is determined by the parameters CJ0, PB, and M. The temperature dependence of the saturation current is defined by the parameters the energy gap, EG, and the saturation current temperature exponent, PT. Reverse breakdown is modeled by an exponential increase in the reverse diode current and is determined by the parameters BV and IBV (both of which are positive numbers). Area-dependent parameters are indicated by (*).

you may not have *all* the equations you wish built-in. For example, this program example includes only *one* of the two exponentials listed in Eq. (3.22). Hence, the user chooses which term dominates by the choice IS and N. It is also apparent that there is a discrepancy of equation notation and CAD specifications. Each program has its idiosyncrasies and the user must determine what exactly is meant by the parameters. For this example, the notation should be relatively transparent—$M \equiv m$, $PB \equiv \phi$, and $TT = \tau_I$ in Eqs. (3.6) and (3.7). One can hope to obtain documentation from the CAD users guide as to exact definitions. In order to pull together all of the above discussion, three device examples and their respective CAD inputs are given next.

3.3.4 Examples

Figure 3.14 shows the topology and section view of three IC diodes—labeled BE, B, and SBD, respectively. All devices have an equal area metal contact so that $A_{BE} = A_B = A_{SBD}$. Shown in Fig. 3.14*b* are the model data for the three devices. A number of things are noteworthy:

1. IS (SBD) > IS (BE) > IS (B), which follows from two facts:

   (a) Figure 3.11.

   (b) The W_{peff}(BE) < $W_{neff}(B)$ in Eq. (3.24).

2. CJ0 (BE) > CJ0 (B) ≈ CJ0 (SBD), which is a result of two facts:

   (a) The n^- region controls the space charge for both the SBD and B devices

   (b) The n^+p space charge of the BE device gives a higher capacitance, ϕ, and m.

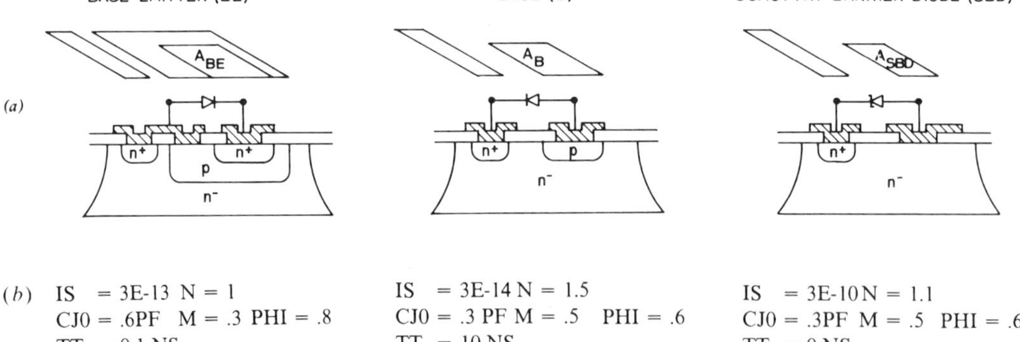

Figure 3.14 Several diode examples including technology view and CAD input: (*a*) the section and topographic views of BE, B, and SBD diodes; (*b*) the representative CAD model specifications for each device.

3. TT (B) > TT (BE) > TT (SBD) = 0, which results again from several factors:

   (a) Charge storage in the n^- epi is large for the B device.

   (b) The charge storage in the p region of the BE diode is much smaller.

   (c) *No* minority charge storage is assumed for the SBD.

The values given in Figure 3.14*b* are typical of diodes with an area of 30 μm $\times$ 30 μm. In the next section similar examples for bipolar CAD models are discussed.

3.4 BIPOLAR MODELING FOR CAD

At this point we have developed a sufficient background so that the model parameters used for bipolar transistor CAD models can be discussed. Figure 3.15 shows the complete bipolar model as implemented in SPICE [6]. The basic topology of the model shown in Fig. 3.5 and 3.15 are the same. The difference, as will be shown, stems from the equations used to describe the dc terminal currents. The equation set, which is shown along with Fig. 3.5, represents the first-order Ebers–Moll [8] relationships. The more-complete equations, which will now be described with reference to Fig. 3.15, are referred to as the Gummel–Poon [9] bipolar transistor model. Table 3.2 lists the set of parameters that completely specify the model along with an indication of which parameters scale with area (*). In addition, the subset of parameters that are unique to the Gummel–Poon model are indicated (#), while the simpler set of Ebers–Moll equations can be obtained by omitting these parameters. The

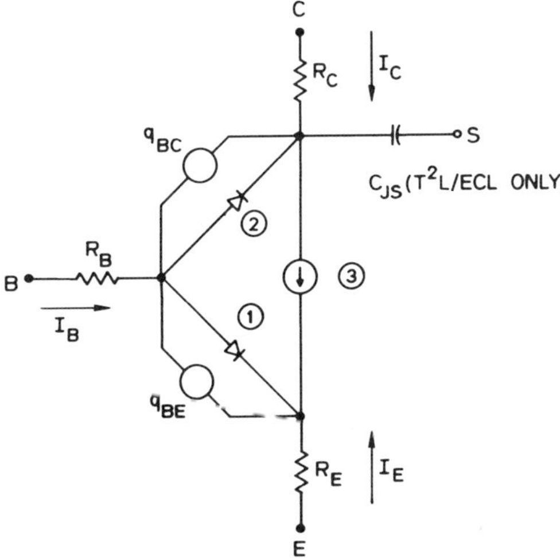

Figure 3.15 The equivalent circuit bipolar device model used for both Ebers–Moll and Gummel–Poon representations. Both dc and charge-storage effects are included as well as the parasitic substrate junction capacitance term.

Ebers–Moll and Gummel–Poon equations will now be discussed using the notation given in Table 3.2. This notation is fairly standard so that for CAD applications it is appropriate. The translation between the technology oriented notation with subscripts u and d will be given explicitly as needed.

3.4.1 The Ebers–Moll Parameter Set

The basic equations for I_C, I_B, and I_E given in Fig. 3.15 are:

$$I_C = \overbrace{I_S\left[\exp\left(\frac{qV_{BE}}{kT}\right) - \exp\left(\frac{qV_{BC}}{kT}\right)\right]\left(1 - \frac{V_{BC}}{V_A}\right)}^{③}$$

$$- \frac{I_S}{\beta_{rm}}\left[\exp\left(\frac{qV_{BC}}{kT}\right) - 1\right] \tag{3.26}$$

$$I_B = \underbrace{\frac{I_S}{\beta_{fm}}\left[\exp\left(\frac{qV_{BE}}{kT}\right) - 1\right]}_{①} + \underbrace{\frac{I_S}{\beta_{rm}}\left[\exp\left(\frac{qV_{BC}}{kT}\right) - 1\right]}_{②} \tag{3.27}$$

where q is the electronic charge, k is Boltzmann's constant, and T is absolute temperature. The four variables, I_S, β_{fm}, β_{rm}, and V_A, are user-specified parameters. The transport current, I_S, is the extrapolated intercept current of $\log(I_C)$ vs V_{BE} in the forward region and $\log(I_E)$ vs V_{BC} in the reverse region as shown in Fig. 3.16. The parameters β_{fm} and β_{rm} are the *maximum* forward and reverse short-circuit current gains, respectively, which are assumed not to vary with operation point. The additional parameter V_A, referred to as the "early voltage," produces a finite value of output conductance, g_0, due to basewidth modulation. The output conductance is given by the equation:

$$g_0 = \frac{\partial I_C}{\partial V_{CE}} = \frac{I_S}{V_A}\left[\exp\left(\frac{qV_{BE}}{kT}\right) - \exp\left(\frac{qV_{BC}}{kT}\right)\right]$$

$$+ \frac{qI_S}{kT}\exp\left(\frac{qV_{BC}}{kT}\right)\left(1 - \frac{V_{BC}}{V_A}\right) \approx \frac{I_C}{V_A} \tag{3.28}$$

Hence, output conductance is proportional to I_C, with the constant of proportionality being $(1/V_A)$. A graphic interpretation of V_A is shown in Fig. 3.17. For most digital applications V_A can be neglected (defaulted). The notation of BE and BC as well as f (forward = BE) and r reverse (reverse = BC) subscripts now refer explicitly to operating conditions. For many ICs where both I^2L and ECL or T^2L devices appear on the same chip, the specific meaning of emitter and collector must be defined. To illustrate, consider the

Table 3.2 Bipolar Model

Name		Parameter	Default
BFM		Sat current/ideal BE sat current	100
BRM		Sat current/ideal BC sat current	1
NE	(#)	BE nonideal emission coefficient	2
NC	(#)	BC nonideal emission coefficient	2
IK	(#) (*)	Forward knee current	Infinite
IKR	(#) (*)	Reverse knee current	Infinite
VA		Forward early voltage	Infinite
VB	(#)	Reverse early voltage	Infinite
PE		BE junction built-in potential	1
PC		BC junction built-in potential	1
PS	(#)	CS junction built-in potential	1
RC	(*)	Collector ohmic resistance	0
RB	(*)	Base ohmic resistance	0
RE	(*)	Emitter ohmic resistance	0
IS	(*)	Saturation current	1.0E − 14
C2	(#)	Nonideal BE sat current/sat current	0
C4	(#)	Nonideal BC sat current/ sat current	0
TF		Forward transit time	0
TR		Reverse transit time	0
CJE	(*)	Zero-bias BE junction capacitance	0
ME	(#)	BE grading coefficient	0.5
CJC	(*)	Zero-bias BC junction capacitance	0
MC	(#)	BC grading coefficient	0.5
CSS	(*)	Zero-bias CS junction capacitance	0
MS	(#)	CS grading coefficient	0
EG		Energy gap	1.1

Note: Parameters indicated by (*) are dependent on area of the device. Parameters indicated by (#) specify the Gummel–Poon model. If default values are used for these parameters, the model reduces to the Ebers–Moll model.

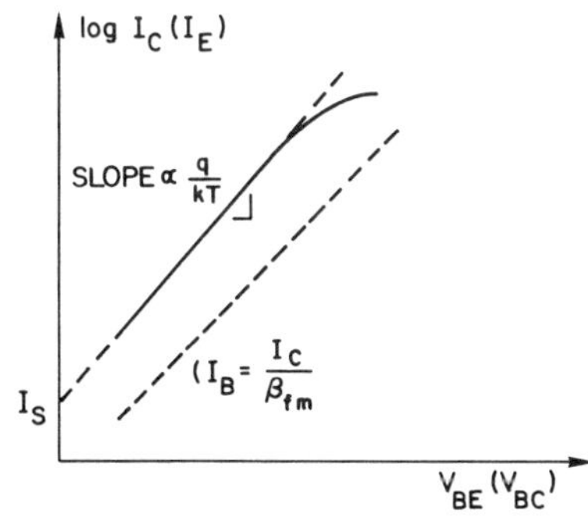

Figure 3.16 Typical log current versus voltage plots to show meaning of I_S and β.

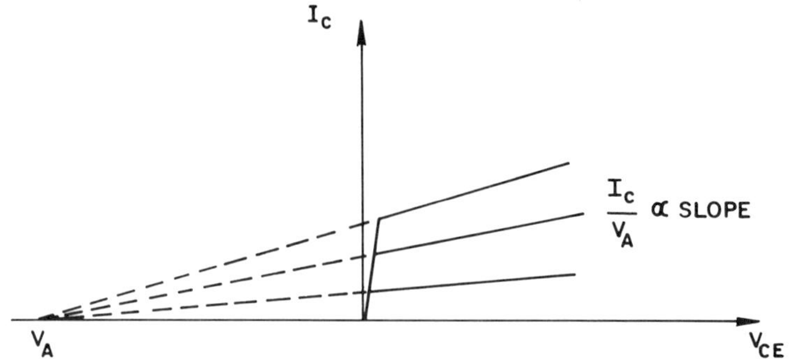

Figure 3.17 The curve-tracer-type plot of I_C vs V_{CE} to show the Early voltage V_A.

case where the same technology fabrication steps are used to create both I^{2}L- and ECL-type devices on the same chip. Figure 3.18 shows the cross section of the devices with BE and BC as indicated. Comparing these structures our basic technology-oriented model shown in Fig. 3.5 finds that:

$$
\begin{array}{ll}
\text{I}^2\text{L} & \text{ECL} \\
V_{BE} = V_u & V_{BE} = V_d \\
V_{BC} = V_d & V_{BC} = V_u \\
\beta_{fm} = \beta_u & \beta_{fm} = \beta_d \\
\beta_{rm} = \beta_d & \beta_{rm} = \beta_u
\end{array}
\tag{3.29}
$$

The subscripts become reversed because the role of the junctions for injection/collection becomes reversed. In particular, note the terminal notation for the quantity V_d. The key point is that the basic device structure (u and d

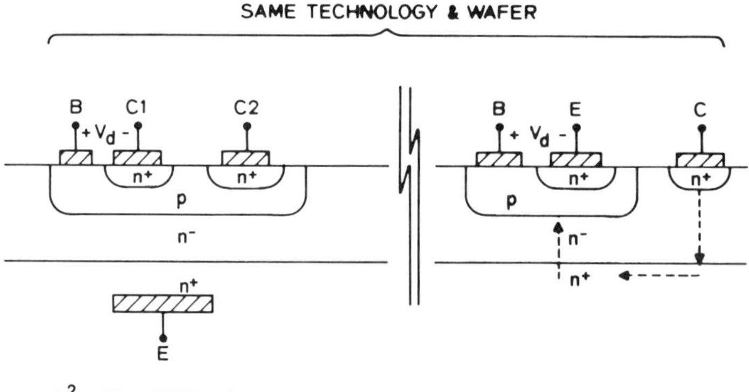

Figure 3.18 Section view of a bipolar technology where both I^{2}L and ECL devices occur on the same chip. Note that the n^+p junction represents a different function on the two "sides" of the chip.

subscripts) has not changed; the circuit use of the devices dictates which junctions should be called the emitter and which the collector.

Now consider the nonlinear charge-storage elements q_{BE} and q_{BC} shown in Fig. 3.15, which are determined by the equations:

$$\begin{aligned} q_{BE} &= q_F + q_{VE} \\ &= \tau_F I_S \left[\exp\left(\frac{qV_{BE}}{kT}\right) - 1\right] + C_{je}\int_0^{V_{BE}} \left[1 - \frac{V}{\phi_e}\right]^{-m_e} dV \end{aligned} \tag{3.30}$$

$$\begin{aligned} q_{BC} &= q_R + q_{VC} \\ &= \tau_R I_S \left[\exp\left(\frac{qV_{BC}}{kT}\right) - 1\right] + C_{jc}\int_0^{V_{BC}} \left[1 - \frac{V}{\phi_c}\right]^{-m_c} dV \end{aligned} \tag{3.31}$$

where the q_F and q_R terms represent injected minority-carrier charge and q_{VE} and q_{VC} represent the space-charge terms. Expressing these charge-storage elements as voltage-dependent small-signal capacitances:

$$\frac{\partial q_{Be}}{\partial V_{BE}} = \underbrace{\tau_F \left(\frac{qI_S}{kT}\right) \exp\left(\frac{qV_{BE}}{kT}\right)}_{\tau_F gm_f} + \underbrace{C_{je}\left(1 - \frac{V_{BE}}{\phi_e}\right)^{-m_e}}_{C_{JE}(V)} \tag{3.32}$$

where gm_f is the forward transconductance given by

$$gm_f = \left.\frac{\partial I_C}{\partial V_{BE}}\right|_{V_{BC}=0} \tag{3.33}$$

$$\frac{\partial q_{BC}}{\partial V_{BC}} = \underbrace{\tau_R \left(\frac{qI_S}{kT}\right) \exp\left(\frac{qV_{BC}}{kT}\right)}_{\tau_R gm_r} + \underbrace{C_{jc}\left(1 - \frac{V_{BC}}{\phi_c}\right)^{-m_c}}_{C_{JC}(V)} \tag{3.34}$$

where gm_r is the reverse transconductance given by

$$gm_r = \left.\frac{\partial I_E}{\partial V_{BC}}\right|_{V_{BE}=0} \tag{3.35}$$

Similar to the case for the dc equations, the relationship of the forward (F) and reverse (R) notation to the technology-oriented form for I^2L and ECL is given below:

$$\begin{array}{cc} I^2L & \text{ECL} \\ \tau_F = \tau_u & \tau_F = \tau_d \\ \tau_R = \tau_d & \tau_R = \tau_u \end{array} \tag{3.36}$$

To summarize the results for time-domain effects, charge storage is modeled by two base-storage terms, which are characterized by the transit times τ_F and

τ_R, and two depletion terms, which are characterized by the parameters C_{je}, ϕ_e, and m_e for the emitter-depletion region and C_{jc}, ϕ_c, and m_c for the collector-depletion region. For the simple Ebers–Moll model, the parameters m_e and m_c are fixed at a value of 0.5. To ensure a finite value for the depletion-layer capacitance with $V > \phi$, a linear approximation is used in SPICE [6] for depletion capacitance in forward bias.

In addition to the intrinsic elements discussed above, the extrinsic elements shown in Fig. 3.15 must be included. In particular R_E, R_B, R_C and C_{js} (for T^2L and ECL devices only) are all constant-valued parameters, which represent passive components to terminal nodes. In the case of C_{js}, the substrate node is assumed to be the most negative supply in the circuit.

The Ebers–Moll transistor model lacks a representation of many of the important second-order effects present in actual devices; the two most important effects are low-current effects and high-level injection. The low-current effects result from additional base current due to recombination which degrades current gain. This effect can be very important for low-current I^2L circuits. The high-level injection effects also reduce current gain and, in addition, cause an increase in τ_F and τ_R. These high-level effects can be important for T^2L and I^2L circuits that operate at high currents and are controlled by minority-charge-storage effects. To properly account for these second-order effects the Gummel–Poon model is described next.

3.4.2 The Gummel–Poon Parameter Set

The Gummel–Poon model [9] makes several important modifications to the dc equations given by Eqs. (3.26) and (3.27). The modified equations are as follows:

$$I_C = \underbrace{\frac{I_S}{Q_B}\left[\exp\left(\frac{qV_{BE}}{kT}\right) - \exp\left(\frac{qV_{BC}}{kT}\right)\right]}_{③} - \frac{I_S}{\beta_{rm}}\left[\exp\left(\frac{qV_{BC}}{kT}\right) - 1\right]$$

$$- C_4 I_S\left[\exp\left(\frac{qV_{BC}}{n_c kT}\right) - 1\right] \tag{3.37}$$

$$I_B = \underbrace{\frac{I_S}{\beta_{fm}}\left[\exp\left(\frac{qV_{BE}}{kT}\right) - 1\right] + C_2 I_S\left[\exp\left(\frac{qV_{BE}}{n_e kT}\right) - 1\right]}_{①}$$

$$\underbrace{+ \frac{I_S}{\beta_{rm}}\left[\exp\left(\frac{qV_{BC}}{kT}\right) - 1\right] + C_4 I_S\left[\exp\left(\frac{qV_{BC}}{n_c kT}\right) - 1\right]}_{②} \tag{3.38}$$

where the normalized base charge, Q_B, models basewidth modulation (BWM), and high-level injection (HLI) is defined by the equation:

$$Q_B = \tfrac{1}{2}\left[Q_1 + \sqrt{Q_1^2 + 4Q_2}\right] \quad \text{where} \tag{3.39}$$

$$Q_1 = 1 + \frac{V_{BC}}{V_A} + \frac{V_{BE}}{V_B} \tag{3.40}$$

$$Q_2 = \frac{I_S}{I_k}\left[\exp\left(\frac{qV_{BE}}{kT}\right) - 1\right] + \frac{I_S}{I_{kr}}\left[\exp\left(\frac{qV_{BC}}{kT}\right) - 1\right] \tag{3.41}$$

The model parameters for the dc portion of the Gummel–Poon model are I_S, β_{fm}, β_{rm}, C_2, C_4, n_e, n_c, I_k, I_{kr}, V_A, and V_B. The charge-storage elements q_{BE} and q_{BC} are identical to the Ebers–Moll model.

The following effects are included in the Gummel–Poon model: *Depletion layer recombination* is included in the model with the two nonideal base-current components determined by the parameters C_2 and n_e for the *emitter*-depletion region and C_4 and n_c for the *collector* region. This is shown in the graph of I_B against V_{BE} shown in Fig. 3.19.

Basewidth modulation and *high-level injection* are introduced into the model via the base charge term, Q_B. The Q_B term, while it is a dimensionless quantity, can be interpreted as the generalized form of denominator in Eq. (3.1). That is, the integrated *base charge* must in fact include effects of both dopant (N_A) and injected charges (I_k) as well as space-charge depletion effects (V_A). This is best understood by considering two limiting cases. First, consider the case when $Q_2 \approx 0$. This is the low-level-injection condition in the forward region of operation and:

$$I_C \simeq \frac{I_S}{1 + V_{BC}/V_A + V_{BE}/V_B}\exp\left(\frac{qV_{BE}}{kT}\right) \approx I_S\exp\left(\frac{qV_{BE}}{kT}\right)\left[1 - \frac{V_{BC}}{V_A}\right] \tag{3.42}$$

This reduces to the Ebers–Moll equation including the basewidth modulation due to the space-charge effects. A similar argument holds for the reverse Early voltage, V_B, in the reverse region.

Now consider the case when $Q_1 \approx 1$ to eliminate effects of basewidth modulation. In the forward region, with low-level injection,

$$Q_2 = \frac{I_S}{I_k}\exp\left(\frac{qV_{BE}}{kT}\right) \ll 1, \tag{3.43}$$

and therefore $Q_B \approx Q_1 = 1$. In this case the collector current follows the "ideal" law

$$I_C \simeq I_S\exp\left(\frac{qV_{BE}}{kT}\right) \tag{3.44}$$

where the exponent gives a slope of q/kT.

For high levels of injection,

$$Q_2 = \frac{I_S}{I_k} \exp\left(\frac{qV_{BE}}{kT}\right) \gg 1, \tag{3.45}$$

and therefore $Q_B \approx \sqrt{Q_2}$. In this case the collector current becomes

$$I_C \simeq \sqrt{I_k I_S} \exp\left(\frac{qV_{BE}}{2kT}\right) \approx I_k Q_B \tag{3.46}$$

which is described as "nonideal" because of the $2KT$ term in the exponent. Hence, the emission coefficient changes from 1 (Ebers–Moll) to 2 as shown in Fig. 3.19. A similar argument holds for the reverse region, where high-level injection occurs when

$$\frac{I_S}{I_{kr}} \exp\left(\frac{qV_{BC}}{kT}\right) \gg 1 \tag{3.47}$$

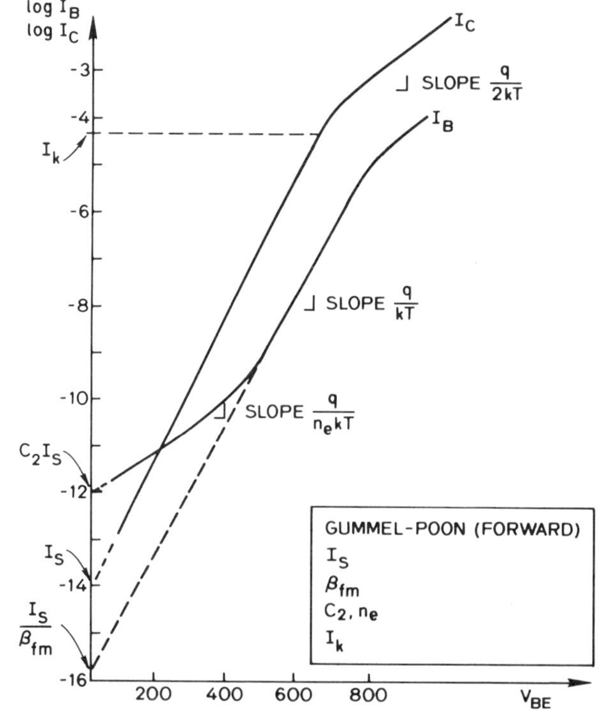

Figure 3.19 A detailed log current versus voltage plot showing the meaning of the dc forward active mode parameters in the Gummel–Poon model.

The parameters I_k and I_{kr} are termed the forward knee current and the reverse knee current, and can be interpreted as the approximate value of collector current (for forward region) and emitter current (for reverse region) where high-level injection becomes significant (see Fig. 3.19).

Additional understanding of the Gummel–Poon parameters is obtained by inspection of the *asymptotic* behavior of the short-circuit current gain as shown in Fig. 3.20*a*. The current gain is essentially constant at a value of β_{fm} for collector currents greater than I_L, falls off with a slope of $1 - 1/n_e$ for collector currents less than I_L, and falls off with a slope of -1 for collector currents greater than I_k; I_L, the low-current breakpoint, is given by the equation

$$I_L = I_S\left[C_2\beta_{fm}\right]^{n_e/(n_e-1)} \tag{3.48}$$

This, of course, is only an asymptotic relation. For cases where low-level and high-level effects overlap, as is true in most transistors, some trial and error is required to obtain the correct parameters.

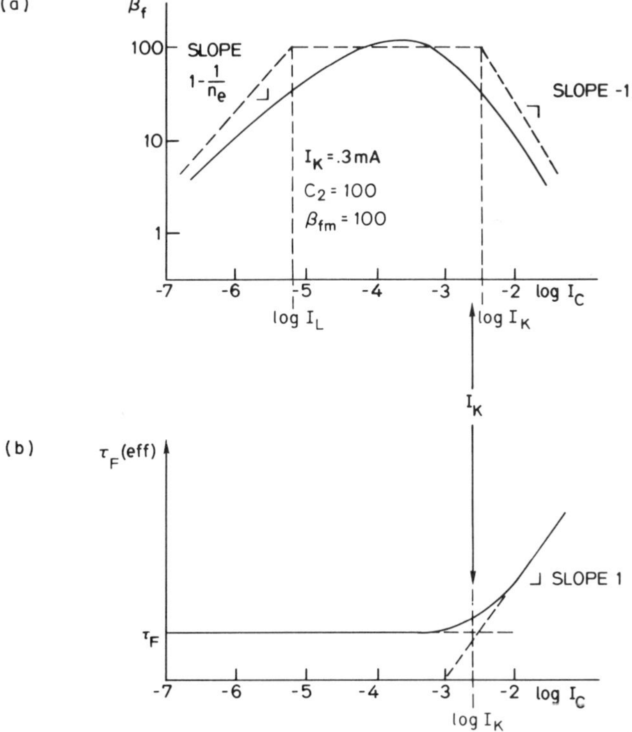

Figure 3.20 (*a*) Log β vs log I_C plot for the Gummel–Poon model to show high- and low-current effects on current gain. (*b*) τ_F versus log I_C to show high-current effects on charge storage.

The effective transit times of the device, and, hence, the f_T, are also dependent on collector current. The effective forward transit time is given by

$$\tau_F(\text{eff}) = \frac{q_{BE}}{I_C} \cong \tau_F Q_B \tag{3.49}$$

Again the normalized charge Q_B enters into charge-storage relationships. Figure 3.20*b* shows a plot of $\tau_F(\text{eff})$ vs log I_C. Above the knee current $\tau_F(\text{eff})$ increases substantially. The results shown in both Figs. 3.20*a* and 3.20*b* represent model effects, which are important as devices are pushed to their high- and low-current performance limits.

The last two sections have defined the Ebers–Moll and Gummel–Poon device equations in the context of notation appropriate for CAD. The following section will present examples of actual parameter calculations for different device types and their use in specifying a circuit topology for CAD.

3.4.3 Examples

Having defined the parameter sets for the Ebers–Moll and Gummel–Poon models, it is useful to show a few examples in the context of current IC technologies. The following examples use I^2L, ECL, and T^2L devices to illustrate typical parameter sets and CAD input. Just as for the diode CAD model examples, separate topology and model parameter input specifications are used. The basic input format as shown in Fig. 3.13 is used for transistors with the exceptions that *three* terminals are required and the parameter names are those shown in Table 3.2.

The first example draws on the I^2L/ECL structure shown in Fig. 3.18. Table 3.3 gives typical values for n^+p and pn^- space-charge capacitance per unit area. To obtain CAD model parameters, the capacitance per unit area must be scaled by the device area, and node numbers must be associated correctly with their appropriate connections. Figure 3.21 shows a topographic view of both the I^2L and ECL devices with terminals and area factors as indicated. Using the correct area factors as determined from Fig. 3.21*a* the CAD inputs for junction capacitance are shown in Fig. 3.21*b*. The areas are multiplied times their corresponding per unit area values. For the I^2L device there are two sets of model inputs. It is useful and appropriate to note that for

Table 3.3 Space-Charge Capacitances

Vertical region	Capacitance/μm^2
n^+p	4×10^{-4} pf/μm^2
pn^-	1.5×10^{-4} pf/μm^2
n^- substrate	1×10^{-4} pf/μm^2

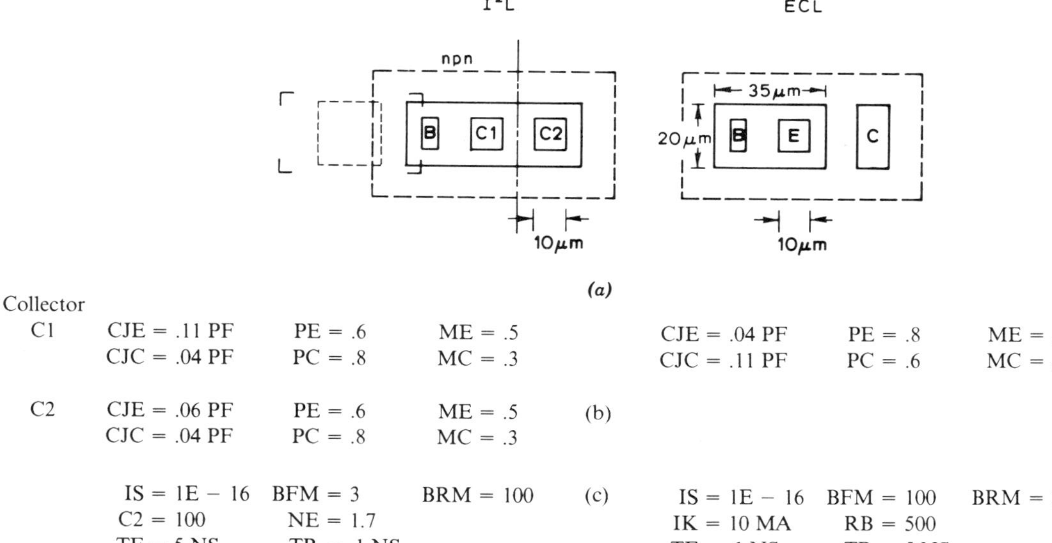

	I²L				ECL		
Collector							
C1	CJE = .11 PF	PE = .6	ME = .5		CJE = .04 PF	PE = .8	ME = .3
	CJC = .04 PF	PC = .8	MC = .3		CJC = .11 PF	PC = .6	MC = .5
C2	CJE = .06 PF	PE = .6	ME = .5	(b)			
	CJC = .04 PF	PC = .8	MC = .3				
	IS = 1E − 16	BFM = 3	BRM = 100	(c)	IS = 1E − 16	BFM = 100	BRM = 3
	C2 = 100	NE = 1.7			IK = 10 MA	RB = 500	
	TF = 5 NS	TR = .1 NS			TF = .1 NS	TR = 5 NS	

Figure 3.21 The topology and model parameter representations for the I^2L/ECL technology example given in Fig. 3.18. (*a*) Topology of the devices where the minimum square is 10 μm × 10 μm and the minimum spacing is 5 μm; for *pnp* see Fig. 3.24. (*b*) The listing of CAD model parameters for the junctions to show differences and correspondence between the devices. (*c*) The dc device model parameters.

the I^2L device labeled collector 1 and the ECL device that although the E and C subscripts are reversed, the junction capacitance values are identical. These values for capacitance ignore contributions of lateral diffusion and sidewall differences in unit area values. By including these effects typically 20% changes occur; however, the fundamental differences from results discussed here can be neglected. However, as geometries are scaled down this statement concerning errors will be blatantly false since sidewall areas dominate in small-geometry devices.

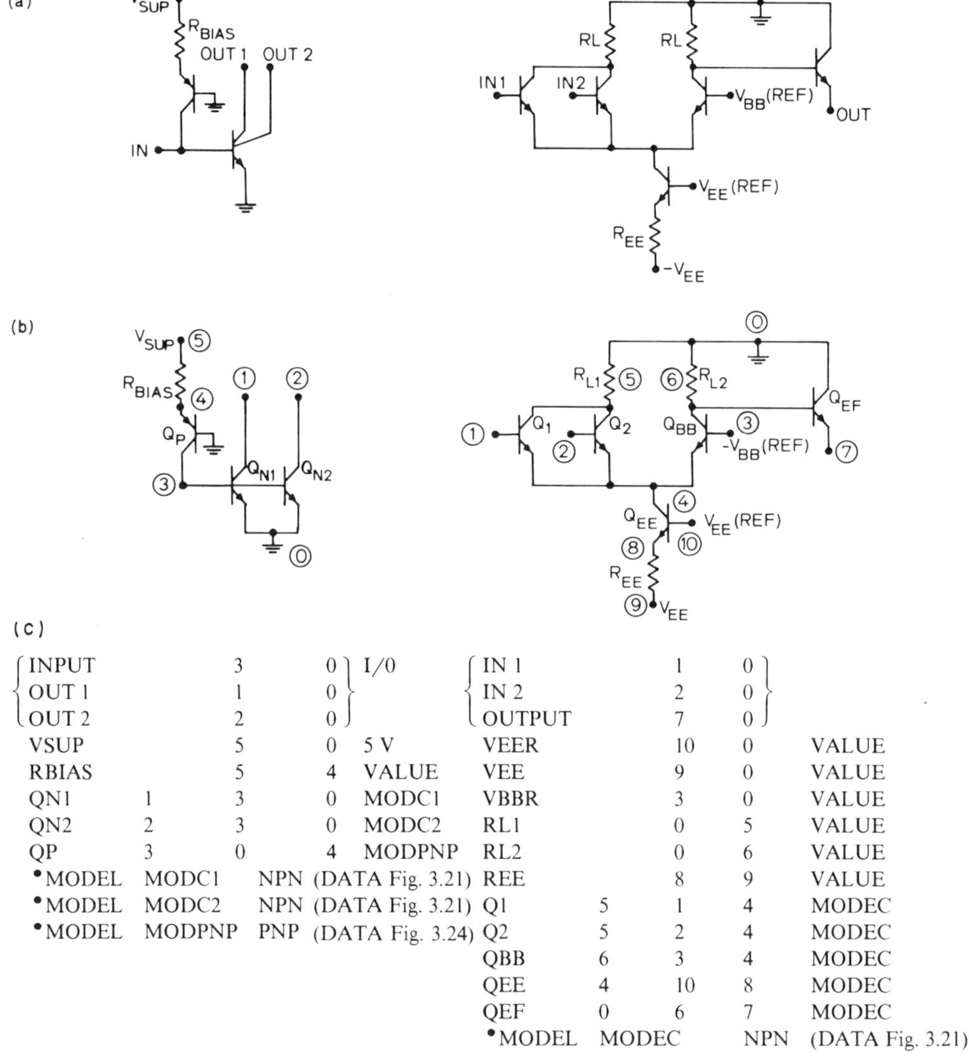

INPUT		3	0	I/0
OUT 1		1	0	
OUT 2		2	0	
VSUP		5	0	5 V
RBIAS		5	4	VALUE
QN1	1	3	0	MODC1
QN2	2	3	0	MODC2
QP	3	0	4	MODPNP
•MODEL	MODC1	NPN	(DATA Fig. 3.21)	
•MODEL	MODC2	NPN	(DATA Fig. 3.21)	
•MODEL	MODPNP	PNP	(DATA Fig. 3.24)	

IN 1		1	0	
IN 2		2	0	
OUTPUT		7	0	
VEER		10	0	VALUE
VEE		9	0	VALUE
VBBR		3	0	VALUE
RL1		0	5	VALUE
RL2		0	6	VALUE
REE		8	9	VALUE
Q1	5	1	4	MODEC
Q2	5	2	4	MODEC
QBB	6	3	4	MODEC
QEE	4	10	8	MODEC
QEF	0	6	7	MODEC
•MODEL	MODEC		NPN	(DATA Fig. 3.21)

Figure 3.22 Equivalent gate representations and CAD specifications for I^2L and ECL gates: (*a*) simple circuit specification, (*b*) carefully labeled CAD schematic, and (*c*) abbreviated CAD input.

Having determined the space-charge-capacitance terms, the other dc and ac model terms are now considered. As discussed in association with Eq. (3.1), the I_S values for both I^{2}L and ECL devices are identical since $A_{\min}$ is the same. Moreover, the forward and reverse parameters for β and charge storage simply reverse roles because of the nature of the terminal connections. Finally, since the I^{2}L is operated primarily at low current, the C_2 and N_E terms are specified whereas for the ECL devices the high-current terms associated with I_k and R_B must be used. The details involved in calculating R_B have been addressed in Chapter 2. For purposes of discussion here, R_B will be taken as an empirical parameter determined from device characterization. Figure 3.21*c* lists pertinent dc parameters for the devices.

To go one step further with this example, consider the two simple circuits shown in Fig. 3.22. These might represent building-block logic components in the combined I^{2}L/ECL technology. The figure sequence goes from the circuit schematic (Fig. 3.22*a*) to the detailed node numbering (Fig. 3.22*b*). The output nodes are connected to appropriate loads. Note that the I^{2}L gate requires two devices to model correctly the multiple collector. Finally, Fig. 3.22*c* gives the CAD input details excluding actual values and necessary I/0 specifications and driving sources. The details of the *pnp* transistor model will be described later.

The second example involves a Schottky T^{2}L device and circuit. Many of the device parameters are the same as for the ECL transistor shown in Fig. 3.18; however, the primary difference is the inclusion of the Schottky clamping diode. Figure 3.23 shows a topology and section view of the Schottky clamped transistor. The emitter is a 10 μm × 20 μm rectangle as is the Schottky clamp

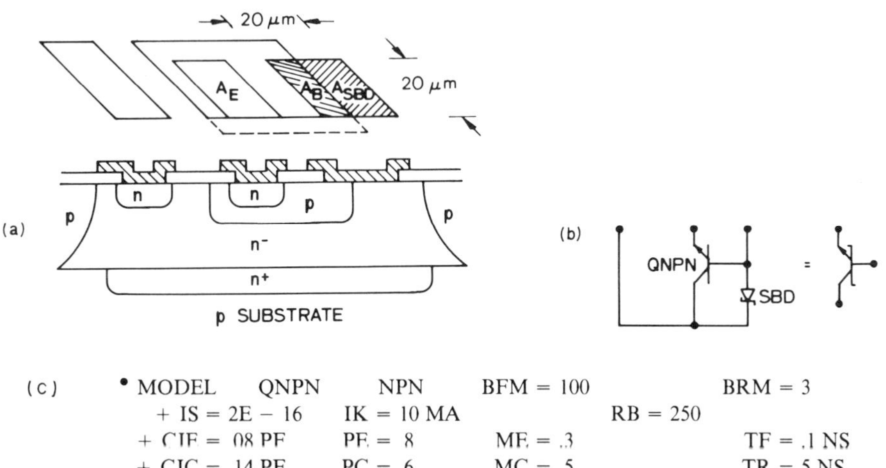

Figure 3.23 Section view, model, and CAD input for a Schottky device: (*a*) the section view with pertinent area factors, (*b*) the two-device equivalent of the clamped transistor, and (*c*) the CAD input.

diode; the base contact measures 5 μm × 20 μm. The nature of CAD modeling requires that two models be used, one for the diode and the other for the transistor. Figure 3.23*b* shows the model as used for CAD as well as its equivalent circuit representation. The device data from Figs. 3.14 and 3.21 can be appropriately scaled to generate the SBD-clamped transistor. Specifically, the emitter area is scaled from 10 μm × 10 μm to 10 μm × 20 μm and the base from 20 μm × 35 μm to 30 μm × 30 μm. The Schottky contact in Fig. 3.14 is 30 μm × 30 μm, while the transistor clamp diode is only 10 μm × 20 μm. The resulting scaled values are shown in Fig. 3.23*c*.

The final example in this section involves the *pnp* device shown in Figs. 3.1*a–c* and 3.22. Figure 3.24*a* shows the section and plan views of a lateral *pnp* transistor. Figure 3.24*b* shows schematically the model topology and the critical factors involving area dependences. As is shown in Fig. 3.24*b*, the two diodes are proportional to the surface areas A_E and A_C, whereas the transport current is proportional to the sidewall area (shaded) as well as the epitaxial thickness (W_{epi}). Using Eq. (3.1) for the *pnp* device:

$$I_{Sp} = \frac{qA_{\min} n_i^2}{\int_0^{W_B} (N_D/D_p)\, dx}$$

where $A_{\min} \leqslant L\, W_{\text{epi}}$. Note that $A_{\min}$ represents the lateral cross section for charge to pass between emitter and collector. The shaded region is a lower bound on this area and the $W_{\text{epi}} L$ is an *upper* bound. The dimensions shown in Fig. 3.24*a* are consistent with the device topology shown in Fig. 3.21. By

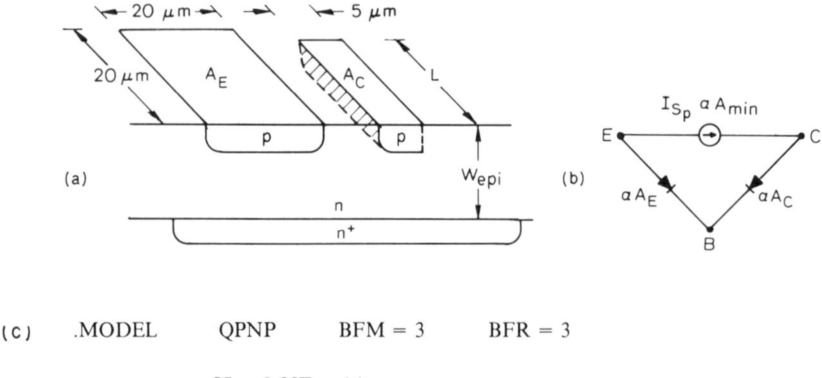

(c) .MODEL QPNP BFM = 3 BFR = 3

IS = 0.55E − 16

CJE = 0.06PF	PE = 0.6	ME = 0.5	TF = 5NS
CJC = 0.015PF	PC = 0.6	MC = 0.5	TR = 5NS

Figure 3.24 Model considerations for the lateral *pnp* device: (*a*) section and plan views (typical dimensions as per Fig. 3.21); (*b*) the model and area dependences are indicated and (*c*) the CAD input.

specifying values for N_D, W_{epi}, and W_B for the *pnp* as well as the counterpart values of N_A and W_B for the *npn* from Fig. 3.21, we can determine a comparative estimate of $I_S(npn)$ and $I_S(pnp)$. The following values result in the calculated transport currents shown:

pnp (Fig. 3.24)	*npn* (Fig. 3.21)
$\overline{N}_D = 4 \times 10^{15}\ \text{cm}^{-3}$	$\overline{N}_A = 10^{17}\ \text{cm}^{-3}$
$W_B = 10\ \mu\text{m}$	$W_B = 0.2\ \mu\text{m}$
$W_{epi} = 3\ \mu\text{m}$	$A_{min} = A_{C1} = 100\ \mu\text{m}^2$
$L = 20\ \mu\text{m}$	
$\overline{D}_p = 11\ \text{cm}^2/\text{s}$	$\overline{D}_n = 15\ \text{cm}^2/\text{s}$
$I_{Sp} \leqslant \dfrac{qn_i^2 L\, W_{epi}}{\int_0^{W_B} (\overline{N}_D/\overline{D}_p)\, dx}$	$I_{Sn} = \dfrac{qn_i^2 A_{min}}{\int_0^{W_B} (\overline{N}_A/D_n)\, dx}$
$I_{Sp} = 0.55 \times 10^{-16}\ \text{A}$	$I_{Sn} = 2.52 \times 10^{-16}\ \text{A}$

As a result the ratio of I_{Sp}/I_{Sn} for this particular example is

$$I_{Sp}/I_{Sn} = 0.22$$

It will be shown in Chapter 5 that a ratio of less than unity is generally desirable for proper operation of I^2L gates. In addition, further details concerning the design options to achieve this result will be discussed.

Continuing with the *pnp* device example, the other device parameters can be computed based on following observations:

$$C_{JC} \alpha A_C$$

$$C_{JE} \alpha A_E$$

$$\tau_F = \tau_R = \tau_u$$

$$\beta_F = \beta_R = \beta_u$$

The first two observations are straightforward. The second two assumptions follow from the symmetry of the device and the fact that to first order the junction for either the emitter of the I^2L *npn* device or the lateral *pnp* involve the same loss and charge-storage mechanisms. Using the above results and the data from Fig. 3.21 the complete model specification for the *pnp* device is shown in Fig. 3.24*c*. One key observation to be made is that the value of I_S is *scaled* relative to the *npn* device (Fig. 3.21) according to the ratio computed above.

3.5 BIPOLAR TRANSISTOR SWITCHING

Sections 3.3 and 3.4 have extended our facility with transistor and diode models to the point where the equations are translated into CAD input. At this point it is appropriate to return to the discussion of bipolar device switching with two objectives in mind:

1. Expand the first-order discussion of switching events introduced in Section 3.2.3. (In addition it is useful to present specific examples.)
2. Benchmark our first-order calculations using CAD simulation results.

To be emphasized throughout this section is the need for *both* CAD and first-order results. Digital device switching events can be very complex when multiple nonlinear devices are involved. Hence CAD is invaluable. On the other hand, first-order models are the *only* viable means to extract transfer-function and rule-of-thumb type approximations. Clearly, in communicating to others the circuit limitations and critical performance factors the first-order models convey information most effectively.

3.5.1 First-Order Switching Approximations

Figure 3.25*a* shows one of the basic building blocks of bipolar digital ICs—a string of inverters (connected in a series chain). Between successive nodes of the inverter chain the signal is of opposite polarity under static conditions. Given an odd number of gates as shown, the output of the last stage can be connected to the input and an oscillator is formed. That is, assuming some state at the input of $Q_1(I)$, the output at the collector of $Q_5(O)$ is the opposite, so that if this new state is inputted at the base of $Q_1(I)$, the change in state must propagate to the output. If this process continues, a ring oscillation is formed with waveforms as shown in Fig. 3.25*b*. Note that for a single period of the voltage at the output node a total of $2N$ coupled switching events have occurred, where N is the number of stages (five in this case). It is the objective of this section to examine in detail the phenomena and modeling that will allow us to determine the switching speed of transistors in circuits such as the ring oscillator of Fig. 3.25. In addition, it is essential that we are able to estimate switching delays for random logic circuits.

Figure 3.26*a* shows a more-detailed view of a portion of the inverter chain and Fig. 3.26*b* shows the first switching couple from Fig. 3.25*b*. The node voltages are subscripted to indicate the collector for each device—for example $V_1(I)$ corresponds to $Q_1(I)$ and V_2 corresponds to $Q_2(D)$. The parenthetic notation of O, I, D, and L correspond to *O*utput, *I*nput, *D*river, and *L*oad.

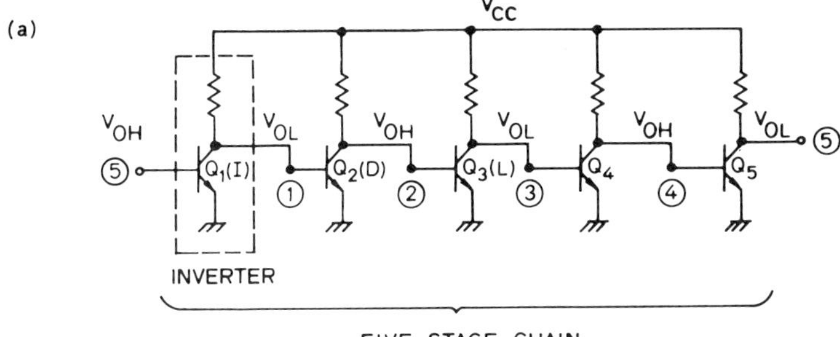

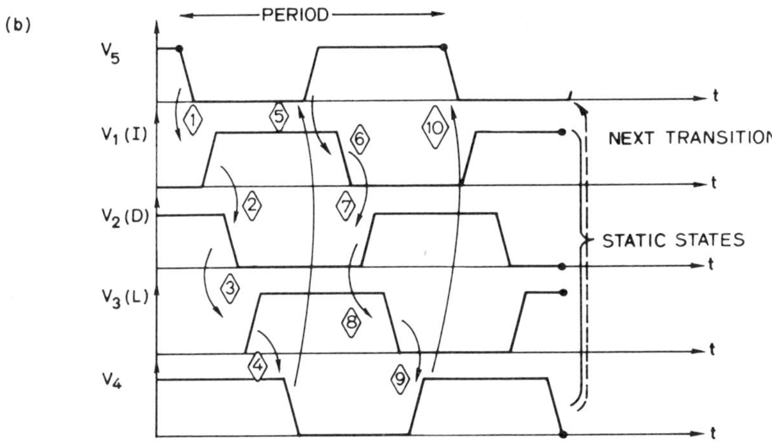

Figure 3.25 Five-stage ring oscillator and voltage waveforms: (a) The basic topology with nodes devices are labeled Q_1–Q_5 and one set of dc levels OL/OH is shown. (b) The time sequence of switching events as the waveform ripples through the circuit. It is assumed that the output is the base of device Q_1.

This notation is used later for analytic derivations in Chapter 5. The numbered nodes are consistent with the SPICE input discussed shortly. The node Ⓝ suggests that this could be the output of *any* odd numbered stages (i.e., 3, 5, 7, ...). The additional detail in Fig. 3.26b shows several key features that are important for first-order switching theory:

1. The subscripted points indicated with t's denote approximately the on-to-off transition point for the base–emitter diodes of each device.
2. The voltage at $V_2(D)$ is approximately constant over the period in which $Q_1(I)$ changes its state.
3. Over the switching couple ◇1 the node voltages at both $V_n(O)$ and $V_1(I)$ are changing.

(a)

(b)

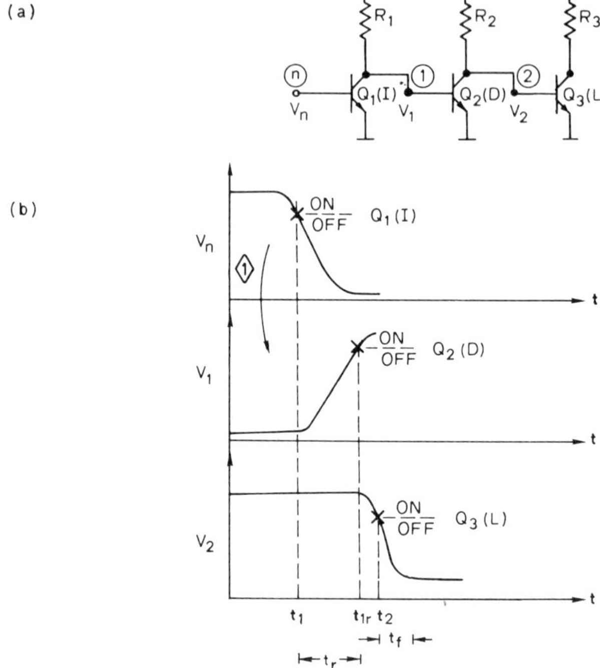

(c)

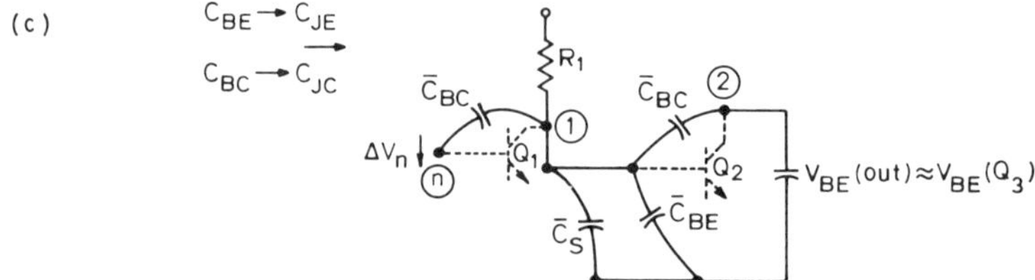

Figure 3.26 A detailed look at nodes n and 1 during switching event 1: (a) shows the circuit topology; (b) shows the detailed waveforms with breakpoints t_1, t_{1r}, and t_2 shown as the "on"–"off" transition point; (c) a first-order model for the charge-up at node 1 assuming that the devices are "off" to first order and that nodes n and 2 have the voltage dependences shown.

The above observations coupled with our discussion from Section 3.2.3 make it possible to give a set of first-order switching approximations:

1. Between t_1 and t_{1r}: $Q_2(D)$ is OFF and $Q_3(L)$ is fully-ON (and saturated) so that all current through R_1 goes to charge the node capacitance at ①.
2. Between t_{1r} and t_2: $Q_2(D)$ is ON and $Q_3(L)$ is coming out of saturation, eventually turning OFF. Voltage swings between nodes ① and ② are relatively small so that charge transfer across junction capacitors is

small. However, there is a substantial change in stored minority charge in transistors Q_2 and Q_3. That is:

(a) The forward base charge in $Q_2(D)$ must build-up.
(b) The saturated charge of $Q_3(L)$ must be removed.
The above two charge components each require a fraction of the current through R_1.

3. After t_2—$Q_3(L)$ is essentially OFF, but the charging current through R_1 now must deal with two competing requirements imposed by transistor $Q_2(D)$:

(a) Since $Q_2(D)$ is ON, there is a dc base current requirement.
(b) Since node ② is falling, the base–collector junction capacitance also requires a fraction of the current through R_1.

The sections which follow give a quantitative, although not rigorous, treatment of the switching events described above. Section 3.5.2 compares these calculations with results obtained using numerical simulation of the coupled phenomena.

The switching event to be discussed first is the rise of node voltage V_1. Observations 2 and 3 above provide the basis for our model. Figure 3.26*c* shows a simplified representation of node ①. Four capacitors are attached to node ①—C_S and C_{JE} are attached to ground, while feedback and feedforward capacitances C_{JC} are attached between ① and nodes ⓝ and ②. Over the time period between t_1 and t_{1r} both the devices shown with dashed lines are assumed off and Q_3 acts as a voltage source as shown, so that only the nonlinear capacitance charging effects need to be considered. Node ① changes through R_1, thus an estimate of the slope at the rising edge of voltage V_1 is

$$\tau \approx R\left[\bar{C}_S + \bar{C}_{JE} + \bar{C}_{JC}(fb) + \bar{C}_{JC}(ff)\right] \tag{3.50}$$

where *fb* and *ff* indicate the connections to nodes ⓝ and ②, respectively and all resistors are assumed to have value R. We will now estimate each of these capacitive contributions. Equation (3.8) provides the basic formula for computing the average capacitor values. The additional piece of information that is needed is the voltage change across each element. These changes are listed below in terms of their nearest integer increment of ϕ:

$$\begin{aligned} &V_{BC}(fb): +\phi \rightarrow -\phi \\ &V_{BE}; V_1: 0 \rightarrow +\phi \\ &V_{BC}(ff): -\phi \rightarrow 0 \end{aligned} \tag{3.51}$$

Using Eq. (3.8) and these values

$$\bar{C}_{JC}(fb) = 1.4C_{JC}$$

$$\bar{C}_{JE} = 2C_{JE}$$

$$\bar{C}_{JC}(ff) = 0.83C_{JC}$$

$$\bar{C}_S = C_S \quad \text{(assumed constant)} \tag{3.52}$$

Hence Eq. (3.50) becomes

$$\tau \approx R(C_S + 2C_{JE} + 2.23C_{JC}) = R\bar{C}_{Tr} \tag{3.53}$$

where it is assumed that devices are identical. To illustrate relative effects of parameter variations, if one assumes initially that $C_S = C_{JE} = C_{JC}$, Eq. (3.53) reveals that a doubling of each component individually results in approximately 19%, 38%, and 43% changes in τ, respectively.

In order to estimate the actual switching time based on this time constant, one must assume a voltage at node 1 of the form

$$V_1(t) = V_{CC}(1 - e^{-t/\tau}) \tag{3.54}$$

and solve for the time t_r when $V_1(t_r) = V_{BE}(\text{on})$, the just-on state for the base–emitter diode of the next stage. Hence,

$$t_r \approx \frac{V_{BE}(\text{on})}{V_{CC}}\tau \tag{3.55}$$

by using only the first-order term of the exponential. For the circuit example used here, V_{CC} will be of the order of 3–5 V. In the next section we will compare results calculated using these first-order approximations with computer simulation.

An equivalent statement to the single RC approximation discussed above is the charge-control approach. Using this second method one can write t_r simply in terms of the transferred charge Δq_{sc} and average current $\bar{I}_{R_1}$ as follows:

$$t_r \approx \frac{\Delta q_{sc}}{\bar{I}_{R_1}} = \frac{V_l \cdot \bar{C}_{Tr}}{\dfrac{V_{CC} - V_{BE}(\text{on})/2}{R}}$$

$$t_r \approx \frac{V_l}{V_{CC} - \frac{1}{2}V_{BE}(\text{on})} \cdot \underbrace{R\bar{C}_{Tr}}_{\tau} \tag{3.56}$$

which is virtually the same as the result given in Eq. (3.55) provided that

$V_l \approx V_1(t_2)$ and that $V_{BE}(\text{on})/2 \ll V_{CC}$. Both results quoted above give a *rise time* that is a fraction of the nodal time constant.

For the time period beyond t_2 the load $Q_3(L)$ is OFF and current through R_1 feeds both the base current $I_{B(Q2)}$ as well as the charging of $\bar{C}_{JC}(fb)$ as shown in Fig. 3.27*a*. The collector current of Q_2 (less the components through C_{JC} and R_2) discharges node ②. Figure 3.27*b* shows the equivalent circuit with Q_3 dashed to indicate its OFF state. To calculate the average values for the capacitances the following voltage changes are assumed:

$$\begin{aligned} V_{BC}(fb) &: 0 \to \phi \\ V_{BE}; V_2 &: \phi \to 0 \\ V_{BC}(\text{ff}) &: +\phi \to -\phi \end{aligned} \tag{3.57}$$

Using Eq. (3.8) and these changes

$$\begin{aligned} \bar{C}_{JC}(fb) &= 2C_{JC} \\ \bar{C}_{JE} &= 2C_{JE} \\ \bar{C}_S &= C_S \quad \text{(assumed constant)} \\ C_{JC}(ff) &= 1.4C_{JC} \end{aligned} \tag{3.58}$$

Writing the node equations for ① and ② and assuming all resistors of value R, one obtains

$$① I_{B(Q2)} = \frac{V_{CC} - V_{BE}(\text{on})}{R} - \frac{\bar{C}_{JC}(fb)V_{BE}(\text{on})}{t_f} \tag{3.59a}$$

$$② \beta_F I_{B(Q2)} = \frac{V_{CC} - V_{BE}(\text{on})}{R} + \frac{\bar{C}_{JC}(fb)V_{BE}(\text{on})}{t_f} + \frac{[\bar{C}_{JE} + \bar{C}_S + \bar{C}_{JC}(ff)]V_{BE}(\text{on})}{t_f} \tag{3.59b}$$

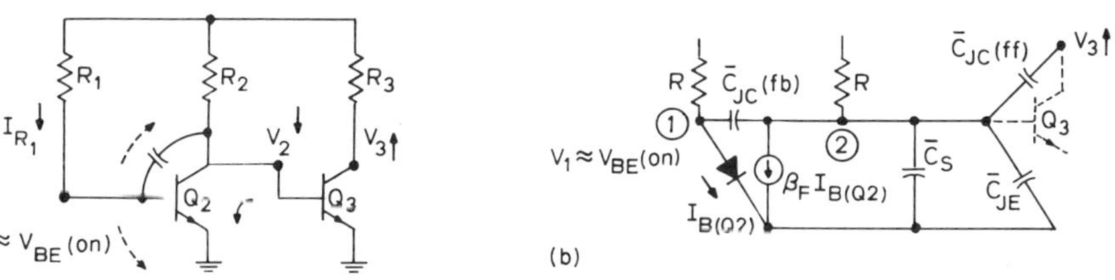

Figure 3.27 The switching model for time beyond t_2: (*a*) the general circuit showing several current components as well as significant voltage changes; (*b*) the pertinent switching model considering dominantly the effects of Q_2.

Solving for t_f it follows that

$$t_f = \frac{V_{BE}(\text{on})}{(\beta_F - 1)[V_{CC} - V_{BE}(\text{on})]} R \overbrace{[(\beta_F + 1)\overline{C}_{JC}(fb) + \overline{C}_S + \overline{C}_{JE} + \overline{C}_{JC}(ff)]}^{\overline{C}_{Tf}}$$

$$= \frac{V_{BE}(\text{on})}{(\beta_F - 1)[V_{CC} - V_{BE}(\text{on})]} R\overline{C}_{Tf} \tag{3.60}$$

where $\overline{C}_{Tf}$ appears to be the Miller capacitance of the Q_2 when it is active. However, making the approximation that the $(\beta_F + 1)$ term in $\overline{C}_{Tf}$ dominates, one obtains:

$$t_f \approx \frac{\overline{C}_{JC}(fb)V_{BE}(\text{on})}{[V_{CC} - V_{BE}(\text{on})]/R} \frac{(\beta_F + 1)}{(\beta_F - 1)} \tag{3.61}$$

which has the form of the resistive charging of $\overline{C}_{JC}(fb)$.

The calculations of t_r and t_f have a simple form and pose no difficulties associated with minority-charge storage. This final set of calculations addresses the problem of saturated switching and the time period between t_{1r} and t_2. The period between the *rise* of V_1 and the *fall* of V_2 marks a *delay* period during which voltages at nodes ① and ② change only slightly; however, substantial internal changes can be transferred. The basic circuit model shown in Fig. 3.26*c* now changes to the model shown in Fig. 3.28*a*, where Q_2 is active but V_1 is nearly constant. The current through R_1 charges the base of Q_2 and discharges the base of Q_3 (via the current gain of Q_2). A simplified equivalent model including these two effects is shown in Fig. 3.28*b* and equal values of R as assumed for R_1 and R_2.

The three charge-storage effects involve q_{F2}, q_{F3}, q_{S3}. The dc current $I_{B(Q2)}$ competes with q_{F2} for the input current, and the output node is driven by $\beta_F I_{B(Q2)}$. Writing expressions for the delay and unknown current $I_{B(Q2)}$ in terms of the known charges and current one obtains

$$\frac{V_{CC} - V_{BE}(\text{on})}{R} - I_{B(Q2)} = \frac{q_{F2}}{t_d} \qquad \text{node ①} \tag{3.62a}$$

$$-\frac{V_{CC} - V_{BE}(\text{on})}{R} + \beta_F I_{B(Q2)} = \frac{q_{F3} + q_{S3}}{t_d} \qquad \text{node ②} \tag{3.62b}$$

Solving for t_d:

$$t_d = \frac{\beta_F q_{F2} + q_{F3} + q_{S3}}{(\beta_F - 1)[V_{CC} - V_{BE}(\text{on})]/R} \tag{3.63}$$

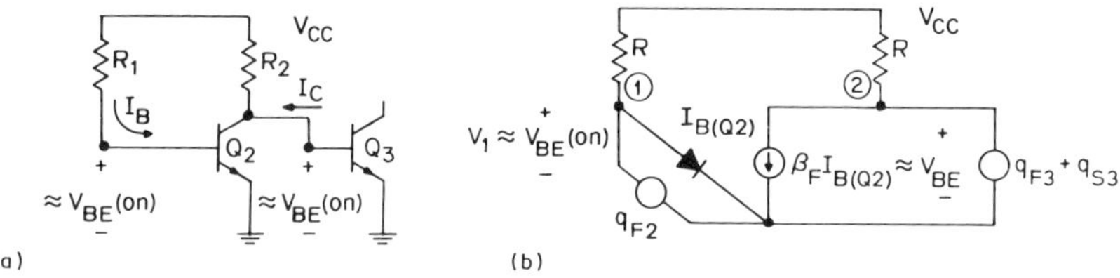

Figure 3.28 Saturated switching model of Q_2 and Q_3: (*a*) the circuit schematic and (*b*) the simplified representation showing competing charges and assumed model for Q_2.

The values for charge can be estimated as follows:

$$q_{F2} = \left[\frac{V_{CC} - V_{BE}(\text{on})}{R} - \frac{V_{BE}(\text{on})\bar{C}_{JC}(fb)}{t_f} \right] \beta_F \tau_F \tag{3.64a}$$

$$q_{F3} = \left(\frac{V_{CC} - V_{BE}(\text{on})}{R} \right) \tau_F \tag{3.64b}$$

$$q_{S3} = \left(\frac{V_{CC} - V_{BE}(\text{on})}{R} \right) \left(1 - \frac{1}{\beta_F} \right) \tau_S \tag{3.64c}$$

and using the simplified expression for t_f from Eq. (3.61) in Eq. (3.64a), q_{F2} can be simplified to the form

$$q_{F2} \approx \left(\frac{V_{CC} - V_{BE}(\text{on})}{R} \right) \left(\frac{2\beta_F}{\beta_F + 1} \right) \tau_F \tag{3.65}$$

Summing the charge terms and computing t_d, the following equation results:

$$t_d \approx \frac{3\tau_F + \tau_S}{\beta_F} \tag{3.66}$$

assuming $\beta_F \gg 1$.

This simplified form suggests that the β_F factor in the denominator, which results from the active-region operation of Q_2, tends to dominate the switching delay of the circuit.

3.5.2 Comparison of Approximate and Computer-Simulated Switching

The results discussed above include many approximations. The purpose of this section is to use computer-aided circuit analysis to validate and in some cases to help refine our first-order bias. Figure 3.29*a* shows the CAD schematic of a three-stage ring oscillator, and the nominal input parameters for SPICE are

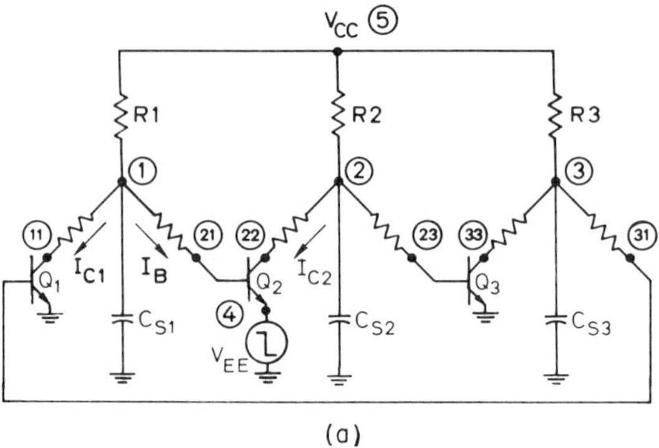

(a)

```
*****10/30/78 ********  SPICE 2D.2 (26SEP76)  ********10:47:34*****

Three Stage Oscillator

         Input Listing                 Temperature = 27.000 DEG C

*******************************************************************


01 11 31 0 NPN
02 22 21 4 NPN
03 33 23 0 NPN
R1 5 1 240K
RC1 1 11 100
RB1 3 31 100
CS 1 0 .5P
R2 5 2 240K
RC2 2 22 100
RB2 1 21 100
CS 2 0 .5P
R3 5 3 240K
RC3 3 33 100
RB3 2 23 100
CS 3 0 .5P
VCC 5 0 DC 3
VEE 4 0 PULSE .57 0 0 10N 500N
.MODEL NPN NPN BF=10 BR=50
+CJE=.5P CJC=.5P
.PLOT TRAN V(1) (-.2,.6) V(1,11) V(1,21) (-.002,.006)
.PLOT TRAN V(2) (-.2,.6) V(2,22) (-.002,.006)
.PLOT TRAN V(3) (-.2,.6)
.PRINT TRAN V(1) V(1,11) V(1,21) V(2) V(2,22) V(2,23) V(3)
.TRAN 10N 500N
.END
```

(b)

Figure 3.29 (*a*) Circuit schematic for a three-stage ring oscillator and (*b*) the CAD input listing used for simulation.

shown in Fig. 3.29*b*. Two CAD circuit "tricks" are employed:

1. 100-Ω current-sensing resistors are added so that I_B and I_C can be monitored.
2. A voltage pulse at the emitter of transistor Q_2 is used to start the oscillator by off-setting the dc bias.

Figures 3.30*a–c* show the voltage and current waveforms at the nodes ①, ②, and ③ = ⓝ, respectively. Several important features can be noted. For the case with no saturation charge storage, the fall at ① and the rise at ② begin precisely at the same time. The current waveforms show a somewhat surprising fine structure, but one key observation is as follows. The net negative I_B over the fall of the voltage at ① (and the mirrored behavior of I_C) can be approximated by averaged box waveforms. This result is consistent with the approximations used in Section 3.5.1. Moreover, a cross check of the waveform

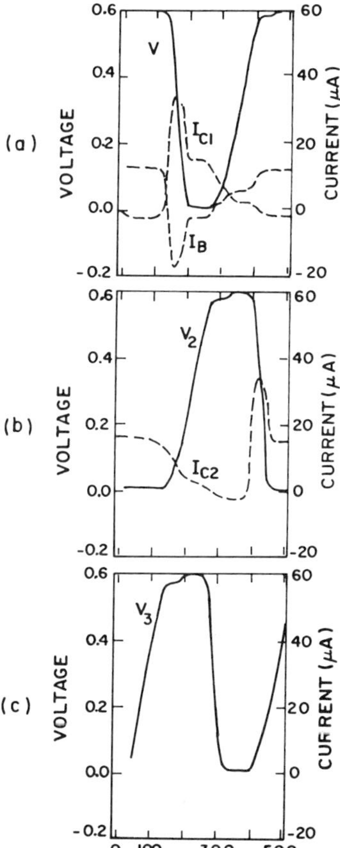

Figure 3.30 The simulated node voltages and selected currents sensed through 100-Ω resistors shown in Fig. 3.28: (*a*) node *A* and driving currents I_{C1} and I_B, (*b*) node *B* and collector current I_{C2}, and (*c*) node *C*.

at ③ = ⓝ over this same period confirms the approximation that $V_{ⓝ} \simeq V_{BE}(\text{on})$. Many of the other details that can be seen are left for the reader to consider further. At this point two sets of calculation and simulation are appropriate. First, a check on nominal values for t_r and t_f. Second, a set of perturbations to test out the applicability of the model results to track correctly element value changes including the introduction of saturated charge.

Based on Eq. (3.53) and (3.55), the estimated t_r is

$$t_r = \frac{0.6}{3\text{ V}}(240\text{K})(2.6\text{ pf}) = 1.25 \times 10^{-7}\text{ s}$$

and the simulated value is 1.2×10^{-7} s which is quite adequate. It is clear from the shape of $V_{①}$ that the solution is not precisely that modeled by Eq. (3.54). An improved estimate between theory and experiment is obtained by using min/max values in the middle of the voltage swing. This approach is used in Table 3.4

Beyond the initial comparisons of switching time a second task of *perturbation* calculations becomes critical. Now we must test our simulations and theory dynamically. Three perturbations have been simulated, which are expected to show a representative set of changes in switching waveforms:

1. $C_S \rightarrow$ twice nominal.
2. $C_{JC} \rightarrow$ twice nominal.
3. Saturated charge is introduced.

For the first two changes we also revise the calculation means for t_r by using $V_B(\text{min}) = 0.13$ and $V_B(\text{max}) = 0.52$ V. The results are summarized in Table 3.4 and show quite favorable agreement. To check the model basis further, consider the comparison of t_f values. Table 3.5 shows both calculated and simulated results based on Eq. (3.61). Although the range of values is not great, the agreement again seems adequate. To push the test still further one can use perturbations, which should alter the results by a factor of *two*. Such examples as well as consideration of saturated charge and active loads (I^2L) will be saved for home problems.

Table 3.4 Calculated and Simulated Values for t_r with $V(\text{min}) = 0.13$ V, $V(\text{max}) = 0.52$ V

	Nominal	$2C_S$	$2C_{JC}$
Calculated t_r ($\times 10^{-7}$ s)	0.8	1.0	1.2
Simulated t_r ($\times 10^{-7}$ s)	0.7	0.9	1.0

Table 3.5 Calculated and Simulated Values for t_f

	Nominal	$2C_S$	$2C_{JC}$
Calculated			
$t_f (\times 10^{-7}$ s)	0.88	0.9	1.6
Simulated			
$t_f (\times 10^{-7}$ s)	0.6	0.6	0.8

3.5.3 An Idealized T^2L Switching Example

Although ring-oscillator circuits and *in situ* experiments involving nonlinear effects provide a realistic measure of logic-circuit performance, they frequently are rather poor examples for studying generic device effects. The following example takes the basic T^2L gate with pass transistor and defines a computer experiment for further analysis of the nonlinear effects.

Figure 3.31*a* shows the idealized circuit used for the experiment. The diode and voltage step simulate the input loading; the pass transistor and diode load

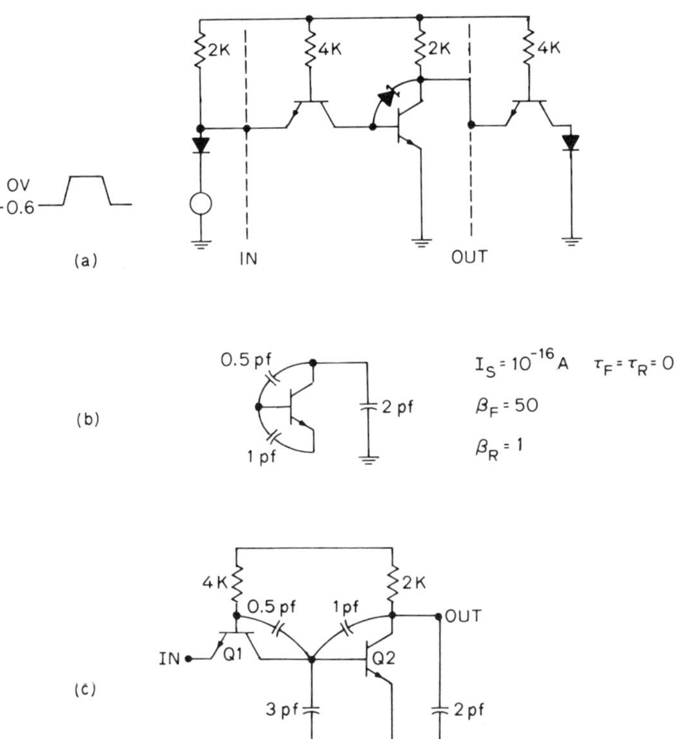

Figure 3.31 An idealized T^2L-type gate analysis: (*a*) the basic gate with input/output loading, (*b*) the simplified device (constant capacitors), and (*c*) the intrinsic gate with total loading shown.

represent the proper effects at V_{out}. Each transistor between IN and OUT is modeled as shown in Fig. 3.31*b* where the capacitance averaging as usually used for nonlinear devices is avoided by using external constant-valued elements. The critical path of the circuit between IN and OUT is shown in Figure 3.31*c*. The 3 pf node-to-ground value is the combination of C_S and C_{JE} for the two devices.

The simulated output waveform is shown in Fig. 3.32. The fall time at the output indicates the dominance of the transistor in its active mode, whereas the rising waveform indicates more typically a passive *RC*-dominated response. Based on a similar approach used in Sections 3.5.1 and 3.5.2 one can determine approximate first-order approximations to compare with results shown in Fig. 3.32.

3.5.4 An ECL Switching Example

The preceding examples have considered switching where the devices saturate and store minority charge in the base and collector. The ECL logic family produces a NOR/OR switching function, but limits the voltage swing at the output (and complements) so that the transistors never saturate. Further

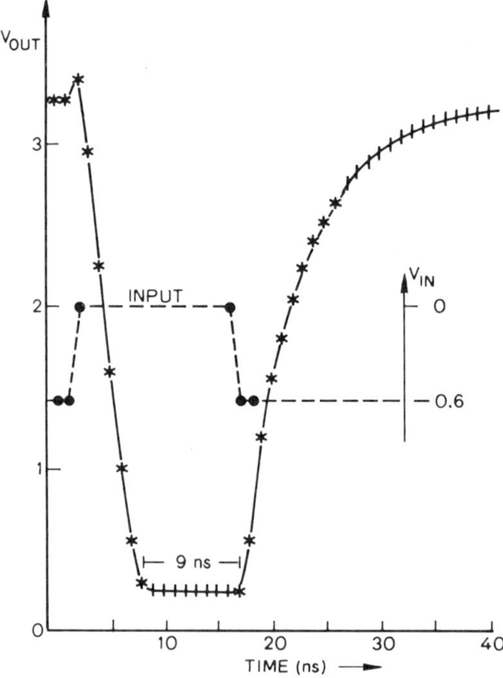

Figure 3.32 Simulated voltage versus time waveform at the output node for the T^2L circuit shown in Fig. 3.31*c*.

details concerning logic design and practical circuit configuration are discussed in Chapter 6. The purpose of this section is to take the basic circuit shown in Fig. 1.6 and demonstrate that many of the switching approximations discussed for I^2L and T^2L circuits can be applied and extended for use with ECL.

Figure 3.33*a* shows the two-stage inverter chain used for this example. The input is driven by a voltage step between the two dc levels of the circuit (-0.7 V and -1.6 V). The current source is replaced by the 1.8K Ω resistor and the V_{BB} reference is a -1.1 V voltage source. All transistors are identical with nominal capacitance values of $C_{JE} = C_{JC} = 1$ pf and $C_S = 3$ pf, which is for purposes of illustration rather than based on any real or postulated technology. In fact, as indicated in Fig. 3.21, these values are typically an order of magnitude too high. A load capacitance of 3 pf is included at the output nodes

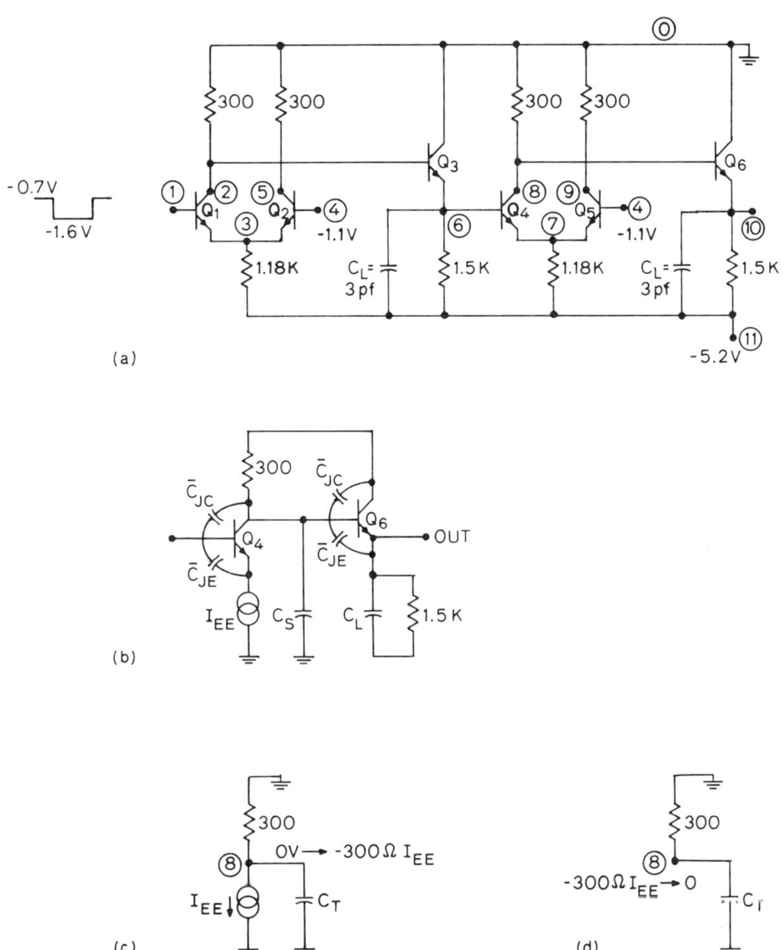

Figure 3.33 An ECL switching circuit example: (*a*) the two-stage circuit, (*b*) a simplified portion of the circuit, and (*c, d*) a simple *RC* analysis of the switching.

⑥ and ⑩ to represent wiring. Figure 3.33*b* shows the equivalent circuit path of Q_4 and Q_6 including the averaged capacitances pertinent for subsequent discussion. The emitter of Q_4 is driven by the equivalent current source shown. However, near the switching threshold the node appears much as a differential ground since Q_4/Q_5 "steer" the current source between their collectors.

Based on the node numbering shown in Fig. 3.33*a*, the outputs ⑥ and ⑩ are used for monitoring switching effects. Figure 3.34*a* shows the output waveforms for the two nodes, and the input steps occur at the times denoted with arrows. The simulation shows that node ⑥ begins to change almost instantaneously with the input pulse, whereas node ⑩, while generally exhibiting the inverse of node ⑥, shows about a 2-ns delay in response. On the transition from -0.7 V to -1.6 V both nodes show a peaked response which occurs, because of capacitive feedthrough as discussed later.

Using node ⑩ and the output path Q_4–Q_6 as the portion of the circuit for analysis, two observations are apparent. First, the "input" swing at node ⑥ is nearly complete before node ⑩ begins to change. Second, the difference in waveforms between node ⑥ with its "ideal" input at ① and the output at node ⑩ is negligible. One can thus infer that an idealized input is a suitable approximation for switching calculations. Based on the circuit shown in Fig. 3.33*b* and a simplified *RC* time-constant model, an approximate set expressions for the observed waveforms at node ⑩ can be determined.

Begin by considering the internal node ⑧ and take the voltage swings across $C_{JC}(Q_4)$ and $C_{JC}(Q_6)$ as $-\phi \rightarrow 0$ and $0 \rightarrow -\phi$, respectively, for the transition where Q_4 is turned ON. As a result, both capacitors have an averaged value:

$$\overline{C}_{JC} = 0.83C_{JC} \tag{3.67}$$

Now since Q_6 is essentially ON over the complete transition, $C_{JE}(Q_6)$ remains constant, and one can assume that its value is large so that node ⑧ sees the effective output loading of C_L, scaled by the current given of Q_6.* Using the above approximations, the total capacitance at node ⑧ is

$$C_T = C_S + \frac{C_L}{\beta_F + 1} + 2\overline{C}_{JC}$$

$$= C_S + \frac{C_L}{\beta_F + 1} + 1.66C_{JC} \tag{3.68}$$

*The series feedback of C_L gives an equivalent input capacitance at node ⑧ "looking into" Q_6 of $C_L/(\beta_F + 1)$. This can be easily derived using a small-signal model and considering only r_π, g_m, and the load C_L.

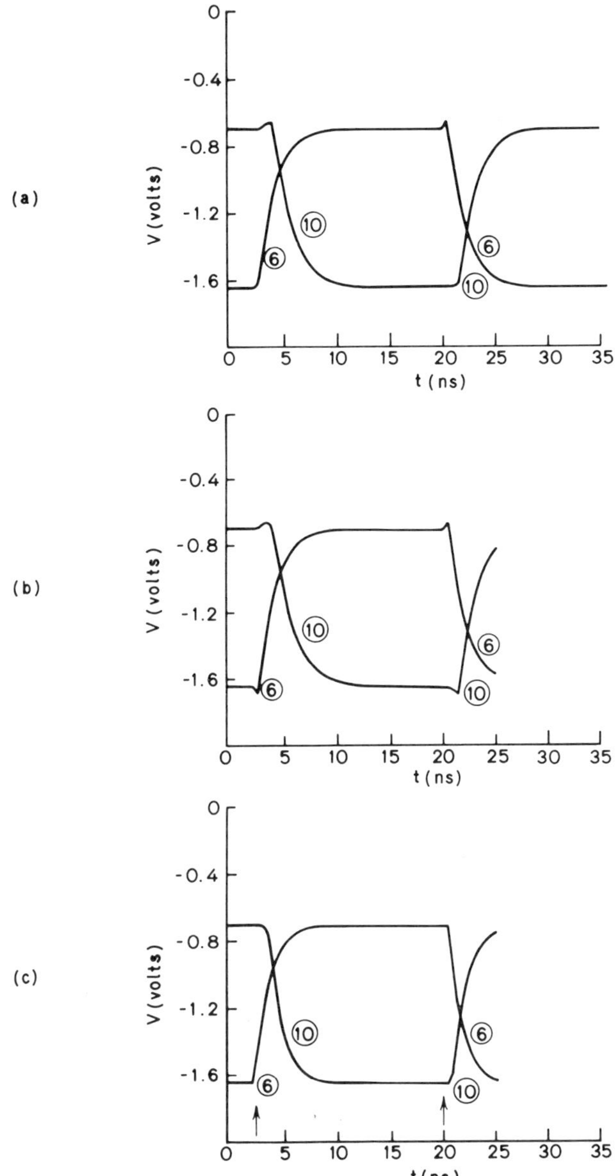

Figure 3.34 Simulated switching waveforms versus time at nodes 6 and 10: (*a*) the nominal switching results, (*b*) results without C_1, and (*c*) results without C_{JC} on Q_4.

Two switching modes are considered for node ⑧:

1. The discharging where Q_4 behaves as a current source.
2. The charging where Q_4 is OFF and the 300-Ω load dominates.

The discharging is then modeled as shown in Fig. 3.33, where

$$I_{EE} = \frac{5.2\text{ V} - 1.1\text{ V} - V_{BE}(\text{on})}{1.18\text{K}} \approx 2.88\text{ mA} \tag{3.69}$$

and the voltage as a function of time is given by:

$$V_{⑧}(t) = -\text{I}_{EE}(300\ \Omega)(1 - e^{-t/\tau}) \tag{3.70}$$

where $\tau = (300\ \Omega)C_T$.

For the charging portion of the waveform the voltage is given by

$$V_{⑧}(t) = -I_{EE}(300\ \Omega)e^{-t/\tau} \tag{3.71}$$

Thus over the linear portion of the switching

$$\frac{\Delta V_{⑧}}{I_{EE}300\ \Omega} \approx \frac{\Delta t}{\tau} \tag{3.72}$$

since $e^{-t/\tau} \approx 1 - t/\tau + \cdots$.

Using typical numbers from the simulator input:

$$C_T = 3\text{ pf} + \frac{3\text{ pf}}{11} + 1.66\text{ pf} = 4.9\text{ pf}$$

$$\tau = (300\ \Omega)C_T = 1.5\text{ ns} \tag{3.73}$$

Over the first 0.5 V swing of *both* the rising and falling waveforms, the simulated time gives 1.5 ns using Eq. (3.72). The result is

$$t = \frac{0.5\text{ V}}{0.86\text{ V}} \cdot 1.5\text{ ns} = 0.9\text{ ns}$$

This 60% apparent difference between the first-order approximation and simulation is misleading since the plot resolution of time is only 0.5 ns. Using the exponential equation to calculate the time required to reach 0.77 V (of the total 0.86 V swing), one obtains 3.4 ns, whereas the simulation gives about 3 ns, which appears to be a much closer agreement. The conclusion to be drawn thus far is that the model is not grossly off, but only perturbation experiments can confirm the correctness of the assumptions.

A partial set of simulations is shown in Figs. 3.34*b* and *c*, which illustrate the sensitivity of two contributions of the total capacitance given in Eq. (3.68). Figure 3.34*b* shows the simulated effect of removing C_L from the circuit. The overall waveforms in Fig. 3.34*b* are nearly identical to those in Fig. 3.34*a*, and the time constant derived from the 10%–90% rise time gives $\tau = 1.74$ ns for both cases. This is consistent with the fact that the $C_L/\beta_F + 1$ contribution is less than 10% of the other factors affecting C_T.

Turning to Fig. 3.34*c* the simulated data are obtained by omitting C_{JC} from the base collector of Q_4. Two things are apparent from the simulation. First, the peaked response and a major part of the associated delay are no longer present. Second, the rise and fall times are much shorter. Again using the 10–90% transition times as a measure of the time constant, the experiments give $\tau = 1.3$ ns. Using Eq. (3.73) and only the Q_6 contribution to the $\overline{C}_{JC}$, one obtains

$$C_T = 3\text{ pf} + \frac{3\text{ pf}}{11} + 0.83\text{ pf} = 4.1\text{ pf}$$

$$\tau = (300\ \Omega)C_T = 1.2\text{ ns}$$

The results are quite comparable. The removal of the capacitive feedback around Q_4 has indeed further reduced the problem to the simple *RC* approximation used for modeling. In fact, in retrospect one can postulate that the difference in measured and calculated τ for the nominal case (i.e., 1.74 ns compared to 1.5 ns) may well reflect the Miller contribution to τ, which is not accounted for by the simple *RC* model.

A final calculation based on the simulated results and a first-order model concerning capacitive feedthrough for Q_4 can now be considered. Figure 3.35 shows the circuit model of Q_4 in the OFF state and the input ramping at rate dv/dt. From the figure one can see that the current is given by $C(dv/dt)$, and the voltage change at the output is then given by

$$\Delta V_{⑧} = RC\frac{dv}{dt}, \qquad R = 300\ \Omega$$

Using the values of $C = \overline{C}_{JC} = 0.83$ pf and $dv/dt = 0.6\text{ V}/1.5$ ns (taken from Fig. 3.2*a*), the calculated ΔV is

$$\Delta V = \frac{(300)(0.83\text{ pf})(0.6\text{ V})}{1.5\text{ ns}} = 0.1\text{ V}$$

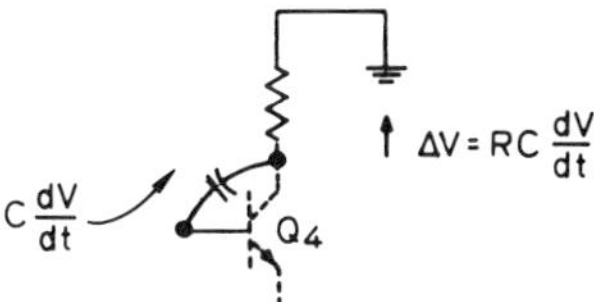

Figure 3.35 A simple circuit schematic to illustrate the effect of feedthrough on voltage peaking at the collector node.

The simulated peak is approximately $\Delta V \approx 0.03$ V. The peak is nearly constant over the time interval before the input at node ⑥ reaches -1.1 V, the threshold for turning ON Q_4. For a comparison, the simulated waveform in Fig. 3.34*c* shows no peaking and node ⑧ begins to fall almost exactly when node ⑥ reaches -1.1 V. As discussed in Chapter 6 one can use node ⑨ to obtain an unpeaked response and, hence, increased circuit speed. The reason node ⑨ shows an effect equivalent to node ⑧ with no C_{JC} is the following. Our basic model for the peaking assumes a base voltage step which feedsthrough to the collector node. The results shown in Fig. 3.34*c* eliminate the effect by removing the capacitance. A *similar result* can be obtained by eliminating the ΔV at the base which is apparent for Q_5, since the base is clamped at the reference.

REFERENCES

1. J. L. Moll and I. M. Ross, *Proc. IRE* **44**, 72 (1956).
2. P. E. Gray, D. DeWitt, A. R. Boothroyd, and J. F. Gibbons, *Physical Electronics and Circuit Models of Transistors*, Vol. 2, SEEC Notes, Wiley, New York, 1964.
3. R. S. Muller and T. I. Kamins, *Device Electronics for Integrated Circuits*, Wiley, New York, 1977.
4. W. J. McCalla and D. O. Pederson, Elements of Computer-Aided Circuit Analysis, *IEEE Trans. Circuit Theory* **CT-18**, 14 (1971).
5. M. Y. Hsieh and D. O. Pederson, An Improved Circuit Approach for Macromodeling of Digital Circuits, *Proc. ISCAS*, 22, *IEEE*, 696–699 (1977).
6. L. W. Nagel, "Spice 2: A Computer Program to Simulate Semiconductor Circuits, PhD. Thesis, University of California, Berkeley, May 1978.
7. V. L. Rideout, A Review of the Theory, Technology and Applications of Metal-Semiconductor Rectifiers, *Thin Solid Films* **48**, 261–291 (1978).
8. J. J. Ebers and J. L. Moll, *Proc. IRE* **42**, 1761 (1954).
9. H. K. Gummel and H. C. Poon, *Bell System Tech. J.* **49**, 827 (1970).

Chapter 4

Logic Families for MSI / LSI: T^2L and ST^2L

4.1 INTRODUCTION

The previous chapters on technology and modeling offer a special background to digital-circuit designers. The features and limitations of a given technology, the relations between circuit parameters and device parameters, and modeling the digital bipolar structure for both "pencil and paper" and CAD are important aspects of circuit design. It is hoped that the reader can follow the main line of thinking used throughout the book by asking questions, for example, How can I take full advantage of a given technology? Can I propose modifications to help me design "better" bipolar digital circuits? What will be an adequate model for my circuit during the first phase of the design and during subsequent phases? In many places in the following chapters the reader will discover that knowing about bipolar integrated technology as well as the different aspects of modeling is essential to improve circuit performance.

The operation of a digital circuit is quite distinct from that of a linear circuit in that the devices are intended to be either "on" or "off" and the "linear-bias condition" occurs only as a transient or to avoid a saturated "on" state. The input and output voltages are thus at two levels, either "high" or "low." These two levels are actually a band of voltages, separated by a buffer-voltage range to ensure that the state of a device can be determined even in the presence of noise.

The *npn* transistor is the primary switching device in bipolar logic families—resistors, diodes, and, in some cases, *pnp* devices are used. As discussed in Chapter 1, these devices can be integrated on a single silicon chip to produce an integrated circuit (IC), and thousands of these can be integrated resulting in a complex MSI (medium-scale-integrated) function (e.g., a counter), an LSI (large-scale-integrated) subsystem (e.g., a microprocessor), or a VLSI (very-large-scale-integrated) system (e.g., a microcomputer).

The purpose of this chapter is to introduce the general characteristics of any logic gate in terms of input, output, and transfer characteristics; logic levels;

noise margins and threshold level; fan-in and fan-out numbers. This will be followed by a discussion of medium performance bipolar logic families suitable for MSI and LSI, in particular, T^2L structures and digital building blocks are discussed. The approach taken here will be to understand logic realizations primarily from the circuit and device technology perspective. Hence, the menu of building blocks will be representative rather than exhaustive. Moreover, the system concepts are left for Chapter 7.

4.2 GENERAL CHARACTERISTICS OF A LOGIC GATE

The following analysis is applicable not only to bipolar gates but to any binary logic gate, including MOS digital circuits.

In a digital system, two basic operations are performed. The first is to realize a logic function: to determine a logic output variable in terms of input logic variables (e.g., A NAND gate) or to determine a logic output variable in terms of the input logic variables and the previous state of the output (e.g., a flip-flop). The second operation preformed in a digital system is transferring a logic signal from one logic block (logic source) to the input of another logic block or blocks [logic detector or detectors] through a transmission system (Fig. 4.1*a*). This transmission system can be either on-chip (e.g., metal lines) or between chips [e.g., on a printed-circuit board (PCB)]. A logic block has usually more than one input and each output is used as an input for a number of other logic blocks. The number of inputs is called the *fan-in* and the blocks driven by one output is called the *fan-out*.

At the input of a logic detector (another logic gate), the voltage signal is detected with respect to a *threshold voltage* V_t. In noninverting logic blocks, if the input voltage is greater than V_t, the output voltage will be at logical "1" and if the input voltage is less than V_t, the output will be at logical "0."

Signal distortion due to attenuation or system noise causes a reduction of the system reliability. Since distortion is primarily introduced by the noise in the system, it is most convenient to refer to the margin allowed for distortion of the logic levels as the noise margins. Two *voltage noise margins* are defined (Fig. 4.1*a*):

$$\begin{aligned}\text{VNM}_0 &= \text{voltage noise margin for a logical "0"}\\ &= V_{t\min} - V_{0\max}\\ \text{VNM}_1 &= \text{voltage noise margin for a logical "1"}\\ &= V_{1\min} - V_{t\max}\end{aligned}$$

The difference between the two levels V_1 and V_0 is called the *logic swing* V_l.

Figure 4.1*b* gives the logic levels for T^2L gates. These levels will be met for the specified load conditions and temperature range. The swing on the output

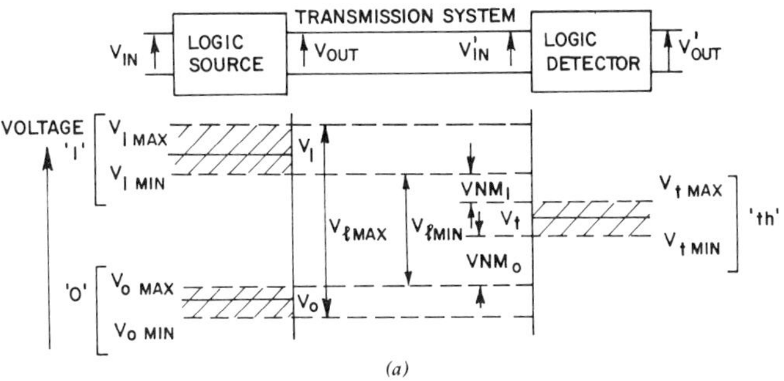

(a)

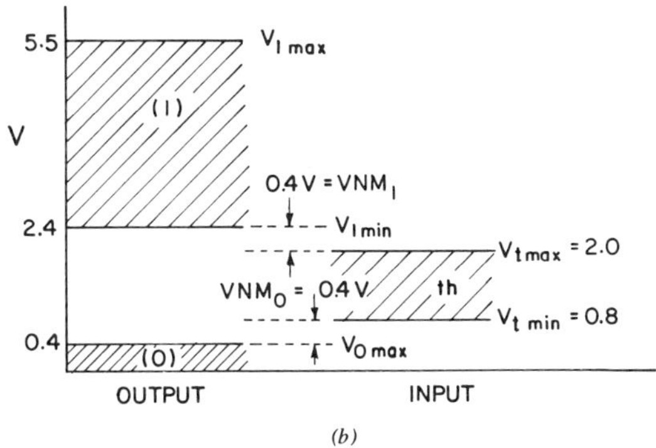

(b)

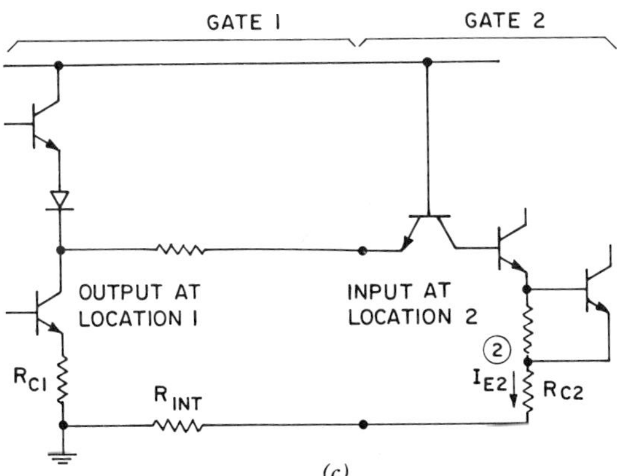

(c)

Figure 4.1 (*a*) System model and logic levels for digital IC circuits. (*b*) T^2L typical logic levels and noise margins. (*c*) Interconnection noise sources in T^2L. (*d*) Cross-coupling noise sources due to stray capacitances between lines.

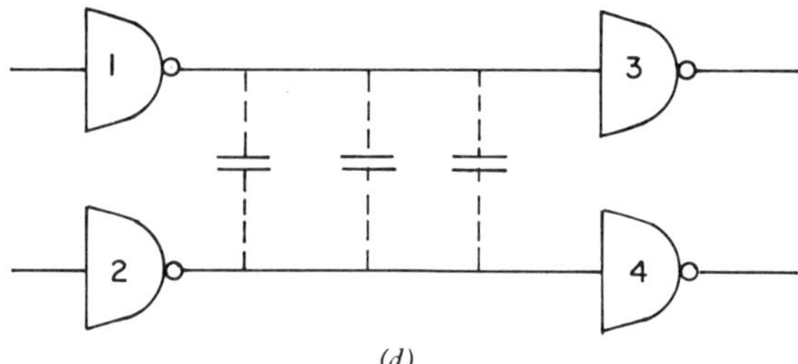

(d)

Figure 4.1 (*Continued*)

of a logic gate is always larger than the threshold limits at the inputs. In Fig. 4.1*b* the noise margin for T^2L is specified as 0.4 V. This margin of voltage allows some noise to be added to different ground potentials or to logic outputs due to coupling from adjacent lines. This 400-mV noise margin is indeed small when a large number of T^2L gates or circuits are used in a logic system. This is especially true in the PCB environment. Care must be used to ensure that the "worst case noise" does not exceed this limit. Noise is inherently difficult to predict since it is statistical, so that maintaining a known level of noise requires careful design. A rule-of-thumb is to keep observable noise below one-half the noise margin for a reliable logic design.

One troublesome factor generating noise in T^2L is the ground potential, since the ground potential is used as a reference for the logic levels. Figure 4.1*c* shows an example of how various noise sources can add to reduce the available noise margin. A logic gate at location 1 is driving the input of a gate at location 2. The power supply is grounded at location 1. I_{E2} is the total current at node ②. This current flows through the connector contact resistance R_{C2} and the ground interconnect resistance R_{INT} causing a voltage drop $I_{E2}(R_{\text{INT}} + R_{C2})$. This voltage drop is in the direction to decrease the noise margin in respect to the output of gate 1. When gate 1 switches, the transient can cause oscillations in the interconnections causing additional loss in the noise margin. The general guideline in producing reliable logic is to minimize the distance between gates and maintain a low-resistance ground reference for all the logic on a chip (as well as in the system). As logic chips and systems become larger, this ground-reference problem becomes a major consideration. It is quite common in logic systems to see large ground buses being used for ground references not only for low resistance but also for low inductance (primarily for PCBs). On the scale of large T^2L ICs, as the conductor sizes are scaled down, the ground references must be considered carefully.

Another source of noise in T^2L ICs is due to the cross coupling of signals. Any two signal lines that are close together (on-chip or off-chip) will have a certain amount of stray capacitance between them, as shown in Fig. 4.1*d*. If the output of gate 2 is switched from high to low, the output of gate 1 is pulled down by ΔV, since the voltage across the coupling capacitance cannot change instantaneously. The change ΔV and its duration are determined by the fall time of gate 2 and the value of the stray capacitance, the output impedance of gate 1 in parallel with the input impedance of gate 3. If ΔV is sufficiently large and depending on the threshold voltage of gate 3, a faulty logic transition may occur at the output of gate 3. Gates with high input impedances tend to be more susceptible to noise sources than ones with low input impedance.

In digital systems, logic signals are specified during time slots to minimize noise effects as well as logical hazards. These time slots are determined using a timing signal called a clock. During these time slots the logic signal can be a *level or pulse signal* (Fig. 4.2*a*), although level signals are mostly used. The level signal is at one of the binary levels representing the bits of information "0" or "1." These two levels can be voltage, current, energy, or impedance. However, for T^2L ICs, *voltage levels* V_0, V_1 are used. The typical voltage representation is a positive logic system, where the logical "1" is positive with respect to the logical "0"; in the negative logic system, the logical "1" is negative with respect to the logical "0."

Even though clocked signals are used, it is important to ensure that transient "glitches"—so-called *hazards*—do not affect the logic operation. For maximum speed it is useful to allow multiple logical operations to be performed between clock cycles—provided that erroneous intermediate results are not clocked-on other blocks. As signals propagate through multiple gates, there exists the possibility that hazards can occur. Two hazard types are identified, *static* and *dynamic*. *Static* refers to an output state that should not change for a new set of inputs; however, momentary spikes occur due to signal races. *Dynamic* refers to a condition where the output should change state, but multiple intermediate output changes occur. Figure 4.2*b* shows a Karnaugh map with both types of hazards indicated. The solid lines indicate the overall output change, and the dashed lines show the intermediate state changes. Figure 4.2*c* shows the bit pattern and "analog" waveform for the static case, and Fig. 4.2*d* shows the dynamic case. When sections of combinational logic feed into latches, there can be potential problems if hazards exist. Two design choices exist. By appropriate logic design one can avoid hazards or one can clock the logic at appropriate time slots to ensure all such transients have died away.

4.2.1 dc Transfer Characteristic

Signal and noise margin definitions have been stated by considering the transmission system, the input and output ports. Alternatively, the input–output gate relationship (the dc transfer characteristic) also provides information about the system [1, 2]. In Fig. 4.3, the unity gain line (with slope of unity) intersects the transfer characteristic at three points, the logical "0," the logical "1," and the threshold level "th." As the parameters of a particular design are varied over their tolerance range, the transfer characteristic (of a noninverting gate) will be shown by the dotted curves *A* and *B*. The shaded region between

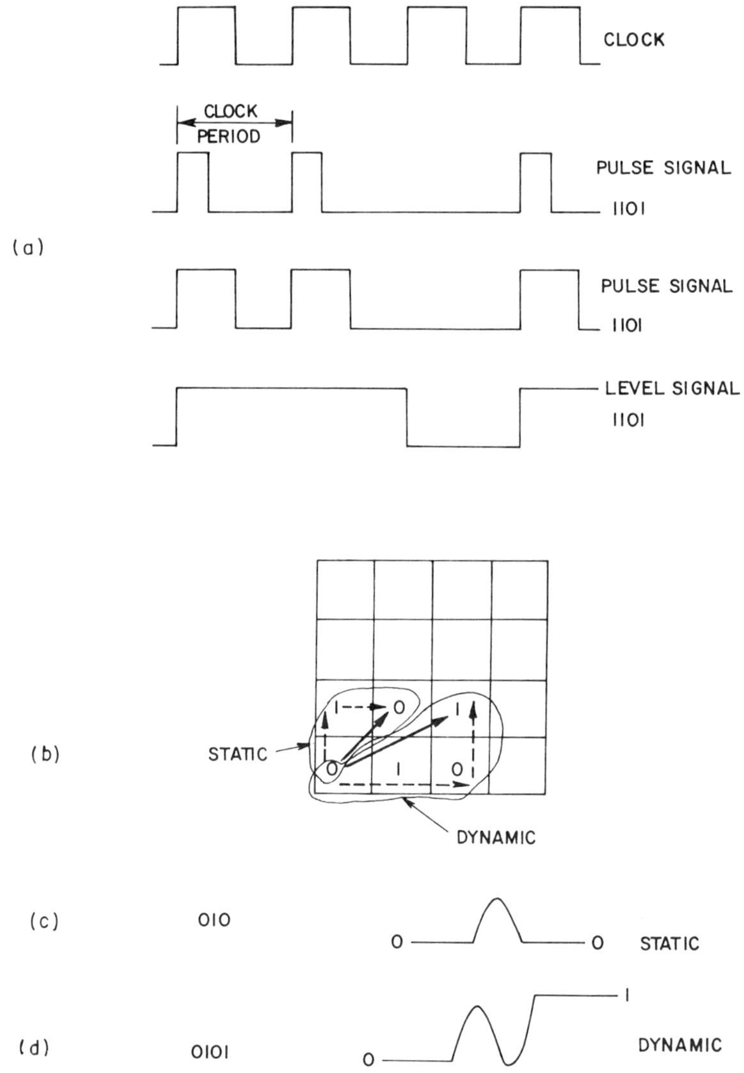

Figure 4.2 (*a*) Pulse and level signals. (*b*) Karnaugh map showing hazards. (*c*) Static hazard waveform. (*d*) Dynamic hazard waveform.

the limit envelopes A and B represents the region of possible transfer characteristics. Since the output does not reach its static value at the same time as the input, a further spreading of the transfer characteristic results. Because of these static and dynamic tolerances, each level becomes a zone as shown in Figs. 4.1 and 4.3. The above definitions of noise margins are shown. It is also shown that (a) when the input is in the region

$$V_{0\min} \leqslant V_{\text{in}} \leqslant V_{t\min}$$

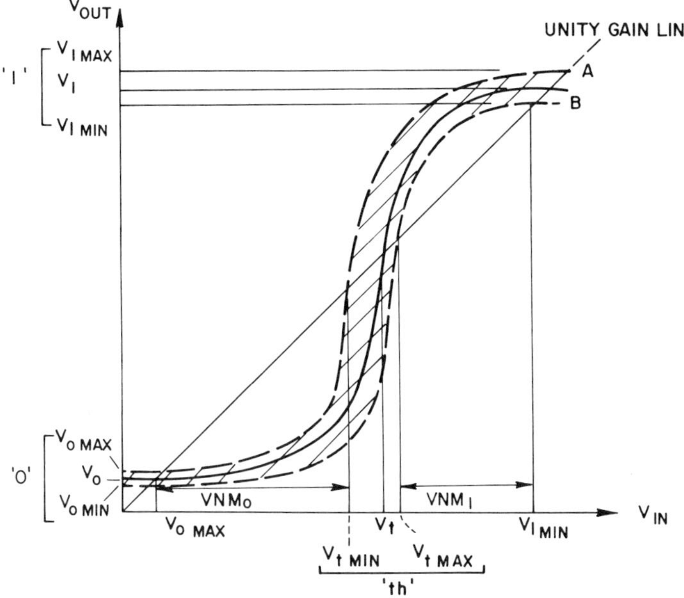

Figure 4.3 Dc transfer characteristic of a noninverting logic block.

the output is in the same region

$$V_{0\,\mathrm{min}} \leqslant V_{\mathrm{out}} \leqslant V_{t\,\mathrm{min}}$$

and (b) when the input is in the region

$$V_{t\,\mathrm{max}} \leqslant V_{\mathrm{in}} \leqslant V_{1\,\mathrm{max}}$$

the output is in the same region

$$V_{t\,\mathrm{max}} \leqslant V_{\mathrm{out}} \leqslant V_{1\,\mathrm{max}}$$

That is, the circuit will recognize and maintain two determined values, "1" above $V_{t\,\mathrm{max}}$ and "0" below $V_{t\,\mathrm{min}}$. The region ($V_{t\,\mathrm{min}} \leqslant V_{\mathrm{out}} \leqslant V_{t\,\mathrm{max}}$) represents an undetermined output transition region, and the logic signal lies in this region only during transitions from one level to the other.

4.2.2 dc Input and Output Characteristics

The dc input characteristic of a logic gate is defined as the I_{in} vs V_{in} relationship. Similarly, the dc output characteristic is defined as the I_{out} vs V_{out} relationship (Fig. 4.4). These two characteristics are useful in determining the

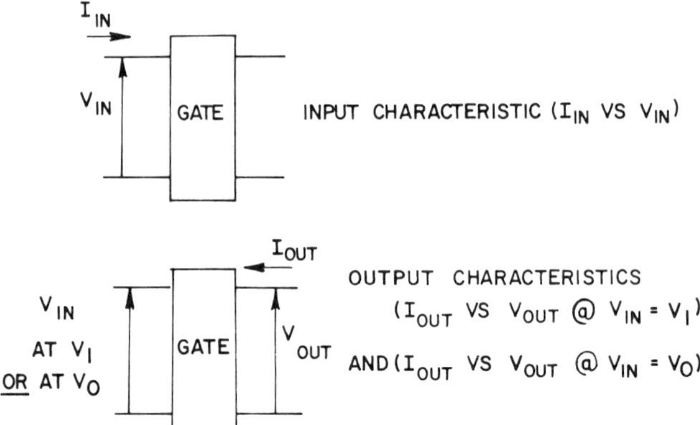

Figure 4.4 Definition of input and output characteristics of a logic gate.

loading conditions, the maximum fan-out, the logic levels, and the noise margins of a logic gate.

4.3 THE BIPOLAR TRANSISTOR AS A SWITCHING DEVICE

Two dc relationships are particularly important in understanding bipolar digital circuits. These are the relationship between the base current (I_B) and the base–emitter voltage (V_{BE}); a diodelike characteristic and the relationship between the collector current (I_C) and the collector–emitter voltage (V_{CE}). In Chapter 3 these characteristics were represented by equations involving exponentials of voltage. In this chapter and subsequent chapters first-order circuit analysis is used, and these relations are simplified to their equivalent piecewise linear representation. For example, the "on"-state base–emitter diode becomes the series combination of a voltage source (V_{on}) and a dynamic conductance (g_π) which is the slope of the $I_B - V_{BE}$ characteristic at a bias point where $I_B = I_C/\beta_F$. The I_C characteristic has two possible "on"-state representations—active and saturated operation modes. As shown in Chapter 3, the active-mode representation is a dependent current generator, which can be represented by either of the following equations:

$$I_C = \beta_F I_B \tag{4.1a}$$

or

$$I_C = I_S e^{qV_{BE}/kT} \tag{4.1b}$$

This family of characteristics is shown in Figure 4.5c for increasing I_B. As the device operates into saturation (decreasing valves of V_{CE} as shown in Fig. 4.5c),

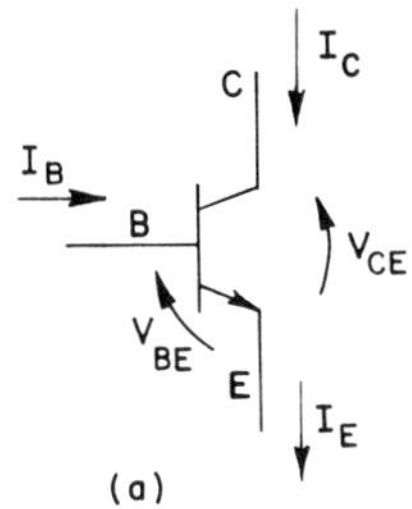

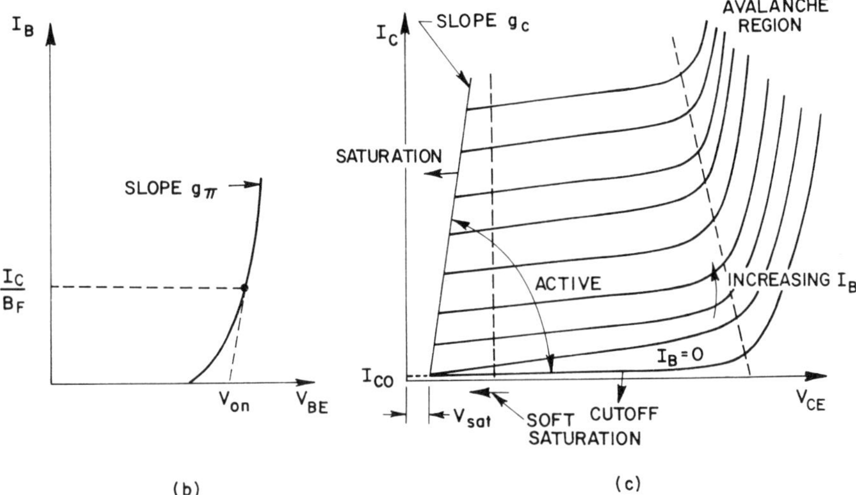

Figure 4.5 (*a*) *npn* transistor voltage and currents. (*b*) I_B versus V_{BE} (input characteristic for common-emitter transistor configuration). (*c*) I_C versus V_{CE} (output characteristic for common-emitter transistor configuration).

the simplified model changes. In particular, the base–collector diode (Fig. 3.9) turns on and allows a shunt path around the dependent current source. In effect the equivalent circuit as viewed from the collector is now a voltage source V_{sat}—the series combination of $(V_{on}(BE)-V_{on}(BC))$, which represents the back-to-back diodes that shunt the dependent generator. In addition to the voltage source, the collector series resistance $(1/g_c)$ is the dominant observable terminal (I_C-V_{CE}) characteristic. A summary of the linearized dc equivalents for both the active and saturated regions—as well as the cutoff mode— are summarized in Figure 4.6.

1. Cutoff Region

In this region both V_{BE} and V_{BC} are reverse biased or (according to the two piecewise approximation of the junctions) forward biased with a voltage less than V_{on}. The collector current is only due to the leakage of the collector junction, and in the ideal case both input and output equivalent circuits are

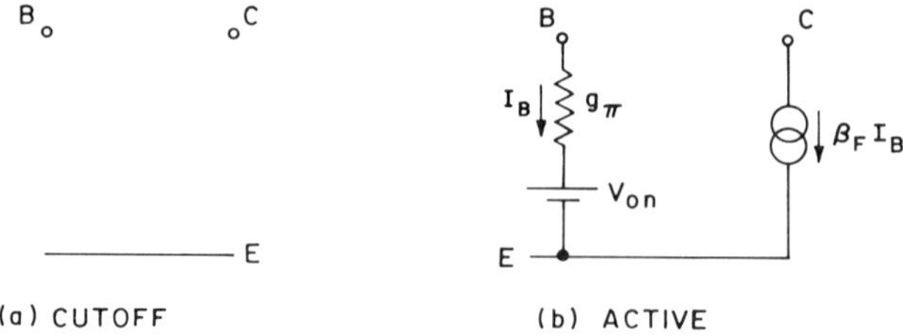

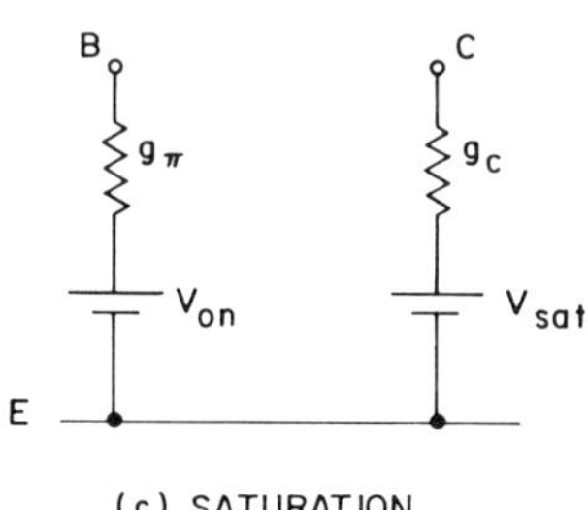

Figure 4.6 Simplified dc equivalent circuits for the bipolar transistor in digital integrated circuits.

presented by open circuits. The total base charge is small (see Chapter 3): $q_{BE} + q_{BC} \simeq 0$, $I_B \approx 0$ and $I_C \approx 0$.

2. Active Region

In this region $V_{BE} \geqslant V_{on}$ and $V_{BC} < V_{on}$. The output circuit is represented by an equivalent current source $\beta_F I_B$. As V_{CE} approaches the soft saturation voltage ($\approx$ 300 mV) the transistor enters into a soft saturation region where the value of β_F starts falling (see Chapter 3). The total charge in the base is mainly the active charge: $q_{BE} \gg q_{BC}$, $q_{BC} \approx 0$, $I_B = I_C/\beta_F$.

3. Saturation Region

In this region both V_{BE} and V_{BC} are $\geqslant V_{on}$. In this case the output circuit is presented by a saturation voltage source V_{sat} in series with a saturation resistance $(1/g_c)$. The total base charge is the summation of an active-mode base charge and an excess storage charge [given by Eq. (3.13)]: $q_{BE} \gg 0$, $q_{BC} \gg 0$, $I_B \gg I_C/\beta_F$.

In addition to the above three regions, the transistor could be operating in the avalanche region where the collector junction is under avalanche breakdown. Because of instability problems, this region is not generally used.

4.4 SATURATED VERSUS NONSATURATED LOGIC FAMILIES

In a digital IC, the bipolar transistor has two steady states. In saturated logic families, these states occur with the transistor is in the cutoff and the

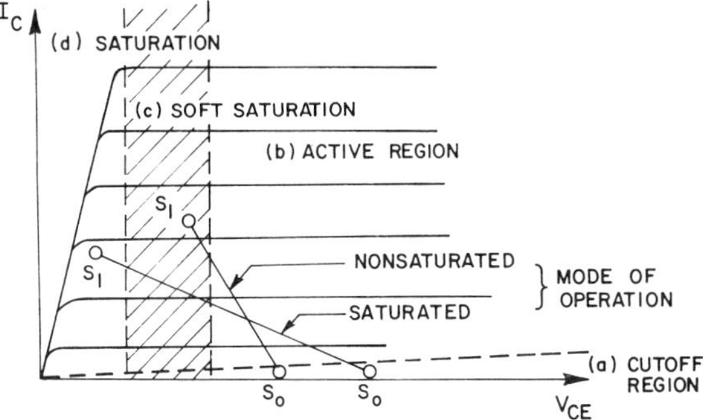

Figure 4.7 Saturated and nonsaturated mode of operation defined on the I_C versus V_{CE} characteristic.

saturation regions. On the other hand, for nonsaturated logic families, the on state occurs with the transistor in the active region. Both cases are shown in Fig. 4.7 on the I_C vs V_{CE} characteristic.

Two main circuit techniques can be used to prevent the bipolar transistor from entering the saturation region:

1. To clamp V_{CE} to a value above the soft saturation voltage ($\approx$ 300 mV).
2. To limit the emitter current (the base or the collector current) to a value corresponding to $I_B \leqslant I_C/\beta_F$, where I_C and I_B are the collector and the base currents determined by the external circuit when the transistor is conducting.

The first technique is used, for example, in Schottky transistor–transistor logic (STTL) where a Schottky diode, with a turn-on voltage of V_{on}(SBD)($\approx$ 400 mV), is connected between the collector and the base (Fig. 4.8). As a result the minimum V_{CE} when the transistor is conducting is $V_{\text{on}}(BE) - V_{\text{on}}(\text{SBD}) \approx 300$ mV.

The second technique is used in current-mode logic families such as emitter-coupled logic (ECL), where a current source I_0 is used to limit the emitter current (and hence the collector current) to I_0. As a result the base current will be limited to $I_0/(\beta + 1)$ provided that $V_{CE} \geqslant 300$ mV.

As shown in Chapter 3, when the transistor enters into saturation, excess charge (q_s) is stored in the device. This charge causes extra delay when the

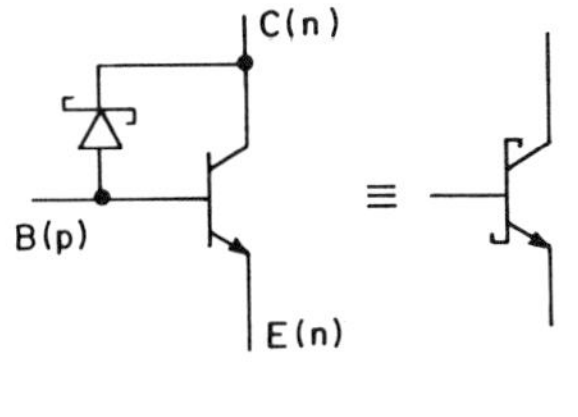

Figure 4.8 Schottky *npn* transistor.

transistor is switched from the on-state to the off-state. As a result nonsaturated logic families are faster than saturated logic families. Saturated logic families include direct-coupled and resistor–transistor logic (DCTL and RTL), diode–transistor logic (DTL), transistor–transistor logic (T^{2}L), and integrated injection logic (I^{2}L). Nonsaturated logic families include emitter-coupled logic (ECL) and emitter-function logic (EFL).

In the following chapters we will present the basic analysis of T^{2}L, ST^{2}L, I^{2}L and other related families, ECL, and EFL. The choice to limit and focus on these families involves two considerations:

1. Today ICs of MSI and even LSI chips use these families almost exclusively.
2. Only with high-packing-density realizations of these logic families can VLSI bipolar chips become feasible.

4.5 THE BIPOLAR TRANSISTOR INVERTER

The analysis of a simple bipolar transistor inverter is given in this section to demonstrate how the calculation of different gate parameters is done. The simple inverter is the basic building block used in the early days of integrated resistor transistor logic (RTL). However, the analysis presented here is more than pedagogical in that the simple transistor inverter is the basic building block for I^{2}L as will be shown in Chapter 5. Moreover, we can treat this minimum inverter gate as the generic unit for on-chip LSI circuits.

Figure 4.9*a* shows the basic inverter with a fan-out n. An input resistor R_b is added to the input to reduce the dependance of the input characteristic of the gate on the diodelike characteristic of the base–emitter junction This makes the circuit less susceptible to “current hogging”, which results from the unmatched base–emitter junction characteristics.

According to the piecewise linear approximation of the input characteristic, the threshold voltage of the inverter V_t is V_{on} of the base–emitter junction, because if $V_{\text{in}} \geqslant V_{\text{on}}$, Q is on and operates in the saturation region and if $V_{\text{in}} < V_{\text{on}}$, Q is off. Figure 4.9*b* shows the dc equivalent circuit of the gate when $V_{\text{in}} = V_1$ and Fig. 4.9*c* shows the dc equivalent circuit when $V_{\text{in}} = V_0$.

With reference to Fig. 4.9, the output voltages can be calculated:

$$\begin{aligned} V_0 &= V_{\text{sat}}\frac{R_L}{R_L + r_c} + V_{CC}\frac{r_c}{R_L + r_c} \\ &\simeq V_{\text{sat}} + V_{CC}\frac{r_c}{R_L} \end{aligned} \tag{4.2}$$

$$V_1 = V_{\text{on}}\frac{R_L}{R_L + (R_b + r_\pi)/n} + V_{CC}\frac{(R_b + r_\pi)/n}{R_L + (R_b + r_\pi)/n} \tag{4.3}$$

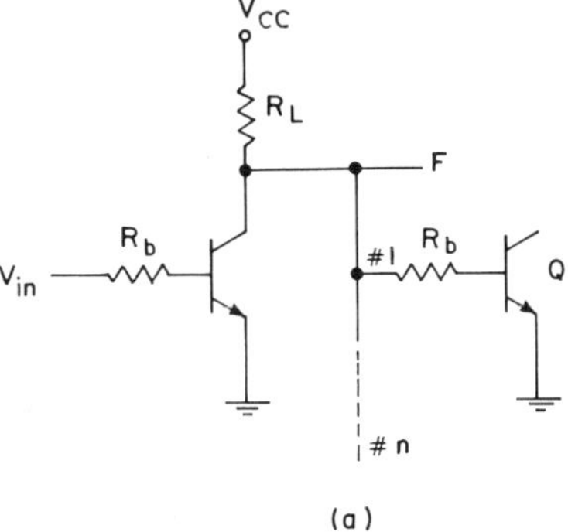

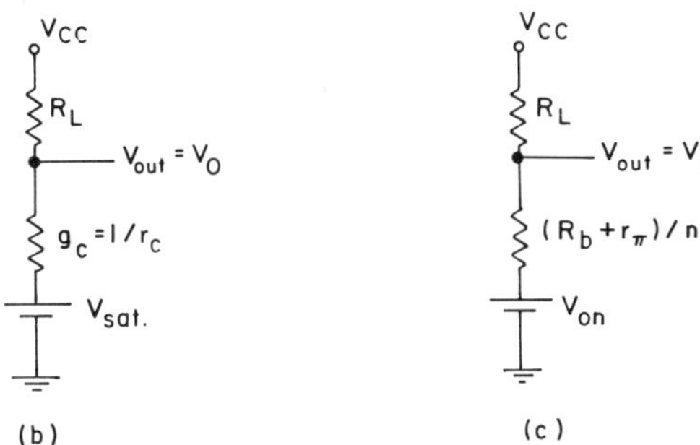

Figure 4.9 The bipolar transistor inverter and dc equivalent circuits.

The approximation given for V_0 is valid, because in general $r_c \ll R_L$. The two noise margins of the gate are given by

$$NM_1 = V_{1\min} - V_{t\max} = V_{1\min} - V_{\text{onmax}}$$

$$NM_0 = V_{t\min} - V_{0\max} = V_{\text{onmin}} - V_{0\max} \tag{4.4}$$

It is clear from Eq. (4.2)–(4.4) that NM_0 is a function of the intrinsic properties of the transistor (r_c, V_{sat}), while NM_1 can be increased by increasing V_{CC}. An increase of the fan-out n reduces V_1. The maximum allowable fan-out can be calculated for a given V_t and a minimum value for NM_1.

The average dc power dissipation of the inverter can be calculated assuming that on the average the inverter stays "on" half of the operating time and "off" the other half. Hence:

$$P_{\text{av}} = (P_{\text{on}} + P_{\text{off}})/2$$

where

$$P_{\text{on}} = V_{CC} I_{\text{on}} \quad \text{and} \quad P_{\text{off}} = V_{CC} I_{\text{off}} \tag{4.5}$$

From the two equivalent circuit of Figs. 4.9*b*, and 4.9*c*

$$I_{\text{on}} = \frac{V_{CC} - V_0}{R_L} \quad \text{and} \quad I_{\text{off}} = \frac{V_{CC} - V_1}{R_L} \tag{4.6}$$

Example. Now let us calculate typical values for the dc parameters of the circuit, given:

$$\begin{aligned}
V_{CC} &= 5 \text{ V}, R_L = 1 \text{ k}\Omega, R_b = 3 \text{ k}\Omega, V_{\text{on}} = 700 \text{ mV}, \\
V_{\text{sat}} &= 50 \text{ mV}, r_c = 50\Omega, \\
r_\pi &\ll R_b, n = 1. \text{ Hence} \\
V_0 &\simeq 300 \text{ mV} \\
V_1 &= 3925 \text{ mV} \\
V_l &= 3625 \text{ mV} \\
V_t &= 700 \text{ mV} \\
V_1 - V_t &= 3225 \text{ mV} \\
V_t - V_0 &= 400 \text{ mV} \\
I_{\text{on}} &= \frac{5 - 0.3}{1} = 4.7 \text{ mA} \\
I_{\text{off}} &= \frac{5 - 3.925}{1} = 1.075 \text{ mA} \\
\frac{I_{\text{on}}}{I_{\text{off}}} &= 4.4 \\
P_{\text{av}} &= 14.4 \text{ mW}
\end{aligned} \tag{4.7}$$

Equation (4.7) shows that this typical gate has a high logic swing, unsymetrical noise margins, a large ratio between I_{on} and I_{off}, and a relatively high power dissipation. The high ratio between I_{on} and I_{off} causes current transient spikes at the power supply line, which can cause a system problem— for example, faulty logic performance.

4.6 TRANSISTOR–TRANSISTOR LOGIC (T²L)

As we mentioned before, in the early days of integrated circuits, the RTL circuit was used as the basic building block. The circuit was chosen because it was a simple discrete component circuit that was already in use and because it was possible to realize it using the state-of-the-art integrated circuits of the day. The basic elements for a NOR gate were a set of *npn* transistors connected in parallel and fabricated on the same isolation land sharing a common collector region and diffused resistors for the base and the collectors.

The shortcomings of RTL lead to diode–transistor logic (DTL)— another discrete component circuit adopted for the integrated form. The DTL has lead to the use of a truly integrated element—the multiemitter *npn* transistor connected to provide multiple *BE* diodes—and eventually to the transistor—transistor logic (T^2L) family. A totem-pole output circuit was added to that basic T^2L circuit if a high driving capability was needed. Hence the family has been widely used in MSI. However, recently with Schottky T^2L (ST^2L), the family has been adopted for LSI applications.

The basic T^2L gate is shown in Fig. 4.10*a* with a fan-out of n. The multiemitter input transistor Q_1 controls the conduction state of the output transistor Q_2. The gate realizes the NAND function since the output F is at logical "0" (low) only if *all* the inputs are at logical "1" (high).

Now let us calculate the dc parameters of the circuit.

Case 1. When one or more of the inputs is low (i.e., at V_0), the base– emitter junction of Q_1 is conducting; with I_1 flowing, no base current is supplied to Q_2, hence, Q_2 is cutoff and the output F is at V_1. Transistor Q_1 is saturated with the base current set at $(V_{CC} - V_{\text{on}} - V_0)/(R_1 + r_\pi)$ and its collector is supplied only by the leakage current I_l of the reverse-biased base–collector junction of Q_2—that is, the condition for saturation ($I_B \gg I_C/\beta$) is satisfied.

In order to calculate V_1 we have to consider the current $n\beta_R I_1'$ supplied by V_{CC} through R_2 to the n output gates. This current results from the fact that the input load transistor(s) Q_1'—corresponding to the next stage(s)—is working in the inverse active mode (the upward mode—see Chapter 3) when the collector–base junction is forward biased. While Q_1' is in the inverse active region, its emitter (which is acting as a collector) current is given by $\beta_R I_1'$, where I_1' is the base current and is given simply by

$$I_1' = \frac{V_{CC} - 2V_{\text{on}}}{R_1 + 2r_\pi}$$

and the total current drawn through R_2 is given by $I_2 = n\beta_R I_1'$. Hence the logical "1" output voltage V_1 is given by

$$V_1 = V_{CC} - n\beta_R R_2 \left(\frac{V_{CC} - 2V_{\text{on}}}{R_1 + 2r_\pi} \right) \tag{4.8}$$

The equivalent circuit model is shown in Fig. 4.10*c*.

From Eq. (4.8) we can conclude that it is advantageous to minimize β_R because the lower β_R is, the higher the maximum allowable fan-out number n. Moreover, the unmatched values of β_R from one loading gate to another results in a current-hogging problem, where more output current is supplied to gates with higher values of β_R. It should be noted that in deriving Eq. (4.8) we have assumed that *all* the emitters of the input transistors of the loading gates are connected to a high-voltage level, that is, all the input transistors of the loading gates are operating in the inverse active mode. This assumption of course gives the lowest possible value for V_1.

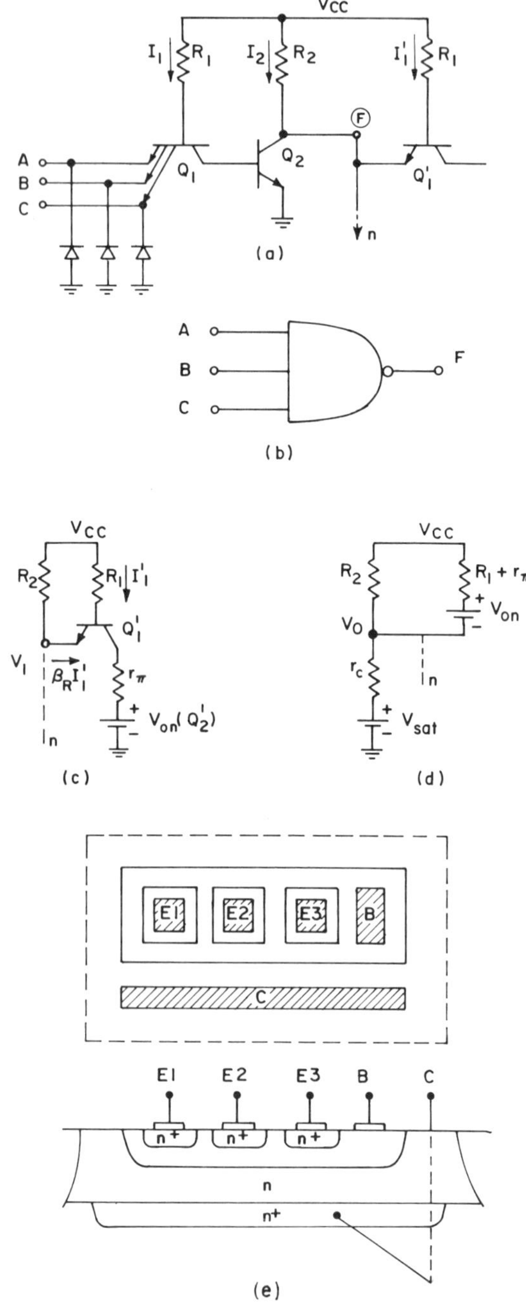

Figure 4.10 (*a*) T^2L basic gate with *n* loading gates. Input diodes are added to reduce input transients. (*b*) Equivalent logic gate. (*c*) Equivalent circuit when the output is high. (*d*) Equivalent circuit when the output is low. (*e*) Layout and cross section.

With reference to Figs. 4.10*a* and 4.10*c*, the currents I_1 and I_2, when the output is V_1, are given by

$$I_1 = \frac{V_{CC} - V_{on} - V_0}{R_1 + r_\pi} \tag{4.9}$$

$$I_2 = n\beta_R \frac{V_{CC} - 2V_{on}}{R_1 + 2r_\pi} \tag{4.10}$$

Case 2. When all the inputs are high (at V_1), the base–collector junction of Q_1 is conducting, supplying the base current to Q_2 through R_1. Hence Q_2 is saturated and the output F is a logical "0," that is, at V_0. The equivalent circuit model for the n loads as well as the collector of Q_2 is shown in Fig. 4.10*d*. With reference to Fig. 4.10*d* V_0 is given by

$$V_0 = V_{sat} \frac{[(R_1 + r_\pi)/n] \| R_2}{\{[(R_1 + r_\pi)/n] \| R_2\} + r_c} + V_{CC} \frac{[(R_1 + r_\pi)/n] \| r_c}{\{[(R_1 + r_\pi)/n] \| r_c\} + R_2}$$

$$+ (V_{CC} - V_{on}) \frac{R_2 \| r_c}{(R_2 \| r_c) + (R_1 + r_\pi)/n}$$

$$\simeq V_{sat} + V_{CC} \frac{r_c}{R_2} + (V_{CC} - V_{on}) \frac{nr_c}{R_1} \tag{4.11}$$

where the symbol $\|$ is used to indicate in parallel with.

It is clear that V_0 is always higher than the saturation voltage of the output transistor Q_2. As the fan-out n increases, V_0 increases, and it is advantageous to reduce r_c.

The currents I_1 and I_2 when the output is V_0 are given by

$$I_1 = \frac{V_{CC} - 2V_{on}}{R_1 + 2r_\pi}$$

$$I_2 = \frac{V_{CC} - V_0}{R_2} \tag{4.12}$$

It should be noted that when the output voltage is V_0, Q_2 is sinking a current equal to $(I_2 + nI_1')$.

If we assume that a typical T^2L gate is at logical "0" state half of its operating time while the other half is operating at "1", the average dc power dissipation of the gate is given by

$$P_{dc} = \frac{V_{CC}}{2}(I_{on} + I_{off}) \tag{4.13}$$

with I_{on} and I_{off} given by

$$I_{on} = [I_1 + I_2] \text{ of Case 2}$$

$$I_{off} = [I_1 + I_2] \text{ of Case 1}$$

The threshold voltage of the gate is the voltage at which the output transistor changes its state from off to saturation. According to the piecewise linear approximation, this is given by

$$V_t = V_{on} - V_{sat} \tag{4.14}$$

where V_{on} is the turn-on voltage of the base–emitter junction of Q_2 and V_{sat} is the saturation voltage of Q_1. This value of V_t can be increased if offset diodes are added between the collector of Q_1 and the base of Q_2. However, these diodes increase the turn-off time of Q_2.

Examining the basic T²L gate of Fig. 4.10*a* from the transient performance viewpoint, we find that Q_1 provides a low impedance path from the base of Q_2 to ground during turnoff transients and the stored charge flows out through Q_1. This provides a faster path for the stored charge than if a simple inverter is used without Q_1. This is one of the key features of T²L. The speed of operation can be further improved if Schottky diodes are added in parallel with the base–collector junctions of Q_1 and Q_2, resulting in a ST²L gate.

Another key feature of T²L is the fact that adding inputs to a single gate requires adding *extra emitters to the multiemitter input transistor* Q_1 (Fig. 4.10*e*). This is a very important feature because the silicon area consumed by a single gate will increase slightly as the number of inputs increases. This explains the suitability of the family for LSI.

Example. Now let us calculate typical values for the dc parameters of the circuit, given $V_{CC} = 5$ V, $R_1 = 4$ kΩ, $R_2 = 1$ kΩ, $V_{on} = 700$ mV, $V_{sat} = 50$ mV, $r_c = 50$ Ω, $r_\pi \ll R_2$, $n = 1$, $\beta_R = 3$. Hence,

$$\begin{aligned} V_1 &= 2300 \text{ mV} \\ V_0 &\approx 353 \text{ mV} \\ V_l &= 1947 \text{ mV} \\ V_t &= 650 \text{ mV} \\ V_1 - V_t &= 1650 \text{ mV} \\ V_t - V_0 &= 297 \text{ mV} \\ I_{on} &= 5.55 \text{ mA} \\ I_{off} &= 3.7 \text{ mA} \\ P_{av} &= 23.1 \text{ mW} \end{aligned} \tag{4.15}$$

4.7 T^2L IN CIRCUIT DESIGN [3–10]

4.7.1 T^2L with Totem-Pole Output

The basic T^2L gate has a limited driving capability because of the relatively high output resistance of the gate when the output voltage is at V_1. If the gate has to be used for off-chip driving, it will have a limited fan-out and slow turn-on time. This can be rectified if Q_2 can be used to drive a totem-pole output circuit as shown in Fig. 4.11. The operation of the circuit is explained as follows:

Case 1. When one or more of the inputs is low (i.e., at V_0), the input transistor Q_1 is saturated and Q_2 is off. As a result Q_3 is on and Q_4 is off. The output voltage level is V_1. The voltage drop across R_2 prevents Q_3 from entering the saturation region, thus it is operating in the active region.

$$V_1 = V_{CC} - 2V_{\text{on}} - R_2 \frac{n\beta_R I_1'}{\beta_F} \approx V_{CC} - 2V_{\text{on}} \tag{4.16}$$

where

$$I_1' = \frac{V_{CC} - 3V_{\text{on}}}{R_1 + 3r_\pi} \approx \frac{V_{CC} - 3V_{\text{on}}}{R_1}$$

$$I_1 = \frac{V_{CC} - V_{\text{on}} - V_0}{R_1 + r_\pi} \approx \frac{V_{CC} - V_{\text{on}} - V_0}{R_1} \tag{4.17}$$

$$I_2 = \frac{n\beta_R I_1'}{\beta_F} \tag{4.18}$$

Case 2. When all the inputs are high (at V_1), Q_1 is operating in the inverse active mode and Q_2 is in saturation. As a result Q_3 is off: its base voltage is held at $(V_{\text{on}} + V_{\text{sat}})$ while its required on voltage is $(2V_{\text{on}} + V_{\text{sat}})$ and Q_4 is saturated.

$$V_0 \simeq V_{\text{sat}} + (V_{CC} - V_{\text{on}}) \frac{nr_c}{R_1} \tag{4.19}$$

$$I_1 = \frac{V_{CC} - 3V_{\text{on}}}{R_1 + 3r_\pi} \tag{4.20}$$

$$I_2 = \frac{V_{CC} - V_{\text{on}} - V_{\text{sat}}}{R_2 + r_c + r_\pi} \tag{4.21}$$

Although for dc conditions the totem-pole has either Q_3 or Q_4 on depending on the logic states, during transient switching both Q_3 and Q_4 are on and the

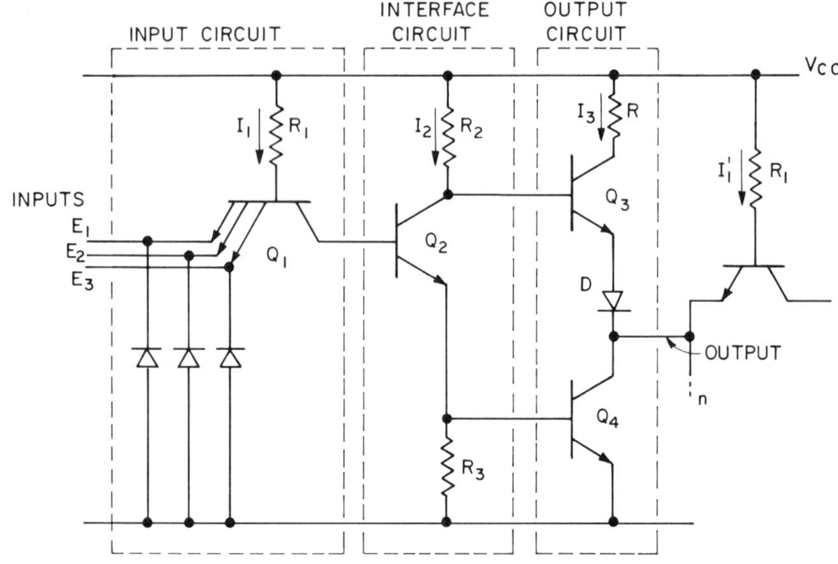

Figure 4.11 T^2L gate with totem-pole output.

current I_3 has to be limited by the resistance R. However, even with this resistor present substantial current spikes between supply and ground do occur.

4.7.2 Modified Totem-Pole Output

The basic circuit of Fig. 4.11 can be modified to meet certain performance requirements. This can be explained by considering as an example the modified circuit of Fig. 4.12. In this circuit the resistor R_3 is replaced by an active pull-down consisting of Q_a, R_a, and R_b. The output transistor Q_3 and the diode D are replaced by a Darlington circuit consisting of Q_A, Q_B, and R_A.

The function of R_3 in Fig. 4.11 is to pull the base of Q_4 down to ground when Q_2 cuts off. This same function is performed by the active pull-down circuit with some advantages. The active pull-down circuit represents a high resistance between the base of Q_4 and ground. As Q_4 is turning on, Q_2 is also turning on. Thus the active pull-down diverts less current from the base of Q_4 allowing Q_4 to turn on faster. During the turning off of Q_4 the active pull-down provides a higher discharging current than the passive pull-down because the collector current of the active transistor Q_a. Thus the turn-off of Q_4 is speeded up. Moreover, the active pull-down allows operation at a wider range of temperature and higher noise margins.

The Darlington circuit offers a lower output resistance when the output voltage is high. This allows higher speed of operation for a given fan-out. It should be noted that R_A is provided for the emitter current to flow because Q_B

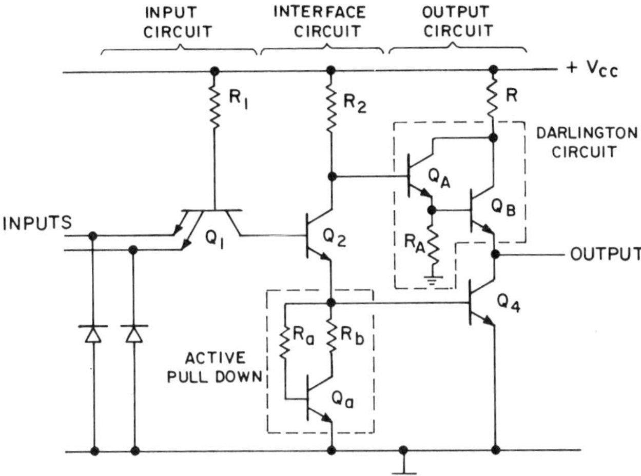

Figure 4.12 T^2L gate with modified totem-pole output.

is always in the active region (its collector junction is reverse biased by the collector emitter voltage drop of Q_A) and its base current is small.

4.7.3 T^2L Circuit Analysis

In analyzing the basic T^2L of Fig. 4.10, we considered two dc steady state cases (cases 1 and 2) and a voltage level (the threshold voltage level) at which a transition occurs from one state to the other. The reader would appreciate the simplicity of the analysis, by examining Table 4.1 where a detailed account is given for the transition from one state to the other. A reference to cases 1 and 2 of the analysis is given. As the input voltage increases from V_0 to V_1, the gate is in transition (between steps 2 and 8 in the table). If detailed (V_{out} vs V_{in}) transfer characteristic is required, break-points corresponding to step 3 up to step 8 could be included. Moreover, in calculating the transient performance of the gate, these break-points could be used in calculating the individual delays. An estimation of the total delay can be obtained by adding these individual delays. A similar approach is applicable to the analysis of T^2L gates with totem-pole output and is left as an exercise for the reader.

4.7.4 T^2L Parasitics

In integrating the basic T^2L gate of Fig. 4.10 there are active parasitic transistors as shown in Fig. 4.13*a*. There is a parasitic *npn* coupling transistor (Q_p) between emitters. Because of the small lateral emitter areas and hence small I_S, this transistor is only important to be considered during transient. There are two *pnp* parasitic transistors associated with Q_1 and Q_2: Q_{p1} and Q_{p2} (see Fig. 4.13*b*). When the inputs are at V_0 and Q_1 is in saturation, Q_{p1} reduces

Table 4.1 Regions of Operation and Breakpoints for the T^2L Gate of Fig. 4.10

			Gate		Load	
		Input	Q_1	Q_2	$\dot{Q}_1$	$\dot{Q}_2$
1	Case 1	V_0	Saturated	Off	Inverse active	Saturated
2	At the threshold	$V_{on} - V_{sat}$	Saturated	Off → active	Inverse active	Saturated
3			Saturated	Active	Inverse active	Saturated
4			Saturated	Active	Saturated	Saturated
5			Saturated	Active	Saturated	Active
6			Saturated	Active	Saturated	Active → off
7			Saturated	Active → saturated	Saturated	Off
8			Saturated	Saturated	Saturated	Off
9	Case 2	V_1	Inverse active	Saturated	Saturated	Off

the sinking current required at the emitters of Q_1. This is because the emitter of Q_{p1} would divert some of the base current of Q_1 to the substrate (S), which is connected to the most negative potential of the circuit (ground in Fig. 4.13). The reduction of the sinking current at the emitters of Q_1 will increase the fan-out capability of the driving gate. However, Q_{p1} has a degrading effect when Q_2 is turning on. This is because it will divert some of the current that would be going to the base of Q_2, thus slowing the turn on time of Q_2. Q_{p2} also diverts current from the base of Q_2 to the substrate, thus reducing the amount of the saturation of Q_2. The effect of both Q_{p1} and Q_{p2} should be considered when the degree of saturation ($I_B - I_C/\beta_F$) is calculated for Q_1 and Q_2.

4.7.5 Wired Logic with T^2L

The basic logic gate realized in T^2L is the NAND gate, and logic functions can be realized using this gate. However, in certain applications it will save silicon area if a wired logic is performed at the output. Figure 4.14 shows a typical example:

$$F = \overline{AB + CD}$$

that is, the gate realizes an AND-OR-Invert (AOI) function.

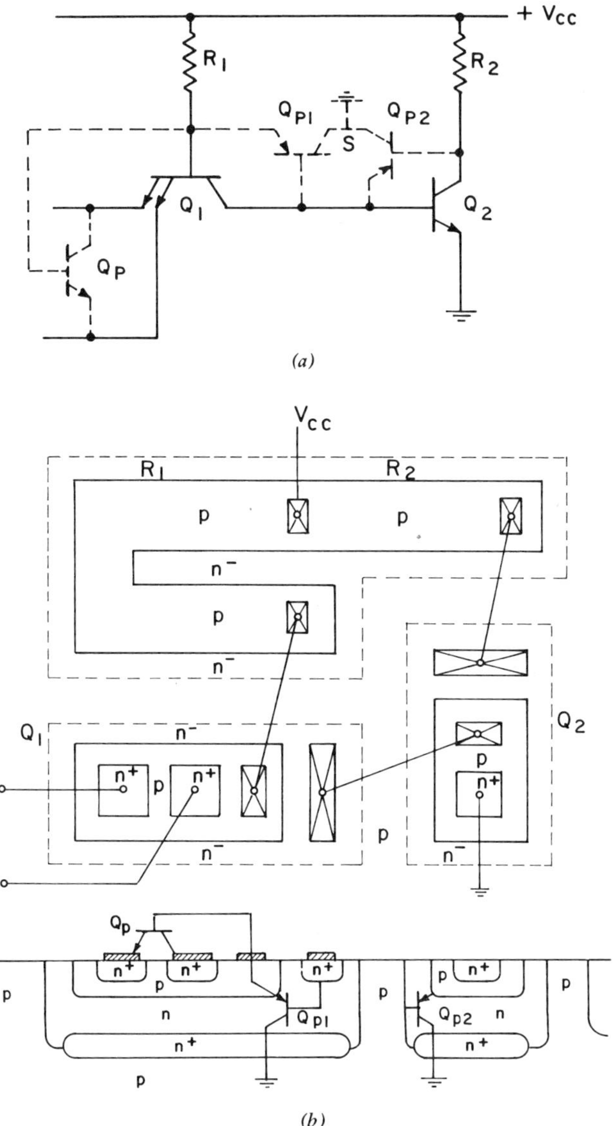

Figure 4.13 (*a*) T^2L gate with active parasitic transistors: Q_p, Q_{p1}, and Q_{p2}. (*b*) Plan and section views of the T^2L gate and the parasitic transistors: Q_p, Q_{p1}, and Q_{p2}.

4.7.6 Open-Collector T^2L

In MSI it is possible to increase the flexibility of the basic T^2L gate by making the collector of the output transistor (Q_2 in Fig. 4.10 and Q_4 in Fig. 4.11) available to the user with no load connected to it (in Fig. 4.10, R_2 is omitted and in Fig. 4.11, R, Q_3, and D are omitted). This allows a choice for a

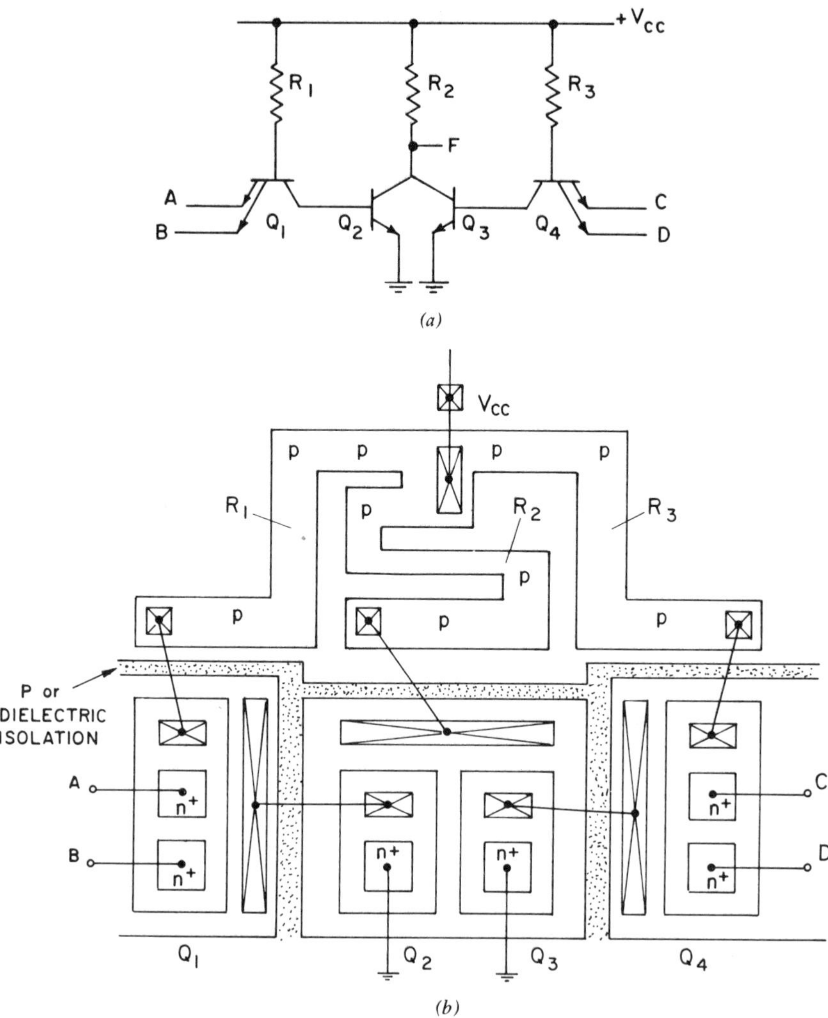

Figure 4.14 (*a*) T^2L gate realizing an AND-OR-Invert logic function. (*b*) Corresponding layout of Fig. 4.14*a*

wired-AND logic at the output, that is, the outputs of several open-collector gates can be tied together to a single pull-up resistor in the case of the basic T^2L gate (Fig. 4.10) or to a single active pull-up in the case of a totem-pole output stage. Moreover, a choice of a suitable V_{CC} and a load value is possible.

4.7.7 Tristate Outputs

When the T^2L totem-pole is used as a driver, it is desirable in many applications that the output, under the control of a disable signal, goes into a

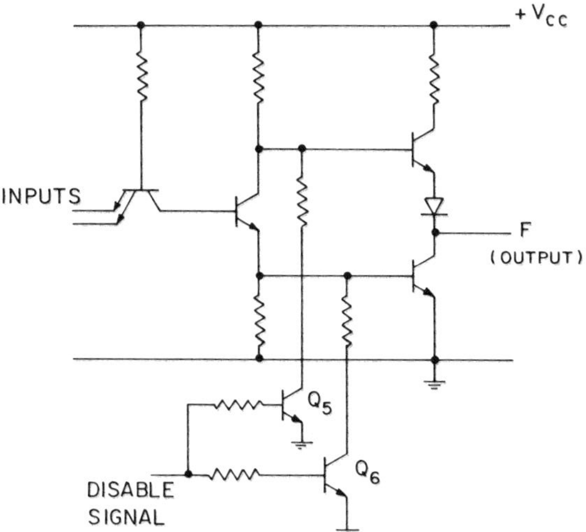

Figure 4.15 T^2L gate with tristate output.

third state where the two output transistors are both turned off. This case arises, for example, where the output is connected to a common signal line shared among many T^2L totem-pole stages. Figure 4.15 shows a typical T^2L gate with a tristate output. The third state occurs when Q_5 and Q_6 are both on, as determined by the disable signal being high.

4.7.8 Low-Power T^2L (LT^2L)

The values of the resistors used in the realization of a T^2L gate determine the level of the power dissipation of the gate, hence its speed. The higher the value of the resistors, the lower the power dissipation and the higher the gate delay. The delay power product stays almost constant. This results in a T^2L logic family (LT^2L) operating at lower speed with lower power dissipation.

4.7.9 Transister Clamped T^2L

In T^2L, the saturation of the output transistor causes an extra delay when it is turning off. Reducing this amount of saturation or preventing it, speeds up the turn-off time. In the next section we shall study a widely used technique: the use of Schottky diodes to clamp the collector junction of the *npn* transistor, resulting in the Schottky T^2L (ST^2L) logic family. In this section an alternative approach is discussed. Similar approach is used in I^2L, the folded-collector I^2L, and is discussed in the next chapter.

The technique uses an extra *npn* transistor Q_3 in Fig. 4.16 to control the amount of saturation of Q_2 [11, 12]. This control is obtained by diverting some of the base current of Q_2 into the collector of Q_3 and by clamping the base–collector voltage of Q_2. As Q_2 tends to saturate a voltage drop occurs

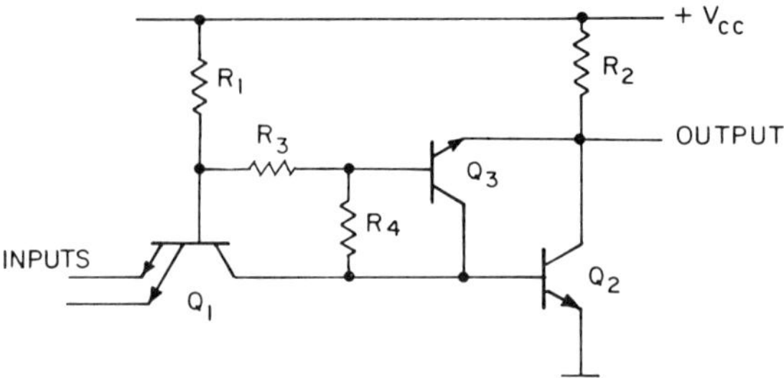

Figure 4.16 Transistor clamped T^2L.

across R_4 and should be enough to turn on Q_3. If the controlling transistor Q_3 is allowed to saturate, then the low logical output V_0 is given by $V_0 = V_{on}$ (of the base–emitter junction of Q_2) $- V_{sat}$ (of Q_3). In laying out the gate of Fig. 4.16, note that Q_1 and Q_3 have a common collector. This method of saturation control requires more silicon area than the Schottky clamp, but does not require any additional processing steps.

4.8 SCHOTTKY T^2L [13–15]

Advances in bipolar technology and the use of Schottky diodes have led to the adoption of T^2L to MSI and LSI. Advances in bipolar technology allow small transistor geometries (see Chapter 2) while Schottky diodes are used to prevent the saturation of the *npn* transistors. Moreover, Schottky diodes are used to replace the input junction diodes in conventional T^2L gates, thus reducing the input transients.

As mentioned in Section 4.4, one method of preventing the transistor from going into saturation is to clamp the base–collector junction with a Schottky diode, which has a lower turn-on voltage than a junction silicon diode (see Fig. 4.8). This reduces the storage charge in the base of the transistor, hence reduces the storage turn-off time. The Schottky clamped transistor (Schottky transistor) is fabricated as a single structure as shown in Fig. 4.17. The diode is built within the same collector isolation region as the transistor and is formed by the interconnection metallization. However, additional metal deposition is sometimes used to reduce the diode leakage.

Figure 4.18 shows the ST^2L versions of Figs. 4.10*a* and 4.12. The transistor Q_B is not replaced with a Schottky transistor because it does not saturate. The high-voltage level is the same as T^2L, while the low-voltage level V_0 is given by $(V_{on} - V_{SBD})$, which is on the order of 300–400 mV. Thus the ST^2L logic swing (V_l) is lower, and this results in even higher speed (delay $\propto V_l$) and lower power dissipation (dynamic power dissipation $= CV_l^2 f$).

ST^2L can offer the same circuit function as that for T^2L. For example, Fig. 4.19 shows a tristate ST^2L driver gate [16]. The transistor Q_1 is an enable device; if Q_1 is off, the gate is acting as a driving inverter. If Q_1 is on

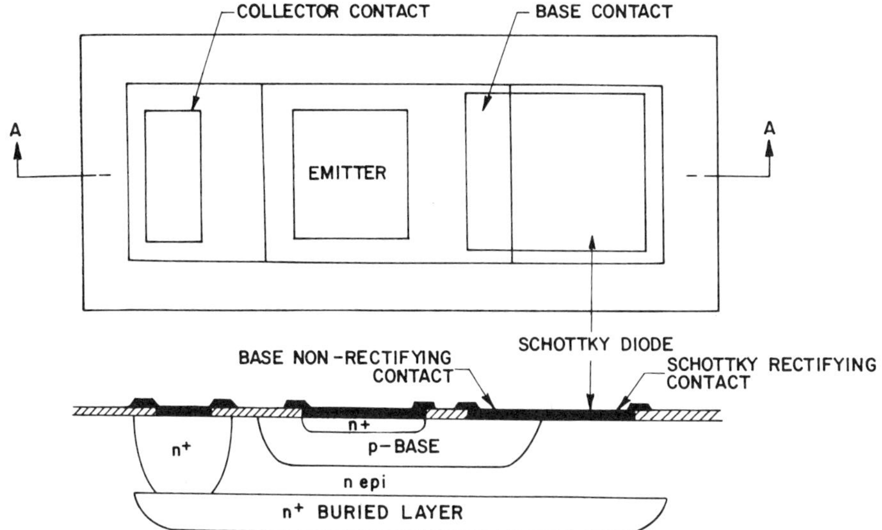

Figure 4.17 Top view and cross section of Schottky transistor.

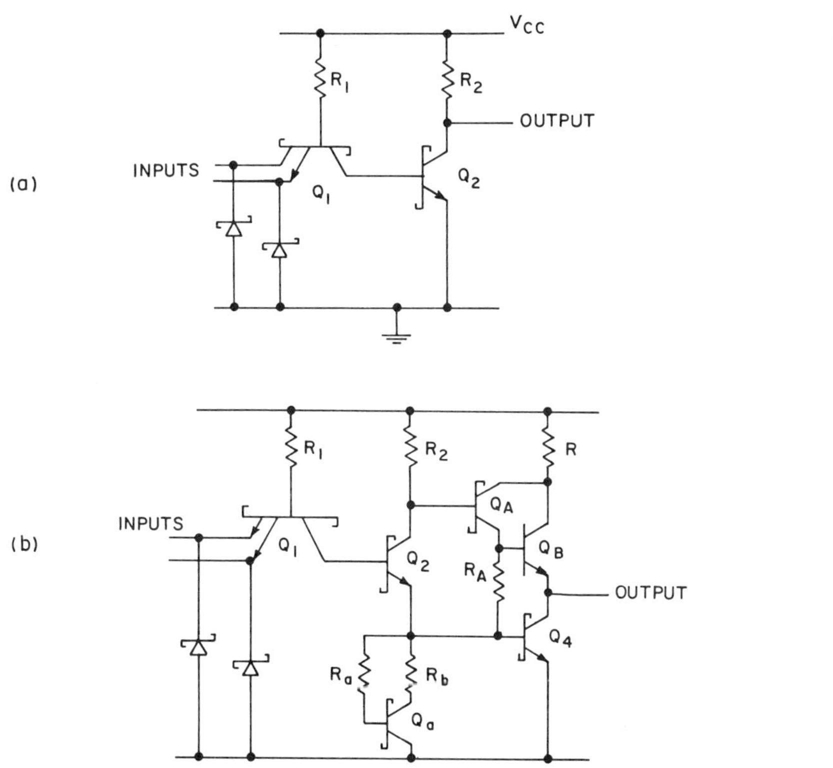

Figure 4.18 ST^{2}L gates: (*a*) basic gate and (*b*) gate with totem-pole output.

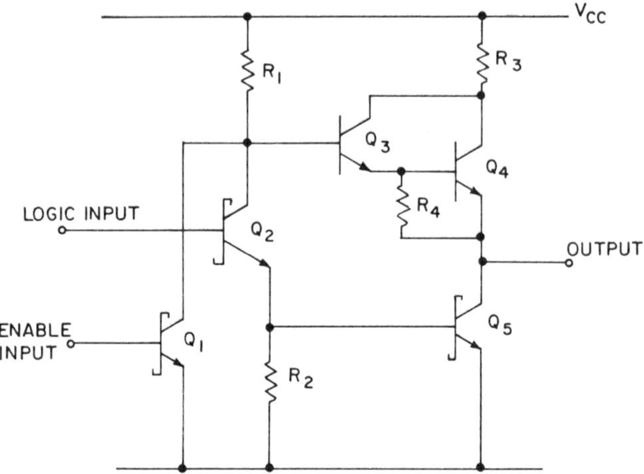

Figure 4.19 Tristate ST²L driver.

(saturated), Q_3–Q_5 are off independent of the logic input signal. The output node then becomes floating, and the output impedance is a function of the output node capacitance and the leakage currents of that node. This allows a parallel interconnection of the outputs of several drivers.

It should be noted that ST²L has a lower logic swing (due to the higher V_0) and lower NMs than T²L. Moreover, the base– collector breakdown voltage is lower because of the Schottky diodes. However, these disadvantages do not affect the performance of the circuit at the chip level, where voltage levels are low and the noise generated is not excessive.

As mentioned in relation to T²L, low-power ST²L is obtained by using higher-value resistors, thus achieving lower power dissipation at lower speed of operation. A modified low-power ST²L gate, LS, is shown in Fig. 4.20. To increase the input impedance of the gate, the multiemitter input transistor is not used and is replaced by an input circuit that consists of resistor R_1 and Schottky diodes D_1 and D_2. The output circuit is similar to that of Figs. 4.12 and 4.18 and consists of Q_5, a Darlington circuit, and a limiting resistor R_3. The interface circuit is identical to that of Figs. 4.12 and 4.18 and consists of Q_1, R_2, and an active pull-down circuit. The threshold voltage of the LS gate is given by

$$V_t = (V_{\text{on}} \text{ of } Q_5) + (V_{\text{on}} \text{ of } Q_1) - (V_{\text{on}} \text{ of } D_2) \sim 1 \text{ V}$$

In order to increase the threshold voltage of the gate and hence its noise margins, and at the same time improve its dynamic performance, the coupling between the input circuit and the interface circuit is modified in a second generation LS gate, LS² [21], as shown in Fig. 4.21. The threshold voltage of the LS² gate is given by

$$V_t = (V_{\text{on}} \text{ of } Q_5) + (V_{\text{on}} \text{ of } Q_1) + (V_{\text{on}} \text{ of } Q_6) - (V_{\text{on}} \text{ of } D_2) \sim 1.4 \text{ V}$$

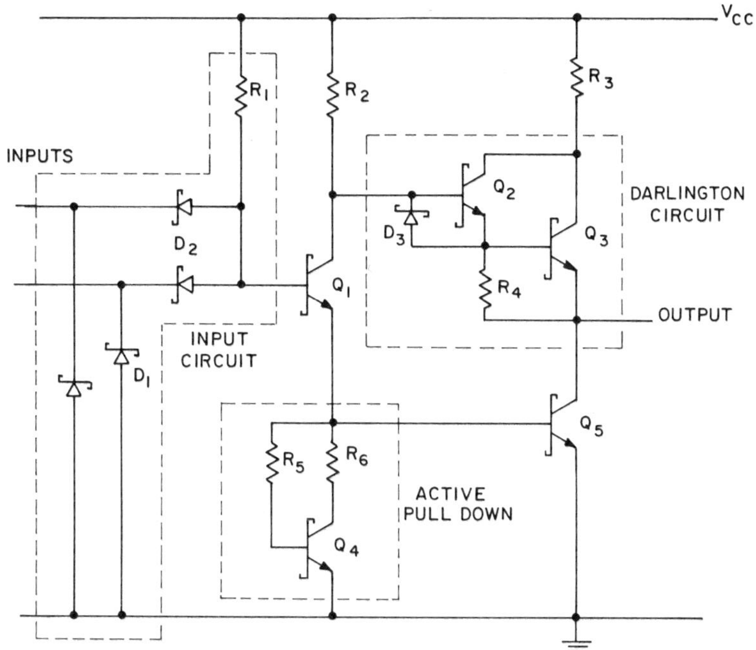

Figure 4.20 A modified totem-pole low-power Schottky (LS) gate.

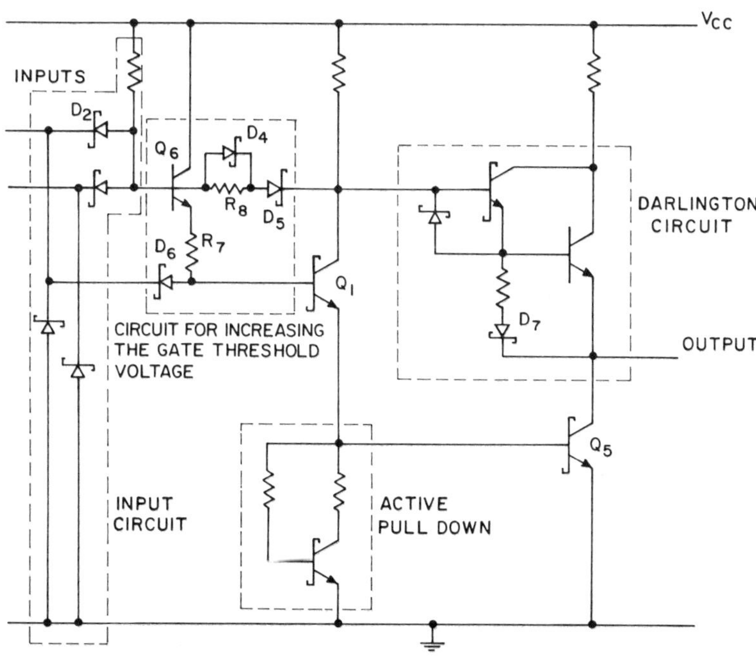

Figure 4.21 Second generation LS gate: LS^2.

The improvement in the dynamic performance is due to the rapid high-to-low transition at the output and the high-capacitance-drive capability. The transistor Q_6 is the key component in this improvement. As the input rises and approaches V_t, Q_6 conducts and supplies a surge of current into the bases of Q_1 and Q_5. This allows the rapid switch of the output from logic "1" to logic "0." This current surge from Q_6 must be controlled or else spiking will result. This control is accomplished by R_7, D_4, D_5, and R_8.

The Darlington circuit of LS^2 is the same as that of LS except for the addition of D_7 to the pull-down resistor R_4. This enables the output to be pulled high while maintaining the "1" logic state, which makes it easy to interface to CMOS logic.

4.8.1 ST^2L Circuits

As mentioned before, ST^2L circuits are based on their T^2L counterparts. However, it is possible to extract the best possible performance from ST^2L by adapting new circuits.

The circuits shown in Figs. 4.18*b*, 4.20, and 4.21 represent conventional gates of the MSI world. On the other hand, the simplified gate shown in Figure 4.18*a* represents a reduced complexity realization, which, although there are noted drawbacks, has substantial density and reduced power advantages for on-chip design purposes. One way to achieve the best of both T^2L worlds is to use on-chip (internal) logic levels and also provide conventional off-chip T^2L buffers. Hodges [17] proposed the gate configuration shown in Fig. 4.22, which meets both these objectives. All the components are directly compatible with a T^2L technology and the speed performance matches that of T^2L quite well. Figure 4.22*b* shows a portion of the layout.

Two sets of parameter values for the circuit in Fig. 4.22 are shown in Table 4.2, together with power consumption and delay for each. The internal supply voltage, V_{CC}, is 3.0 V at room temperature. This is provided from 5.0 V by an on-chip regulator. Nominal logic levels at the output are 0.4 and 1.4 V.

When the input is open-circuited, Q_1 and Q_3 conduct and V_{in} is about 1.5 V; Q_2 is off. V_{out} is about 0.4 V. When the input is grounded, Q_1 and Q_3 are off and Q_2 conducts at a low-current level, fed by the base current flowing via R_2, R_3, and D_1. V_{out} is clamped at about 1.4 V by the feedback through Q_2. The transition point, at which $V_{in} = V_{out}$, is about 1.0 V, resulting in 0.4 V or more of noise immunity in each state. This is entirely adequate for on-chip interconnections.

The diodes D_1–D_4 are all formed on the *n* epitaxial collector region of Q_3. Alternative diode or multiemitter transistor gating arrangements result in substantially more total capacitance. Q_1 and Q_3 are Schottky-clamped transistors. The clamped saturation voltage of Q_3 determines the low output level. The output high level is clamped by action of Q_2, R_2, D_1, R_5, and R_4; the voltage at node 3 is about 0.5 V. Turn-on delay of Q_3 is thereby reduced compared to the situation in which this voltage goes to zero. Circuit perfor-

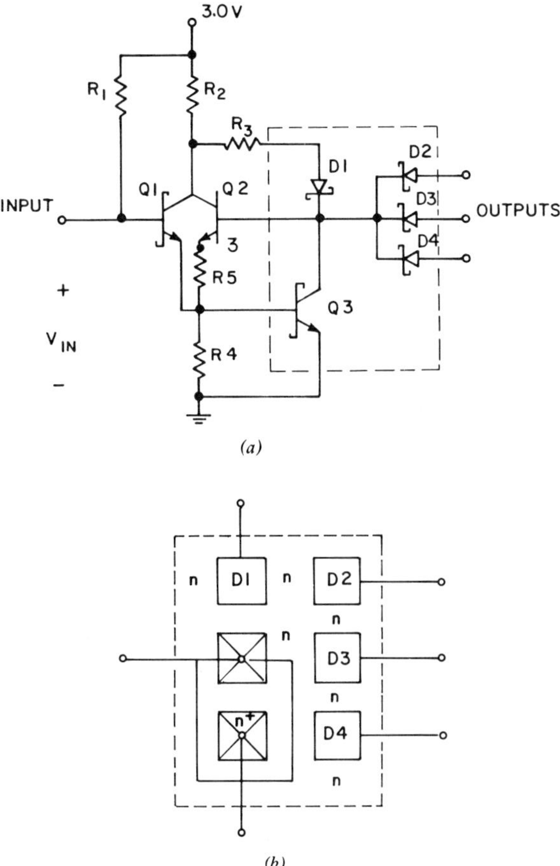

Figure 4.22 Basic double-clamped NAND gate: (*a*) circuit diagram and (*b*) layout of Q_3 and diodes $D1$–$D4$.

Table 4.2 Parameters for Double-Clamped Gates

	Low Power	High Speed
R_1	12 kΩ	2.4 kΩ
R_2	5 kΩ	1 kΩ
R_3	8 kΩ	1.6 kΩ
R_4	2.5 kΩ	0.5 kΩ
R_5	.5 kΩ	0.1 kΩ
dc power	1.4 mW	7 mW
Fan-out	3	3
Propagation delay	3.5 ns	1 ns
Power × delay	5 pJ	7 pJ

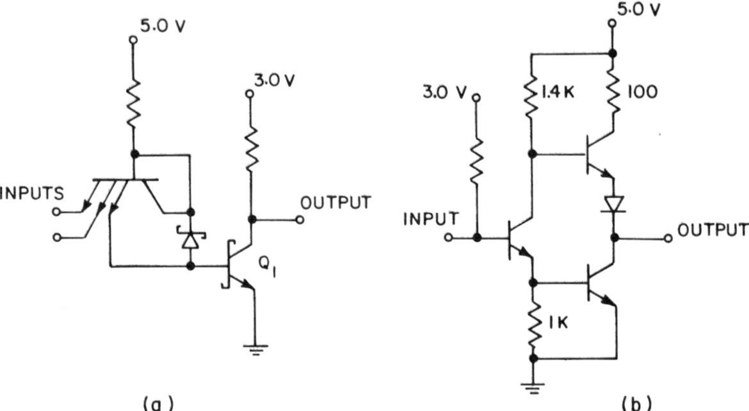

Figure 4.23 (*a*) Input interface circuit: 5-V T^2L to 3-V double-clamped ST^2L. (*b*) Output interface circuit: 3-V double-clamped ST^2L to 5-V T^2L.

mance is summarized in Table 4.2. A fan-out of three identical circuits was assumed, with one on-chip loading included. These are intended for use as internal gates in MSI or LSI logic components. Chip area per gate is about 0.02 mm^2, or about one-third that of a standard Schottky TTL.

Figure 4.23 shows interface circuits to and from standard T^2L signal levels (Transition voltage $\simeq$ 1.5 V at room temperature.) These circuits would be used at the inputs and outputs of an MSI or LSI component. Simple interface circuits are essential if any overall advantage is to be obtained in MSI circuits, since these usually have only three to six gates in any path from input to output.

4.8.2 ST^2L and LSI

Since the introduction of ST^2L in 1970, the family has found its applications first in MSI and then in LSI. This is mainly due to the smaller silicon area taken by ST^2L and the almost half speed-power product of ST^2L compared to that of T^2L. ST^2L was originally designed as an extension to standard T^2L. ST^2L MSI functions were available for one-to-one replacement with T^2L MSI [15]. In fact, a circuit designer can mix ST^2L and T^2L for any subsystem, using the former where speed is important and using the later where NMs are important. For example in many minicomputers, ST^2L is used for computation registers—adders, multipliers, accumulators, etc.—while T^2L is used in the input–output stages, memory drivers, and other interface circuits.

ST^2L found its way to LSI not only because of the lower area and lower power-speed product, but also because of other factors. One important factor is that no level shifting is required to interface ST^2L with any conventional logic families (T^2L, DTL, or low-threshold MOS) except for ECL. Another factor is that the Schottky-clamped input diodes suppress the effects of line noise (ringing) because the transfer characteristics are sharper than T^2L. The fact that the designer is flexible in choosing between ST^2L, low-power ST^2L, and T^2L is another advantage. As a result, ST^2L LSI functions are available. For example, a single chip microcontroller is used in association with other ST^2L chips in minicomputers [18] and field-programmable arrays can replace random logic functions [19].

4.9 SWITCHING MATRIXES AND PLAs

The impact of ST^2L on logic design is enhanced by using regular logic arrays. This section deals with the concepts of switching matrixes and programmable logic arrays (PLAs). Although the discussion here is focused on ST^2L by using Schottky diodes, the approach is more universal and is applicable to bipolar and MOS logic families.

A switching matrix is a circuit that gives multiple output logic functions for mutiple inputs. A multiple-output circuit can be described by a suitable Boolean matrix function. Figure 4.24 shows a two-input switching matrix; providing four outputs for two input variables. The particular matrix shown uses logical AND to give all the product combinations P_0, P_1, P_2, and P_3. The output functions (f_0, f_1, f_2, and f_3) can be related to the products in a matrix form; in our example, the switching matrix is rectangular and diagonal, and it transforms the input minterms (products) into the output functions:

$$\begin{bmatrix} f_0 \\ f_1 \\ f_2 \\ f_3 \end{bmatrix} = \begin{bmatrix} 1 & 0 & 0 & 0 \\ 0 & 1 & 0 & 0 \\ 0 & 0 & 1 & 0 \\ 0 & 0 & 0 & 1 \end{bmatrix} \begin{bmatrix} P_0 \\ P_1 \\ P_2 \\ P_3 \end{bmatrix}$$

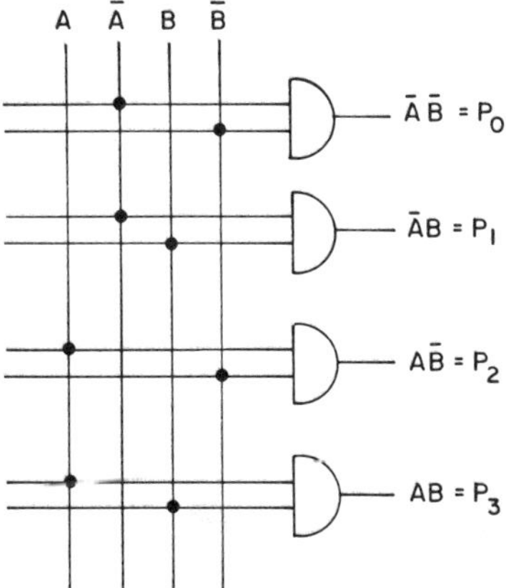

Figure 4.24 Two-input switching matrix.

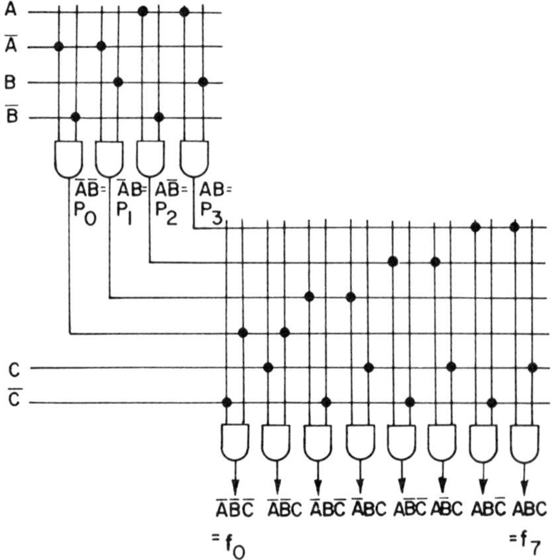

Figure 4.25 Two-level switching matrix for three input variables.

If the AND gates are replaced by OR gates, the switching matrix remains the same, and the products are replaced by logical sums S_0, S_1, S_2, and S_3, namely, the maxterms.

Other switching functions can be obtained for two inputs (or more) either by changing the number of outputs, by altering the interconnection nodes, or by replacing the AND or OR gates by different logic elements.

The above concept can be extended to multiple-level logic circuits. Figure 4.25 shows a two-level switching matrix for three input variables [20]. The first level is a four-output matrix, as the one shown in Fig. 4.24, which drives, in turn, a second switching matrix. Both matrixes use AND gates as logic elements. Each input pass through two AND elements before emerging at the output lines (f_0–f_7).

One of the simplest ways to implement the matrix functions is to use diodes. In ST^{2}L, these diodes are Schottky diodes. As an example, Fig. 4.26 shows a three-input *decoding matrix*. A circuit similar to that of Fig. 4.26 can be used as a *selection matrix* (Fig. 4.27) for eight inputs I_0–I_7 giving the output:

$$f = I_0P_0 + I_1P_1 + I_2P_2 + \cdots + I_7P_7$$

i.e., the output f is function of the eight inputs and the three control inputs A, B, and C ($P_0 = \bar{A}\bar{B}\bar{C}$, etc.). In general, similar circuits can be constructed to give *encoders*; giving n output pairs for 2^n inputs and *decoders*; giving 2^n outputs distributed from one input according to the state of n input pairs.

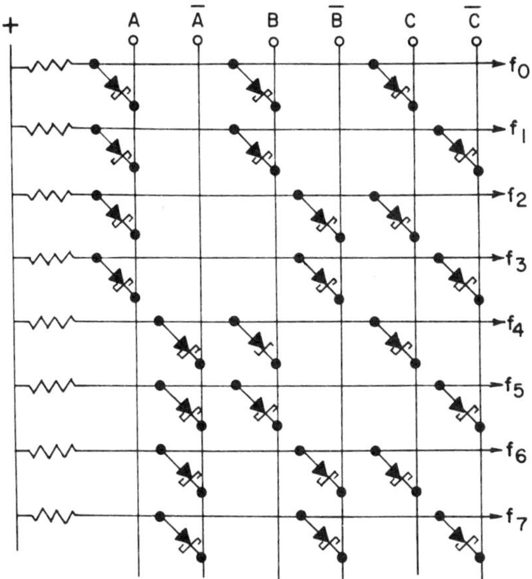

Figure 4.26 A three-input rectangular Schottky diode (SD) switching matrix.

As an example let us consider using a SD matrix to realize the sum (S) and the carry (C) of a full adder:

$$S = \overline{X}\,\overline{Y}C_0 + \overline{X}Y\overline{C}_0 + X\overline{Y}\,\overline{C}_0 + XYC_0$$

$$C = \overline{X}YC_0 + X\overline{Y}C_0 + XY\overline{C}_0 + XYC_0$$

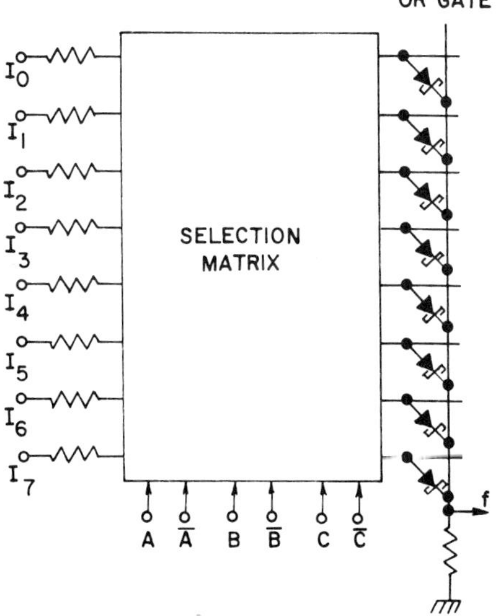

Figure 4.27 An SD selection matrix.

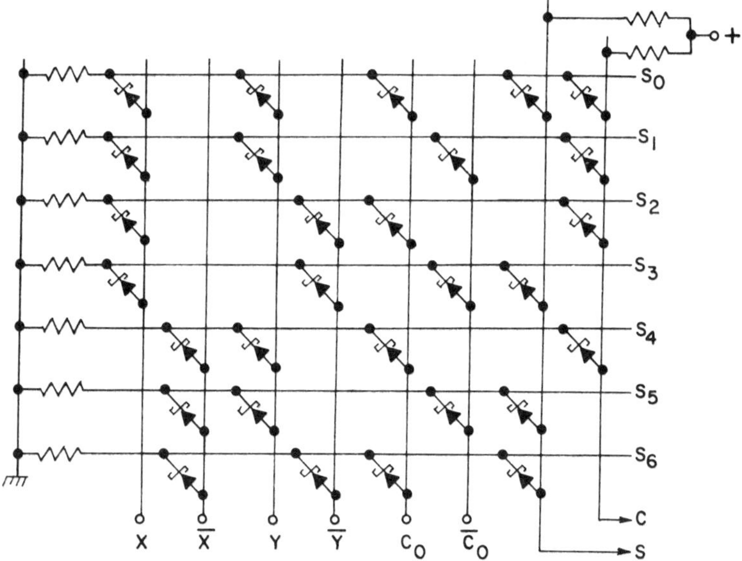

Figure 4.28 Full adder realization using SD matrix.

where X, Y are the inputs and C_0 is the previous carry. The realization is shown in Fig. 4.28 where an OR gate matrix is followed by an AND gate matrix: $S = S_0S_3S_5S_6$ and $C = S_0S_1S_2S_4$.

The matrix of Fig. 4.28 could be of a universal nature if diodes are allocated on the silicon chip at the signal line intersections and the user programs the matrix to obtain the required logic function. This is referred to as programmable logic arrays (PLAs). A PLA has an AND-matrix (logical product generator) and an OR-matrix (logical summer generator), and can be used to realize combinational logic functions. If a sequential circuit is to be realized with PLAs, the output of the OR-matrix can be fed via time delay elements (e.g., flip-flops) back to the input of the AND-matrix (Fig. 4.29).

Read-only-memories (ROMs) can also be used to implement logic functions in a fashion similar to PLAs. A ROM is generally realized in a matrix form. n rows representing the number of words and m columns representing the number of bits in a word. Thus, a ROM can store the contents of a switching matrix; for example, a switching OR-matrix of two variables requires two-word two-bit per word ROM.

It should be noted that PLAs are universal logic blocks and can be underutilized if the function to be realized is simple. Figure 4.30 illustrates this point. The silicon area and the delay for a given PLA is constant, independent of the function to be realized. Thus, there is an optimum complexity below which it is uneconomical to use a given PLA. This optimum complexity is a function of the size of PLA and the LSI technology used for its realization.

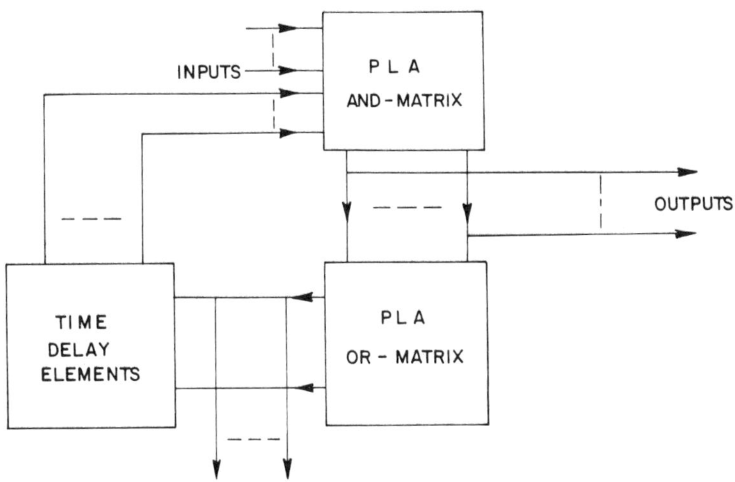

Figure 4.29 Sequential logic realization using PLAs.

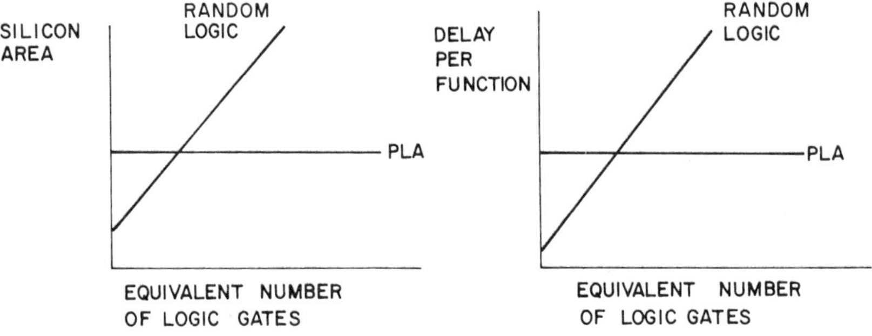

Figure 4.30 PLAs performance curves.

REFERENCES

1. C. S. Meyer, D. K. Lynn, and D. J. Hamilton, *Analysis and Design of Integrated Circuits*, McGraw-Hill, New York, 1968.
2. T. R. Blakeslee, *Digital Design with Standard MSI and LSI*, Wiley, New York, 1975.
3. L. S. Garrett, Integrated-Circuit Digital Logic Families: TTL Devices, *IEEE Spectrum* **7**, 63–72 (Nov. 1970).
4. A. B. Glaser and G. E. Subak-Shorpe, *Integrated Circuit Engineering: Design, Fabrication and Applications*, Addison-Wesley Reading, MA, 1977.
5. D. J. Hamilton and W. G. Howard, *Basic Integrated Circuit Engineering*, McGraw-Hill, New York, 1975.
6. H. Taub and D. Schilling, *Digital Integrated Electronics*, McGraw-Hill, New York, 1977.
7. C. A. Holt, *Electronic Circuits: Digital and Analog*, Wiley, New York, 1978.

8. V. H. Grinich and H. G. Jackson, *Introduction to Integrated Circuits*, McGraw-Hill, New York, 1975.

9. R. L. Morris and J. R. Miller, *Designing with T^2L Integrated Circuits*, McGraw-Hill, New York, 1971.

10. B. T. Murphy and V. J. Glinski, Transistor-Logic with High Packing Density and Optimum Performance at High Inverse Currents, *IEEE JSSC* **SC-3**, 261–267 (1968).

11. S. K. Wiedmann, A Novel Saturation Control in T^2L Circuits, *IEEE JSSC* **SC-7**, 243–250 (1972).

12. F. Capocaccia, An Integrated Circuit Technique for Saturation Control of Switching Transistors, *IEEE JSSC* **SC-3**, 267–270 (1968).

13. Y. Tarui *et al.*, Transistor Schottky-Barrier Diode Integrated Logic Circuit, *IEEE JSSC*, **SC-4**, 3–12 (1969).

14. R. N. Noyce, R. E. Bohn, and H. T. Chua, Schottky Diodes Make IC Scene, *Electronics* **42**, 74–80 (July 21, 1969).

15. L. Altman, *Large-Scale-Integration*, Electronics Book Series, McGraw-Hill, New York, 1976.

16. J. C. Barrett *et al.*, Design Considerations for a High-Speed Bipolar Read-Only Memory, *IEEE JSSC* **SC-5** 196–202 (1970).

17. D. A. Hodges, Department of Electrical Engineering and Computer Sciences, University of California, Berkeley, CA, personal communication.

18. R. J. Clayton *et al.*, Evaluation Breeds a Microcomputer that Can Take On Its Big Brothers, *Electronics* **44**, 62–66 (October 11, 1971).

19. N. Cavlan and S. J. Durham, Field Programmable Arrays: Powerful Alternatives to Random Logic, *Electronics* **52**, 109–114 (July 5, 1979).

20. B. Zacharov, *Digital Systems Logic and Circuits*, American Elsevier, New York, 1968.

21. P. Hoffpair, LS^2 Family Squeezes More Speed from Existing T^2L Sockets, *Electronics* **53**, 149–153 (February 28, 1980).

Chapter 5

Logic Families for LSI / VLSI: I^2L and Related Families

5.1 INTRODUCTION

As we mentioned in Chapter 4, T^2L is primarily used in MSI, whereas ST^2L can be adopted for LSI because of the lower logic swing and lower power dissipation as well as smaller gate area. Bipolar logic families can be used further in LSI provided they can continue to decrease power dissipation and shrink gate area. Integrated injection logic (I^2L) represents the first bipolar family that meets these design objectives and is the subject of this chapter.

In addition to the importance of the I^2L family in LSI circuit design, the development of the family offers a unique opportunity for the reader to examine closely such LSI circuit techniques as merging of semiconductor regions and repartitioning of logic circuits. In fact these same techniques are used in developing high-speed LSI realizations of ECL, as treated in Chapter 6. It is demonstrated that the circuit and logic design functions can be tightly coupled as shown in Section 5.6.1 by examples of folded-collector I^2L applied to threshold logic and in multiple-level stacked I^2L. Finally, the I^2L family offers another opportunity to examine, model, and characterize the upward mode of operating the *npn* integrated transistor.

Before we study the family, we would like to draw the reader's attention to the discussion of Section 4.2 related to the definition of noise margins. We have stated that in logic circuits, two levels of an electrical quantity—e.g., voltage, current, energy, or impedance—can be used to represent the logic binary information "0" and "1." We have used the voltage-level representation in defining V_0 and V_1, and then we have defined two voltage noise margins VNM_0 and VNM_1. However, as it will become clear in this chapter, the I^2L gate is current operated and current noise margins are more appropriate to use and are defined in Section 5.2.1. In fact, the VNM_1 of I^2L is close to zero. However, it should be pointed out that although the I^2L gate is current operated, it still produces two distinct voltage levels at the output: V_0 and V_1.

5.2 INTEGRATED INJECTION LOGIC (I^2L)

Integrated injection logic (I^2L), also known as merged transistor logic (MTL), is a low-power, high-density bipolar logic family suitable for LSI. It was introduced in 1972 by Berger and Wiedman [1] and Hart and Slob [2]. It is based on repartitioning the direct-coupled transistor logic (DCTL) and merging of a vertical *npn* transistor with a lateral *pnp* transistor. The vertical transistor acts as a switching transistor, while the lateral one is used as a current source.

Figure 5.1 shows the development of I^2L from DCTL. If the load resistance of DCTL is considered to be associated with the input circuit as shown in Fig. 5.1*b*, rather than the output, the input circuit becomes a current source I_0 steered by the input level and the output circuits becomes a multicollector transistor as shown in Fig. 5.1*c*. In I^2L this current source I_0 is realized using a lateral *pnp* transistor as shown in Fig. 5.1*d*. The *npn* transistor is allowed to operate in the active upward mode with the epitaxial layer and n^+ buried layer acting as an emitter and the surface n^+ diffusions acting as collectors. This is basically the same structure as used for the T^2L input device (Fig. 4.11*a*), but the collector–emitter terminals are functionally reversed. Figure 5.2*a* shows a section view of the device. Inspection of the basic gate topology shown in Fig. 5.1*d* reveals that the *pnp* and *npn* devices share a common ground n^+ region and the collector of the *pnp* is interconnected to another *p* region, which is the base of the *npn*. As a result, it is possible to merge the *pnp* and *npn* transistors to save silicon area. In this case the *pnp* collector and *npn* base is one region (p_2 in Fig. 5.2*b*) and the *pnp* base and the *npn* emitter is also one region (*n* in Fig. 5.2*b*). The emitter (p_1) of the *pnp* is called the injector of the I^2L structure. The section view of the overall merged structure is shown in Fig. 5.2*c*. The n^+ isolation collars are added to confine hole injection from the *p* regions under bias as will be discussed shortly. The *pnp* is forward biased using a power supply ($+V_{CC}$) and a biasing resistance R_B. The supply voltage ($+V_{CC}$) could be as low as 1 V, since all that is required is for the $p^+ - n$ injector to be forward-biased enough for the desired injector current to flow. Hence the structure can operate at low-power levels with only microamperes as bias.

On the other hand, the supply voltage could be as high as 10–15 V, depending on the allowed power dissipation in R_B. The higher the operating injector current is, the larger the power dissipation of the structure and the faster the operating speed. This represents an advantage for I^2L because of the flexibility in operating the circuit at different current levels and hence with different speed and power dissipation levels.

The isolation between structures is made using either shallow or deep n^+ diffusions or using a dielectric isolation. These isolation regions are usually referred to as collars and can touch the *p* diffusion walls—the p_1 and p_2 regions—since only low voltages appear across these junctions. The dielectric isolation provides the best approach since it reduces the leakage currents and

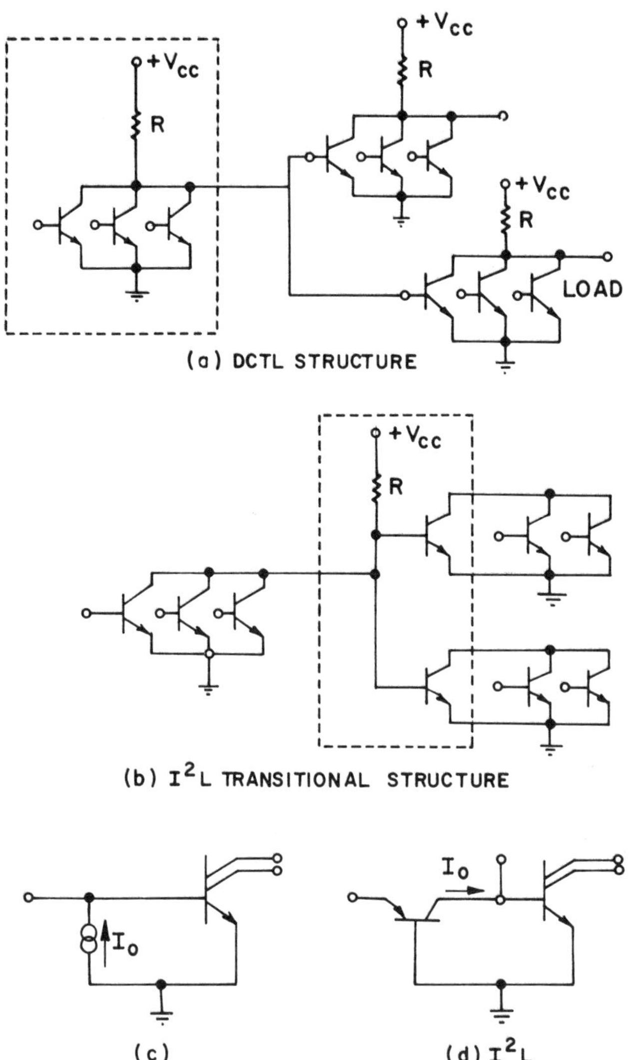

Figure 5.1 From direct-coupled transistor logic (DCTL) to integrated injection logic (I^2L).

parasitic capacitors typical of junction isolation. The second best solution is the deep n^+ isolation collar followed by the shallow n^+ collar. Figure 5.3 shows typical delay versus power dissipation curves for the three cases where the speed-power product ranges between 0.1–1 pJ depending on the technology used [3]. It is appropriate to note that the relation between stage delays and power dissipation becomes nonlinear at increased power levels. In fact a delay minima occurs so that diminished returns in stage delays are expected as one approaches the bottom of the curve. This minimum delay point occurs when the saturated charge storage in the n regions shown in Fig. 5.2c dominates over

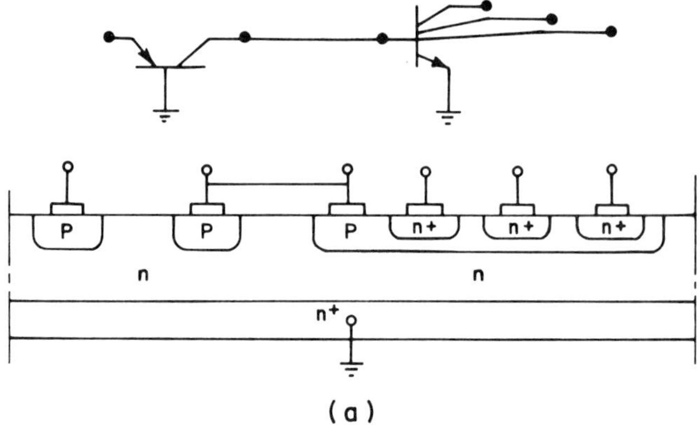

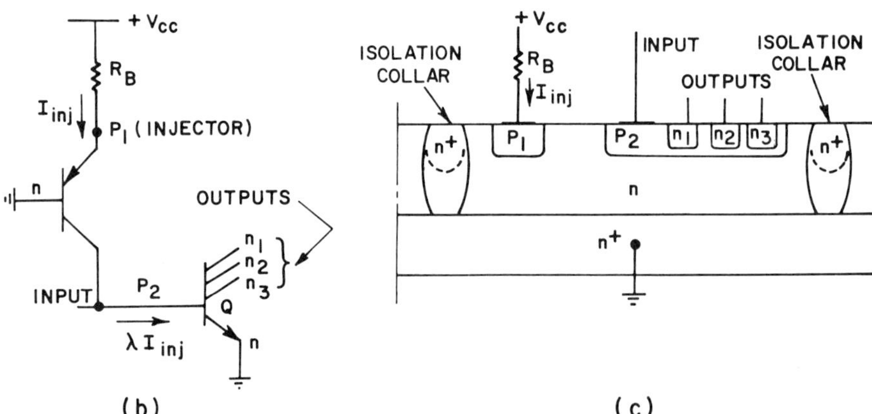

Figure 5.2 I^2L: (*a*) *npn* and *pnp* transistors before merging; (*b*) and (*c*) are the merged structure.

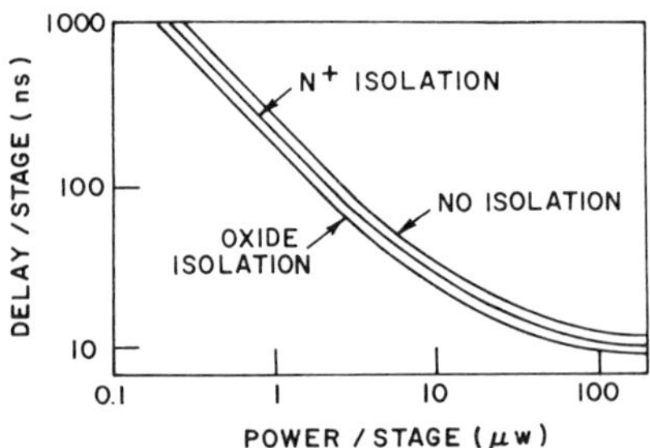

Figure 5.3 Power-delay curves of I^2L inverters with n^+ and oxide isolation structures [3].

space-charge-storage effects. Owing to the device topology with the *npn* emitter having the largest area, it is difficult to reduce the stored charge and hence the minimum delay of the basic gate.

5.2.1 The Basic I^2L Gates

Before discussing further the details of device performance effects let us consider how basic I^2L gates are used to form more complex logic. Figure 5.4 shows three I^2L logic gates. Each gate realizes a wired-AND function at the input, while the output is a multicollector inverter. The operation of the gate is simply described by considering device Q_3. The injector current I_{inj} supplies the *npn* base with an input current $I_B = \lambda I_{\text{inj}}$, where λ is the injector efficiency (λ is defined in Section 5.5.1). If one of the inputs (F_1 and F_2 represent collectors from Q_1 and Q_2, respectively), is at a low-voltage level, the input current is routed to that collector, Q_3 is cutoff, and the output voltage is determined by the junction voltage V_{on} of the next stage. If, on the other hand, all the input transistors (Q_1 and Q_2) are cutoff—equivalent to a higher voltage level of a junction voltage V_{on} on the base of Q_3—then the input current is routed to the base of Q_3, which allows the device to saturate. Thus the low-voltage level of an I^2L corresponds approximately to the saturation voltage of the *npn* transistor (V_{sat}), while the high-voltage level corresponds to $\sim V_{\text{on}}$. Thus the logic swing is $\sim (V_{\text{on}} - V_{\text{sat}})$. The threshold voltage of the gate is slightly less than V_{on}. Thus $VNM_1 \sim 0$ and $VNM_0 \sim (V_{\text{on}} - V_{\text{sat}})$. However, the I^2L gate is current operated and current noise margins are appropriate to use. They can be defined similar to voltage noise margins as follows.

For a given I^2L gate, the input base current *required* to turn-on the *npn* transistor is nI_0/β_u, where n is the fan-out and β_u is the upward β of the *npn* transistor. The input current supplied to the gate to turn it on, from the previous stages when their logical output is "1," is the leakage current ~ 0. Thus the current I_0 supplied by the injector to the *npn* transistor is *available* to turn on the *npn* transistor. Thus the current noise margin is, assuming an ideal λ value $= 1$,

$$INM_1 = I_0 - \frac{nI_0}{\beta_u} = I_0\left(1 - \frac{n}{\beta_u}\right)$$

Similarly, the input sinking current supplied to the gate, from the previous stages when their logical output is "0," is $m'I_0$, where m' is the number of turned-on stages (m' has range between 1 and m, where m is the fan-in). Thus the net current supplied to the gate is $I_0 \quad m'I_0$, and the current noise margin is

$$INM_0 = \frac{nI_0}{\beta_u} - (I_0 - m'I_0) = \frac{nI_0}{\beta_u} - I_0 + m'I_0$$

At a worst value of m' equal to 1, $INM_0 = nI_0/\beta_u$. It is interesting to note that increasing n improves INM_0 and degrades INM_1.

It should be pointed out that, as shown in Fig. 5.4, the I^2L fan-out requires *separate* collectors. This is because fanning-out from the same collector to different bases will affect other inputs when that collector is a logical low. Figure 5.4*c* illustrates this point.

5.2.2 I^2L in Circuit Design

Having discussed the basic I^2L gate in terms of an inverter and a wired-AND now let us build a representative set of digital subblocks. The emphasis is to show examples at the technology layout level and to provide the basic circuit know-how so that the reader can construct more general designs. This section will be subdivided as follows. First several general logic realizations will be considered. Standard building blocks include: decoders, half and full adders, and bus structures. Next the flip-flop will be discussed with emphasis toward

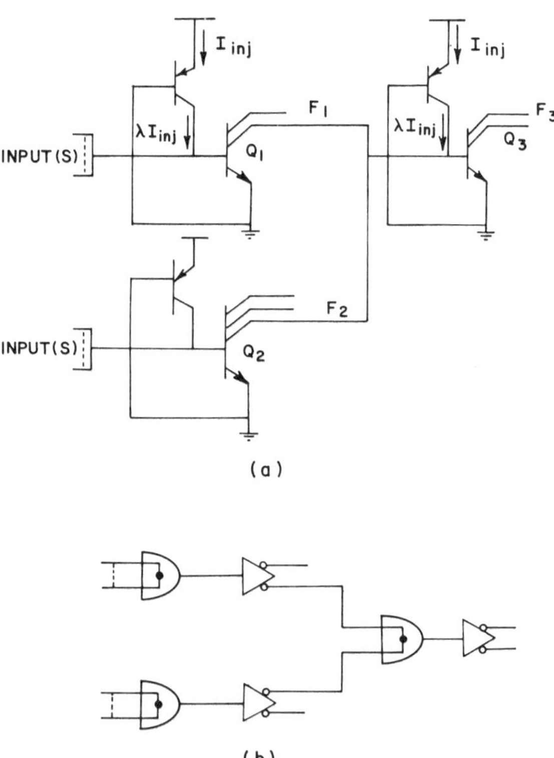

Figure 5.4 I^2L gates: (*a*) circuit diagram and (*b*) logic diagram. (*c*) I^2L wired logic: (*i*) fanning-out from the same collector: faulty logic operation; (*ii*) fanning-out from different collectors: correct logic operation.

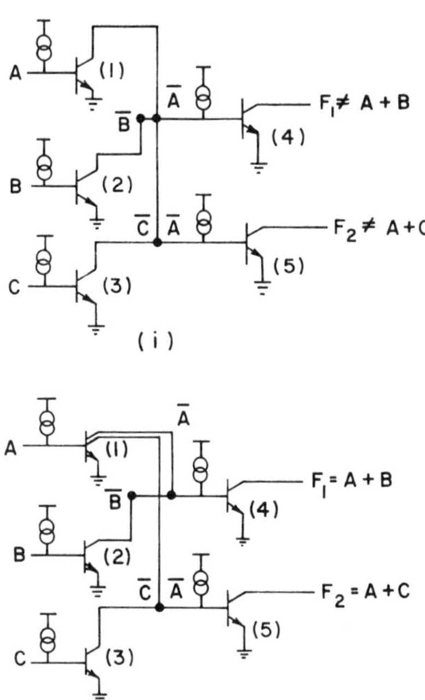

Figure 5.4 (*Continued*)

the memory function. Finally the more general considerations of flip-flops for registers and shift registers will be discussed.

I^2L Logic Structures. The basic need for logically equivalent gate realizations stems from the constraints of a given technology. For T^2L circuits the NAND gate is the basic unit, so it is well to translate all logic to a NAND form. For

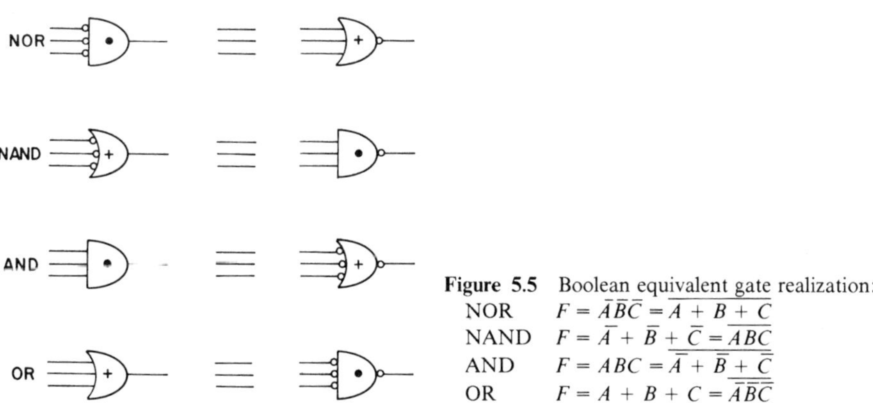

Figure 5.5 Boolean equivalent gate realization:
NOR $F = \bar{A}\bar{B}\bar{C} = \overline{A + B + C}$
NAND $F = \bar{A} + \bar{B} + \bar{C} = \overline{ABC}$
AND $F = ABC = \overline{\bar{A} + \bar{B} + \bar{C}}$
OR $F = A + B + C = \overline{\bar{A}\bar{B}\bar{C}}$

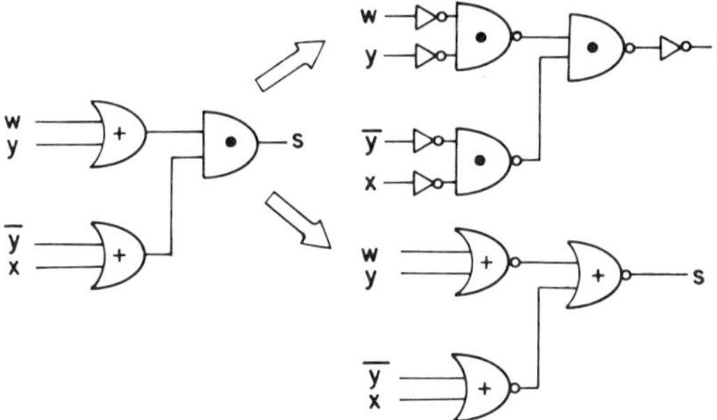

Figure 5.6 Product-of-sums gate implementation using all-NAND or all-NOR gates:

$$S = (w + y)(\bar{y} + x) = \overline{\overline{(\bar{w}\bar{y})(\bar{y}\bar{x})}}$$

$$= \overline{\overline{(w + y)} + \overline{(\bar{y} + x)}}$$

I^2L the situation is somewhat more flexible as will be discussed. However, the need is to pick a notation for convenience. Figure 5.5 shows a set of Boolean equivalent logic realizations. Figure 5.6 shows the translation of a desired Boolean function into its respective all-NOR or all-NAND realizations. One might note that the "gate-count" from an MSI viewpoint looks unfavorable for the NAND realization as compared with the NOR realization. However, as we look at the problem at an IC technology level we may find the difference to be very small. Frequently, it may be more convenient to start directly from the truth tables or Boolean functions and determine the logic realizations. Figure 5.7 shows the sum-of-products form equivalent to Fig. 5.6. Also shown is the NAND realization, which is considerably more compact than that shown in Fig. 5.6.

The basic I^2L gate configuration, when connected to realize logic functions, can be interpreted from either the NAND or NOR perspective. Figure 5.8 illustrates the point by focusing on the two interpretations of the same set of

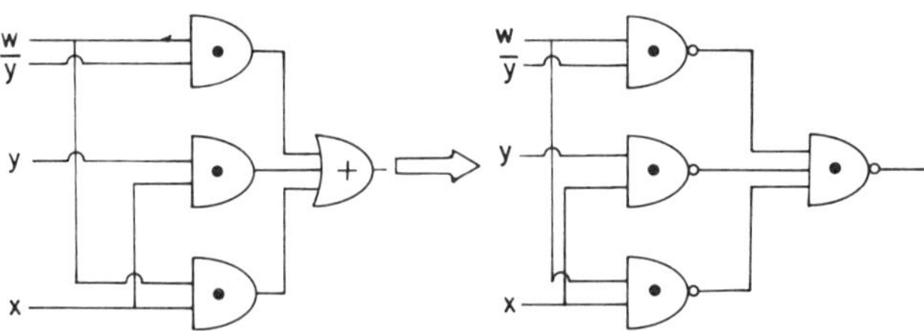

Figure 5.7 Sum-of-products gate implementation and NAND realization of function shown in Fig. 5.6:

$$S = w\bar{y} + yx + wx = \overline{\overline{(w\bar{y})}\,\overline{(yx)}(yx)}$$

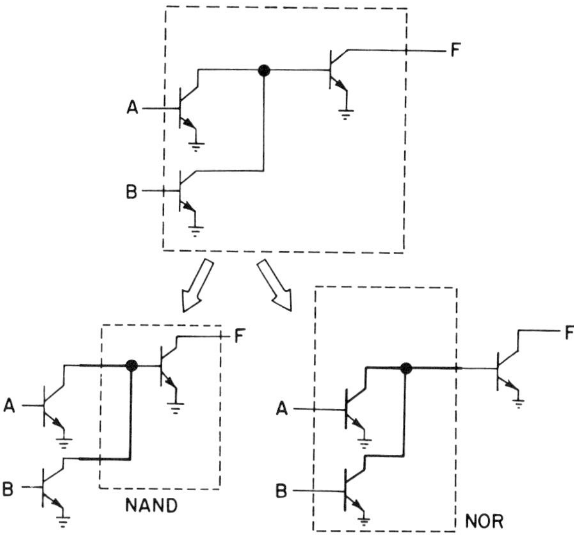

Figure 5.8 The basic connection of three transistors and two equivalent logic interpretations:

$$F = \overline{\overline{A}\,\overline{B}}$$

$$= \overline{\overline{A + B}}$$

devices. The full set of three devices would in turn be characterized as shown in Fig. 5.9. The implementation of a given logical function in I^2L technology then translates into the designer preference. Algorithms for each generic-logic-type NAND/NOR is given below.

Consider first the NOR realization. The algorithm for I^2L realization is as follows:

1. Assign a number to each logic gate.
2. Each gate represents a *node*.
3. Inputs to the gate are transistor collectors.
4. Outputs from the node define the number of collectors on the following transistor.

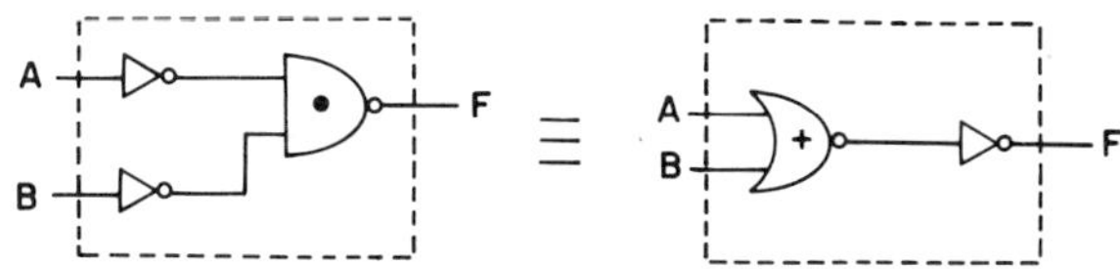

Figure 5.9 The gate equivalents to the partitionings shown in Fig. 5.8.

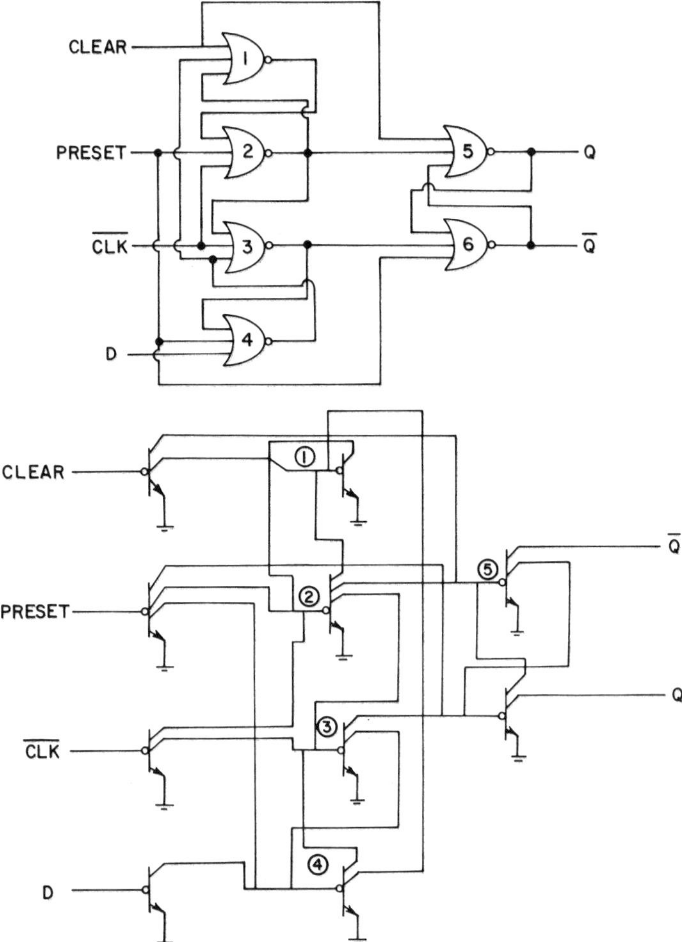

Figure 5.10 The gate equivalent and I^2L realization using NOR gates for a *D*-type flip-flop. The injector connection is not shown for simplicity.

Figure 5.10 illustrates the procedure in going from a *D* flip-flop realized with NOR logic to the corresponding I^2L implementation. Note that each node has a fan-in of three, which matches the original gate input counts.

The equivalent set of guidelines for NAND realization are as follows:

1. Assign a number to each NAND gate.
2. Each gate represents a transistor and multiple outputs correspond to multiple collectors.

Figure 5.11 shows the NAND gate and I^2L realizations of the *D*-type flip-flop.

For the examples shown in Fig. 5.10 and 5.11 the merits of the particular gate logic should be considered a minor point, and attention should be focused primarily on the device media for implementing a logical connection of gates.

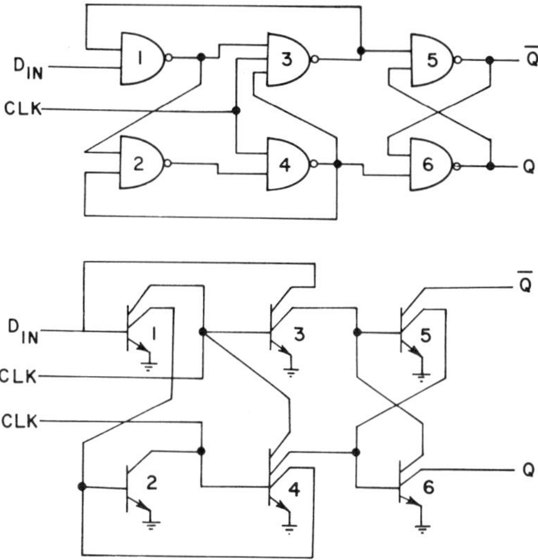

Figure 5.11 The gate equivalent and I^2L realization using NAND gates for a *D*-type flip-flop. The clear and the preset signals have been excluded.

At this point we can go one step further and consider the actual gate layout. When considering I^2L one can imagine a series of *p*-diffused regions, each of which represents a potential gate. Figure 5.12*a* shows a series of these devices all driven by a single injector diffusion. The symbolic layout/circuit equivalent, assuming each *npn* has three collectors, is shown in Fig. 5.12*b*. Two parts should be emphasized before actually creating a logic gate. In terms of relative positions of base and collector regions for each *npn* device, the designer is free to make interchanges to facilitate gate interconnects. Second, although identical *npn* gates are shown, the designer is also free to use fewer collectors and customize the size of the *p* regions to best accommodate the design. Having stated these two points, the final layout shown in Fig. 5.12*c* reflects both design options and corresponds to the layout of the flip-flop shown in Fig. 5.11.

The examples given above illustrate the principles of logic conversion to the layout technology realization for I^2L. Several typical logic structures will now be used to construct a sample menu of digital building-block elements. The first example is that of a simple decoder structure. The decoder serves an important function in activating one specific output line based on an input word. For example, an address word of 3 bits must be decoded to activate one of the eight possible address lines to a storage location in a memory. Figure 5.13*a* shows the logic realization for a one-of-eight decoder, and Fig. 5.13*b* shows the technology level layout for such a structure. One can note from the layout that gate "placement" is important to layout and some of the wasted space in terms of long gates results from the need to "route" metal *over* devices in order to make connections to other devices.

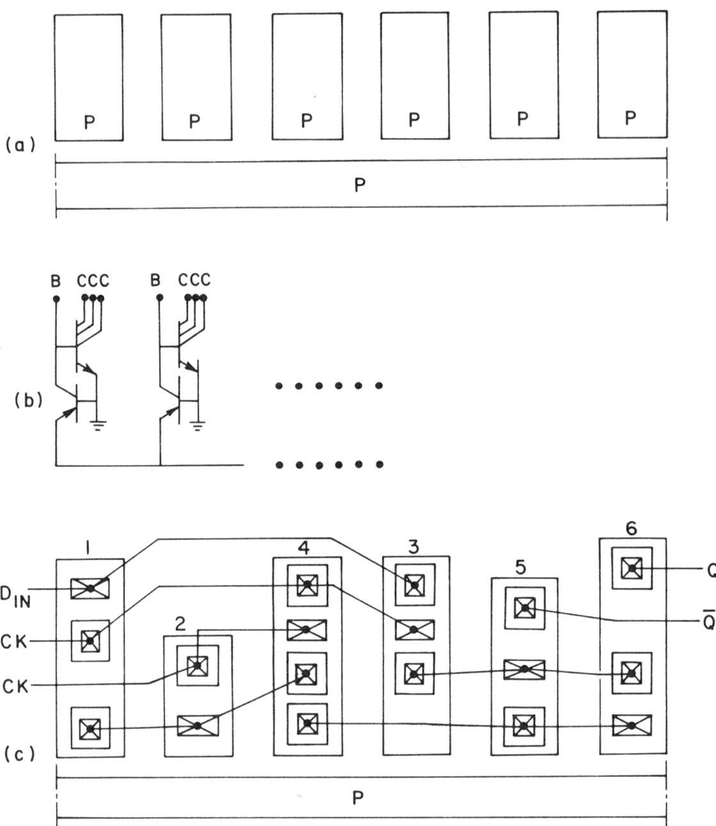

Figure 5.12 I^2L layout: (*a*) base diffusions and injector without collectors, (*b*) layout/circuit equivalent showing three collectors, and (*c*) layout and interconnection for the *D*-type flip-flop of Fig. 5.11.

Another important example of logic for digital ICs involves half- and full-adders. Figure 5.14*a* shows the three-gate NOR equivalent for a half-adder. Figure 5.14*b* shows the I^2L schematic realization and Fig. 5.14*c* shows the gate layout. The injectors for each base are not shown, except in the layout. It is clear from Fig. 5.14*c* that the gate layout is extremely dense with interconnects and crossovers (over the injection rail) having a natural flow. To go from the half-adder to a full-adder several approaches can be followed. Figure 5.15*a* shows a realization using half-adders. This would be a direct use of the results from Fig. 5.14, however, more efficient designs are possible. Figure 5.15*b* shows the truth table and simplified Boolean form for a full-adder. Figures 5.15*c* and 5.15*d* in turn show the gate schematic and layout for the optimum full-adder. It is clear that at the technology level the designer has substantial flexibility in reducing effective gate count and layout by using well-designed gate structures.

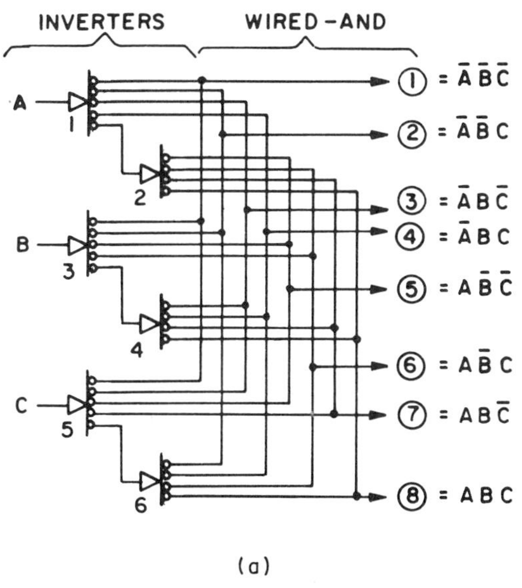

(a)

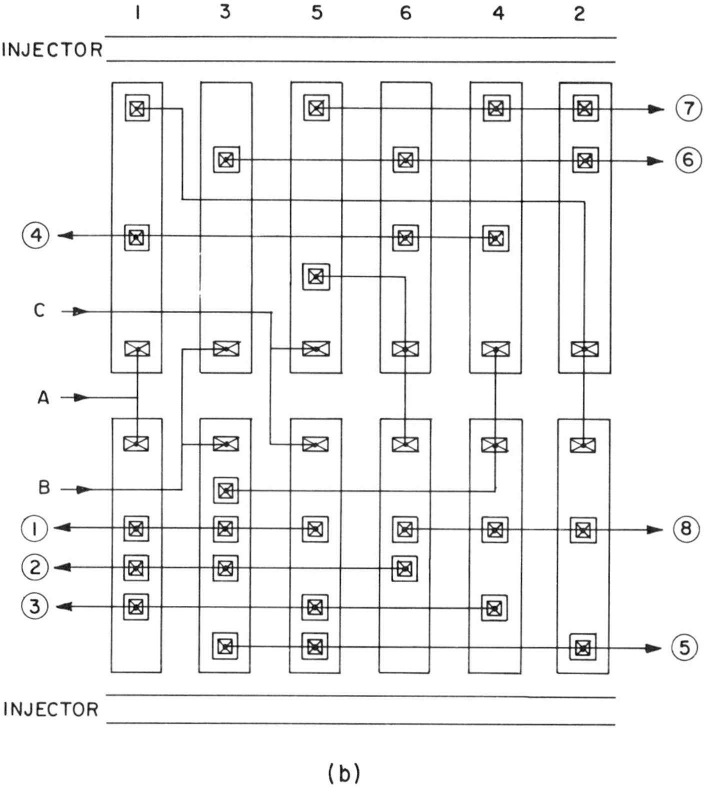

(b)

Figure 5.13 One-of-eight decoder: (*a*) inverter-wired AND logic representation and (*b*) layout.

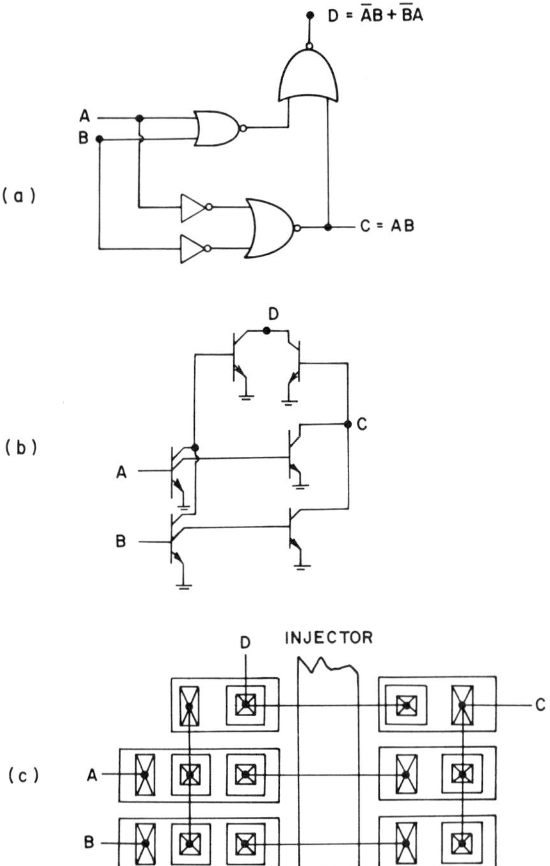

Figure 5.14 Half-adder: (*a*) logic representation, (*b*) I^2L gate implementation, and (*c*) layout.

I^2L Memory Structures. The above discussion of logic structures was concerned with ways to manipulate and effectively transmit digital information. The purpose of this section is to consider digital storage of information. The fundamental storage element is the flip-flop or bistable latch. The flip-flop from an I^2L perspective is realized as shown in Fig. 5.16*a*. The layout is dense and interconnections are straightforward as shown in Fig. 5.16*b*. The primary issue from the functional circuit point of view is how to read and write the contents of the cell. A number of techniques have been proposed [4, 5] and only two will be presented here for purposes of demonstration.

Figure 5.17*a* shows one configuration for read/write where extra collectors are added to the *npn*s to provide bit line (BL) and $\overline{\text{BL}}$. Also note that the cell voltage is variable so that standby power can be kept low in addition to the requirements for cell read/write. The basic operation is as follows. Once the

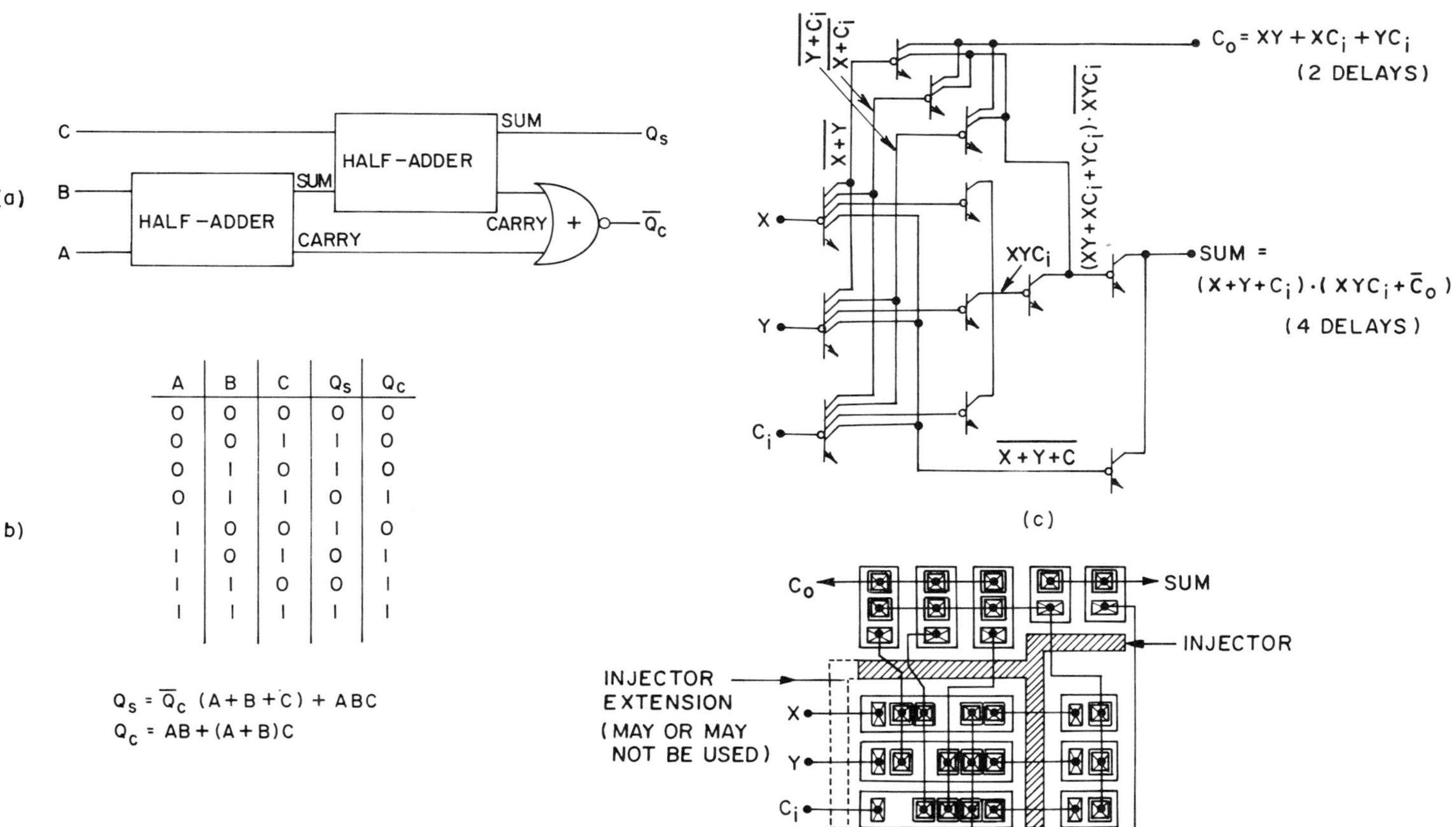

A	B	C	Q_s	Q_c
0	0	0	0	0
0	0	1	1	0
0	1	0	1	0
0	1	1	0	1
1	0	0	1	0
1	0	1	0	1
1	1	0	0	1
1	1	1	1	1

Figure 5.15 Full-adder: (*a*) using half-adders, (*b*) truth table and Boolean expression, (*c*) a direct I^2L gate implementation, and (*d*) layout

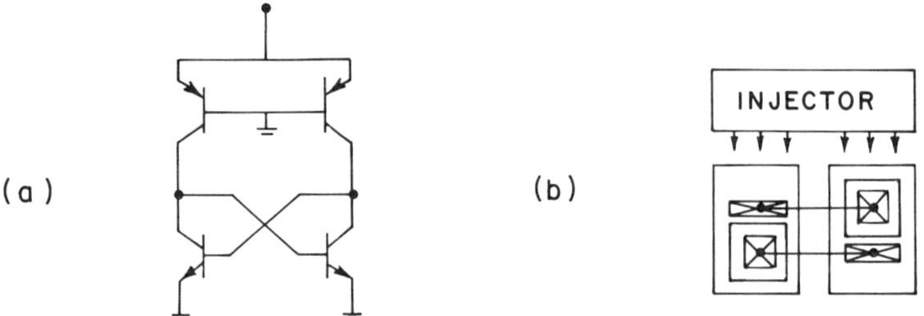

Figure 5.16 The bistable latch: (*a*) an I^2L gate realization and (*b*) layout.

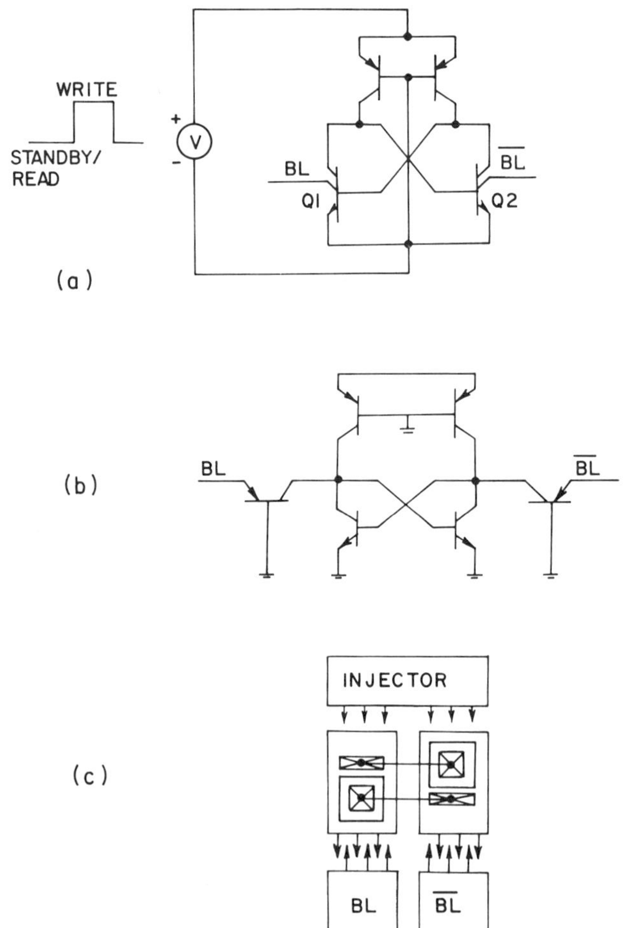

Figure 5.17 Read/write configurations for an I^2L flip-flop: (*a*) a multiple collector realization, (*b*) a multiple injector realization, and (*c*) layout for (*b*).

flip-flop state is set, the *pnp* voltage is at its standby/read level. Current sense amplifiers attached to the BL/$\overline{BL}$ lines can detect the state of the devices since the "on" *npn* transistor will draw current on the BL (assuming an initial state where Q_1 is "on"), whereas the "off" device will not. To achieve a write operation, where Q_2 becomes saturated, first one sets $\overline{BL} = 0$ and BL = 1. Next the write pulse is applied to the supply.

In the initial state Q_1 has one collector grounded and therefore requires a larger V_{BE} drive to draw current in the second collector and keep Q_1 out of conduction. As a result Q_2 starts conducting first and goes into saturation, which keeps Q_1 out of saturation. In essence the extra current of the write pulse is robbed by Q_2 and, in turn, Q_1 loses base drive. The gate structure involves multicollector devices and the support circuitry is somewhat complex. A simpler arrangement is shown next.

Figure 5.17*b* shows another I^2L memory arrangement. The read/write function is performed by the two lateral *pnp*s. One can consider the operation in terms of the addition of independent controlled injectors. By applying either BL high or BL low and the opposite to $\overline{BL}$, the *pnp* "current source" forces the *npn*s to their proper state by injecting extra charge or robbing it from the *npn*. By separating the control injectors one reduces capacitance of the *npn* as compared with the structure shown in Fig. 5.17*a*. This can be seen quite clearly by inspection of the layout shown in Fig. 5.17*c*.

As one can see from the two memory examples shown, the basic cells are quite simple, and it is in fact the support circuitry that can be a major factor in the destiny of a given memory chip. Moreover, architecture, sense amplifiers, and well-controlled clocks and supplies are critical factors. In the next and final subsection on I^2L subblocks we will consider storage from the perspective of registers and shift registers. While the basic flip-flop circuit will be apparent, it is the issues concerning support circuitry that will be addressed.

I^2L Registers. Registers are a basic and essential component in digital IC architectures. The varieties in implementation and application are beyond the scope of this book and are well described elsewhere [6, 7]. The simplest realizations involve set–reset (SR) flip-flops with the aid of logic and timing support circuitry. Other registers use flip-flops based on the *D*, *T*, or JK flip-flops. The trade-offs in choice for a particular application involve both timing and support logic to ensure proper operation. Figure 5.18 shows two registers realized with JK flip-flops. Figure 5.18*a* shows a basic 4-bit shift register, and Fig. 5.18*b* shows a multifunction register design including both load and shift operations. Table 5.1 shows the set of control inputs to the multifunction register as shown in Fig. 5.18*b*. Both designs shown in Fig. 5.18 make optimum use of the JK features, which involve master–slave (MS) operation and explicit timing constraints. In essence the JK represents a potent building block, which we will now consider in detail as a representative digital component.

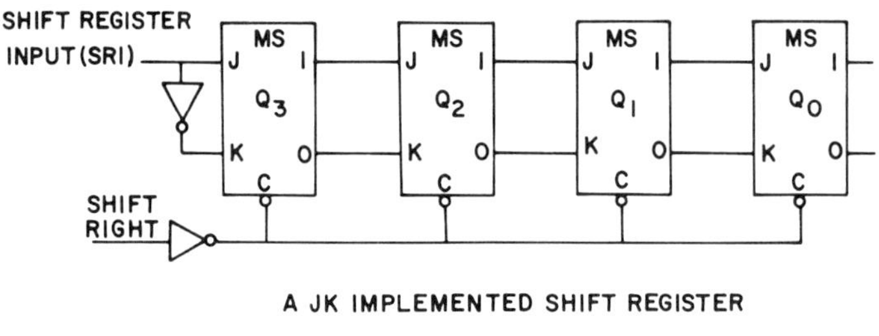

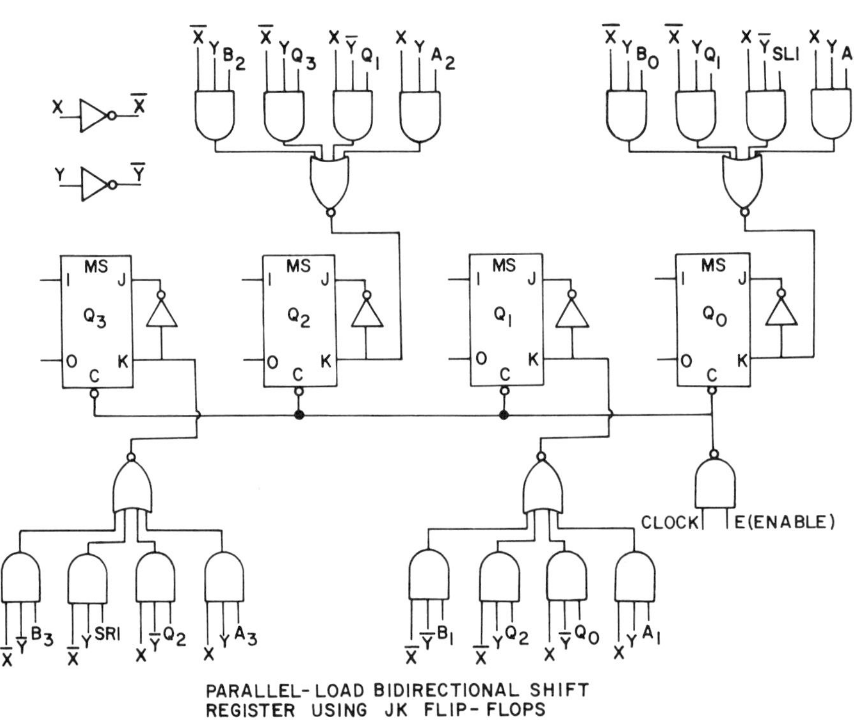

Figure 5.18 Applications for JK flip-flops: (*a*) a shift register and (*b*) a multifunction register.

The JK flip-flop is characterized by the logic table given in Fig. 5.19*a*. The time-dependent next-state output equations are

$$Q(t+\Delta) = \bar{J}(t)\bar{K}(t)Q(t) + J(t)\bar{K}(t) + J(t)K(t)\bar{Q}(t)$$

$$= J(t)\bar{Q}(t) + \bar{K}(t)Q(t)$$

Table 5.1 Multifunction Register Function Specifications

Control Inputs			Register Function To Be Done on Next Clock Pulse	Next Stage of the Register			
$E(t)$	$X(t)$	$Y(t)$		$Q_3(t+\Delta)$	$Q_2(t+\Delta)$	$Q_1(t+\Delta)$	$Q_0(t+\Delta)$
1	0	0	Load register with the data on the parallel B input lines.	$B_3(t)$	$B_2(t)$	$B_1(t)$	$B_0(t)$
1	0	1	Shift register data right one place with SRI (shift right input) shifted into Q_3.	SRI(t)	$Q_3(t)$	$Q_2(t)$	$Q_1(t)$
1	1	0	Shift register data left one place with SLI (shift left input) shifted into Q_0.	$Q_2(t)$	$Q_1(t)$	$Q_0(t)$	SLI(t)
1	1	1	Load register with the data on the parallel A input lines.	$A_3(t)$	$A_2(t)$	$A_1(t)$	$A_0(t)$
0	d[a]	d	Chip disabled, retain old data.	$Q_3(t)$	$Q_2(t)$	$Q_1(t)$	$A_0(t)$

[a]d = don't care.

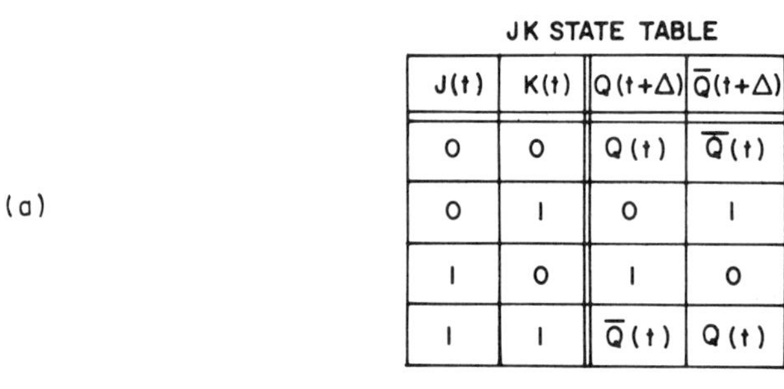

(a)

JK STATE TABLE

J(t)	K(t)	Q(t+Δ)	$\overline{Q}$(t+Δ)
0	0	Q(t)	$\overline{Q}$(t)
0	1	0	1
1	0	1	0
1	1	$\overline{Q}$(t)	Q(t)

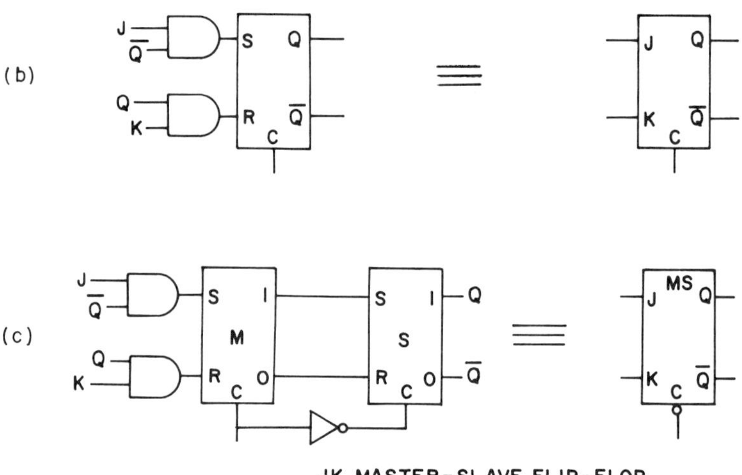

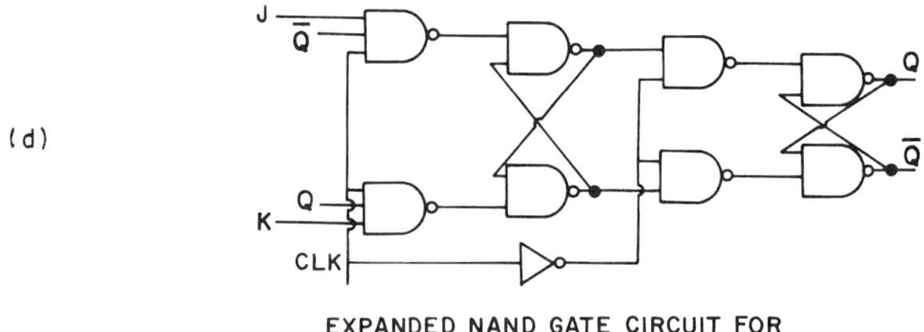

Figure 5.19 The JK flip-flop: (*a*) state table, (*b*) RS equivalent, (*c*) master–slave, and (*d*) NAND realization of (*c*).

and equivalently,

$$\bar{Q}(t + \Delta) = \bar{J}(t)\bar{Q}(t) + K(t)Q(t)$$

Figure 5.19*b* shows the realization of the JK using the basic RS flip-flop configuration, which was discussed in relation to I^2L memory cells. The major difficulty with this realization is the race conditions, which may occur at the clock period if the S and R inputs change and the feedback is complete before the end of the clock. To overcome this problem a "master–slave" realization is used as shown in Fig. 5.19*c* where an inverted clock signal is sent to the "slave" SR section, which locks-out the signal until the end of the clock. Figure 5.19*d* shows the expanded NAND gate realization of the circuit. In essence we have now generated the building block used in Fig. 5.18.

Beyond the basic master–slave JK there are several useful modifications that can be added to realize a more-general element. *Preset* and *preclear* inputs are especially useful when considering general purpose registers where it is desirable to perform these operations without creating the all-0's or all-1's words. Figure 5.20*a* shows how the preset (PS) and preclear (PC) can be simply added to the output stage of the basic SR latch, and Figure 5.20*b* shows the needed implementation changes to create the PS and PC functions in our overall master–slave JK. Figures 5.20(*c*)–5.20(*e*) show the I^2L version of the master–slave JK flip-flops with S and R inputs.

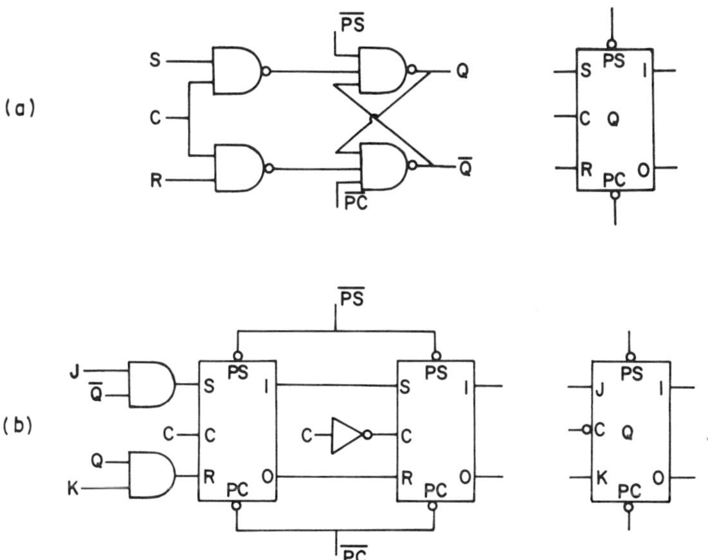

Figure 5.20 Implementation of the preset/preclear functions: (*a*) for the RS flip-flop and (*b*) for the master–slave JK flip-flop; (*c*) logic gate realization, (*d*) I^2L gate implementation, and (*e*) layout, single layer of metallization is used.

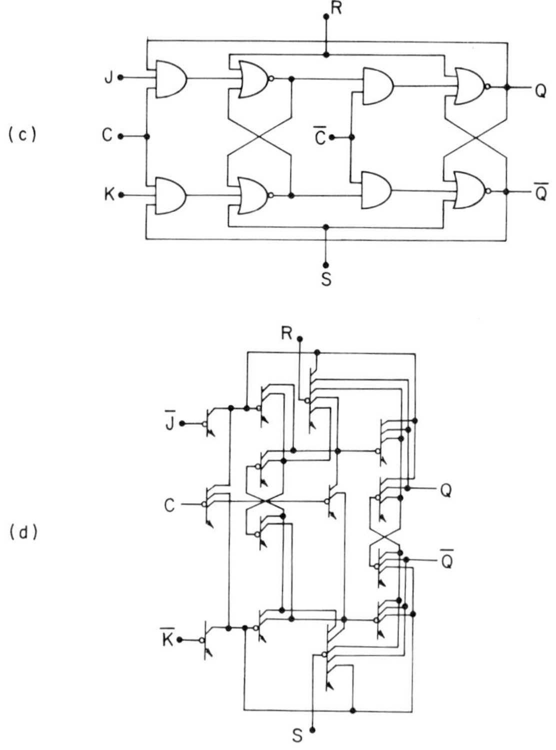

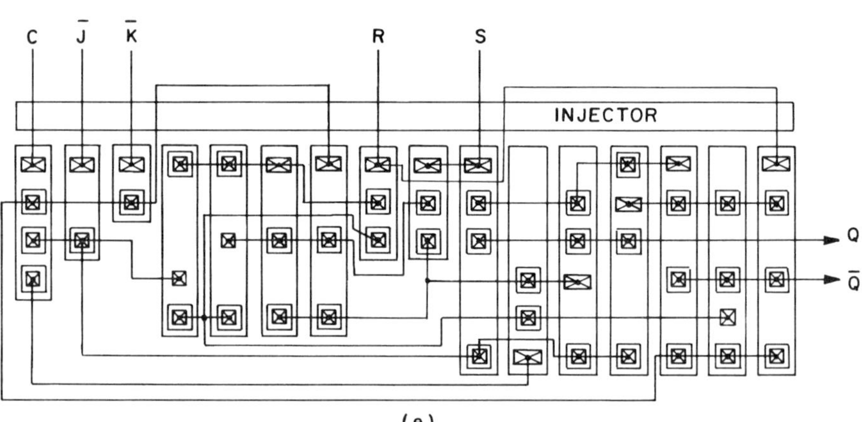

Figure 5.20 (*Continued*)

5.3 I^{2}L CURRENT COMPONENTS

Having discussed the gate-level considerations for I^{2}L, it is useful to return to the device level to determine performance limits. Figure 5.21 shows the different current components associated with a typical I^{2}L structure [8, 9]. Three *terminal* currents can be identified. These are the injector current I_{inj}, the input (base) current I_{B}, and the output (collector) current(s) I_C. The internal current density components can be classified as follows:

1. The *useful* current density components J_n and J_{pL}. J_n is the upward electron current density, which is injected from the emitter toward the I^{2}L collector(s). Thus $I_C = J_n A_{BC}$, where A_{BC} is the collector–base area. J_{pL} is the lateral hole current density component injected from the injector toward the *npn* base(s).
2. The *parasitic upward* current densities J_{nm} and J_{no}, which results from the electron injection from the emitter toward the surface; J_{no} is the upward current density which faces an oxide–semiconductor interface and J_{nm} is the upward current density facing a metal-contact-semiconductor interface.
3. The *parasitic downward* current density J_p, which results from the hole injection from the *p* region toward the emitter region. J_p is the main current component contributing to the base current I_B and to the injector current I_{inj} and thus should be minimized.
4. The *parasitic lateral* current densities J_{pl1}, J_{pl2}, and J_{Bbi}. These components are injected laterally by the *npn* base; J_{pl1} designates the component facing another *npn* base (in the case of using a shallow n^+ isolation collar) and J_{Bbi} designates the back base injection component, that is, facing the injector. If a dielectric isolation is used, J_{pl1} and J_{pl2} are

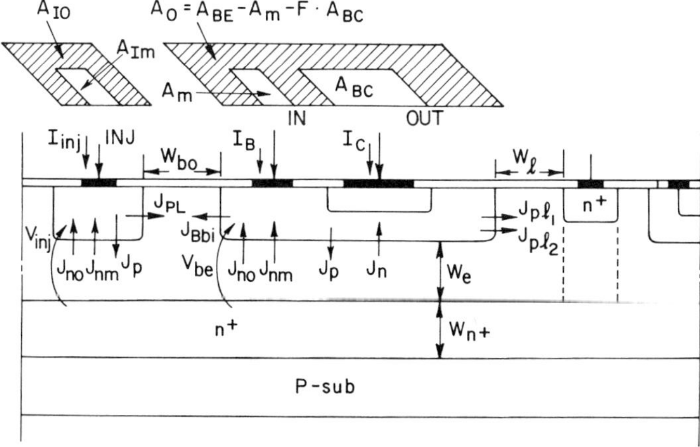

Figure 5.21 I^{2}L current components.

eliminated. J_{Bbi} can be used to an advantage to control the degree of saturation of the *npn* transistor as we shall see in Section 5.6.1.

As suggested in the discussion above, it is desirable to optimize the *useful* current components and to minimize *parasitics* where possible. Isolation techniques and impurity-profile tailoring offer several options. However, the area factors, which scale components such as J_{no}, J_{nm}, J_p, and J_{Bbi}, are clearly set by minimum feature sizes and junction depths. To see more precisely how these geometry effects control the performance of I^2L gates, the dc and transient analysis is presented next.

5.3.1 dc Analysis and Scaling of I^2L

The dc performance of an I^2L structure is determined by area-scaled parameters relative to the one-dimensional intrinsic controlling current densities (e.g., J_n). The intrinsic upward beta of the *npn* transistor (β_{ui}) is the ratio of the collector current to the base current under the I^2L collector, that is,

$$\beta_{ui} = \frac{J_n A_{BC}}{J_p A_{BC}} = \frac{J_n}{J_p} \tag{5.1a}$$

On the other hand, the terminal upward β of the *npn* transistor, β_u, is given by

$$\beta_u = I_C / I_B \text{ at a given } V_{\text{inj}} \ (\text{e.g., } V_{\text{inj}} = 0) \tag{5.1b}$$

Both Eqs. (5.1a) and (5.1b) make reference to Fig. 5.21 and the area scaling factors are as shown. The actual components in Eq. (5.1b) are given by

$$I_C = J_n A_{BC} \tag{5.1c}$$

$$I_B = A_{BE} J_p + A_o J_{no} + A_m J_{nm}$$
$$+ A_{l1} J_{pl1} + A_{l2} J_{pl2} + A_{IL} J_{Bbi} \tag{5.1d}$$

and the area factors subscripted with l and L represent lateral contributions.

It is worth noting at this point that the intrinsic β_{ui} is a one-dimensional best-case value, whereas the list of contributions to I_B given by Eq. (5.1d) all degrade this maximum performance and result in the actual β_u [as used in Chapter 3, Eq. (3.5)]. In order to project how circuit performance scales with technology and gate fan-out, it is useful to write normalized equations for the actual β_u in terms of the area factors shown in Fig. 5.21.

It is desirable to maximize β_u, that is, minimize β_{ui}/β_u. Based on Eq. (5.1b) and (5.1d) it is appropriate to minimize all area and current density factors for the *lateral parasitic* contributions. Assuming this to be true, the approximate form β_{ui}/β_u is

$$\frac{\beta_{ui}}{\beta_u} \approx \frac{A_{BE}}{A_{BC}} + \frac{A_o}{A_{BC}} \frac{J_{no}}{J_p} + \frac{A_m}{A_{BC}} \frac{J_{nm}}{J_p} \tag{5.2}$$

Hence there is a trade-off between β_u and the area ratio A_{BE}/A_{BC}, which also reflects the proportionality $\beta_{ui}/\beta_u \propto$ fan-out (F). It is clear that β_u is increased by either increasing A_{BC} or decreasing A_o and A_m.

In a manner similar to the discussion for base-current components, the injector current is given by

$$I_{\text{inj}} = J_{pL} A_{IL} + J_{no} A_{IO} + J_{nm} A_{Im} + J_p (A_{Im} + A_{IO}) + A_{Il1} J_{pl1} + A_{Il2} J_{pl2} \tag{5.3}$$

where again the I notation refers to the injection and the l and L terms indicate lateral area factors. The useful injected current is I_0 given by

$$I_o = A_{IL} J_{pL} \tag{5.4}$$

Now assuming the parasitic lateral contributions can be minimized and hence neglected, the relationship for injector efficiency is given by

$$\lambda = \frac{I_0}{I_{\text{inj}}} = \frac{1}{1 + \dfrac{J_{no}}{J_{pL}} \dfrac{A_{IO}}{A_{IL}} + \dfrac{J_{nm}}{J_{pL}} \dfrac{A_{Im}}{A_{IL}} + \dfrac{J_p}{J_{pL}} \dfrac{A_{Im}}{A_{IL}} + \dfrac{J_p}{J_{pL}} \dfrac{A_{IO}}{A_{IL}}} \tag{5.5}$$

Clearly to achieve maximum efficiency J_{pL} should dominate J_{no}, J_p, or J_{nm}. This is also consistent with the design objectives for maximum β_u. One must also seek to minimize the area ratios.

Figure 5.21 has defined many of the key features that govern I²L performance. However, when the basic gate with multiple collectors is embedded in a logic structure, the effects are not as simple as shown by the single-collector-device effects of Fig. 5.21. The following discussion will elaborate on the multiple-collector I²L gate.

Consider the circuit shown in Fig. 5.22a. This is the basic multicollector I²L gate equivalent to Fig. 5.2*a*. For the discussion of this section, we will focus on the circuit performance effects.

The corresponding cross section is shown in Fig. 5.22*b*, and the junctions have now been identified with a number which will appear as a subscript in describing device parameters. The transport models for the different transistors within the gate are shown in Fig. 5.22*c*. The form of the transistor models is that of the CAD format used for circuit simulation. As will become clear in a moment, the diodes associated with the *p*-type region of the multicollector *npn*

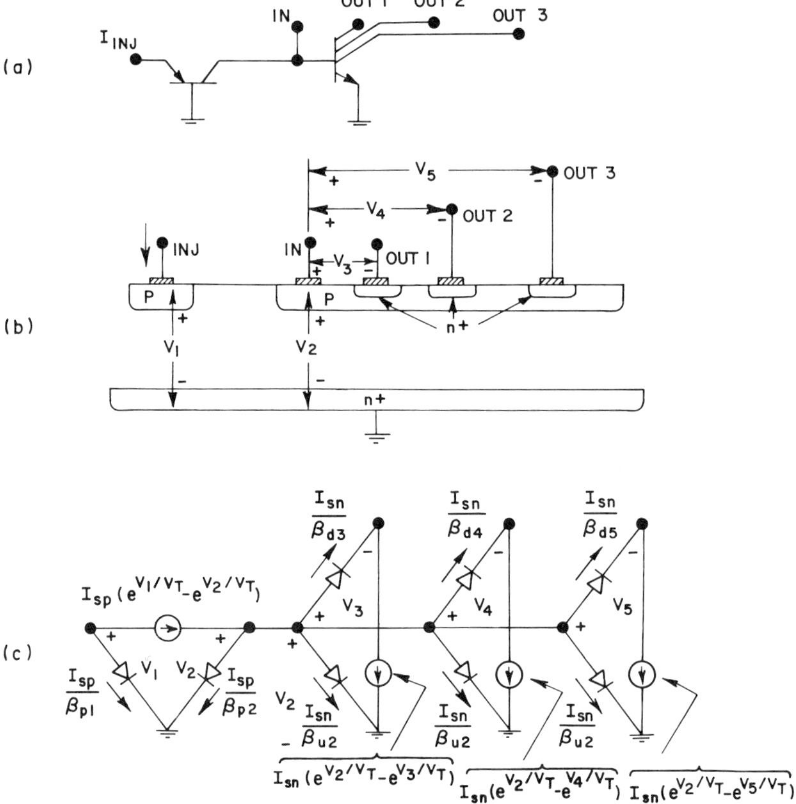

Figure 5.22 An I²L gate and circuit equivalent model: (*a*) the basic gate with three collectors, (*b*) cross section with junction voltage notation, (*c*) dc model circuit equivalent—both junction diode and transport currents shown relative to pertinent exponential-voltage relationships. I_{sp} and I_{sn} are the transport current components associated with the lateral *pnp* and the vertical *npn* transistor, respectively. I_{sp}/β_{p1} and I_{sp}/β_{p2} are base-current components associated with the lateral transistor. I_{sn}/β_{u2}, I_{sn}/β_{d3}, I_{sn}/β_{d4}, I_{sn}/β_{d5} are the base-current components associated with junction voltages V_2, V_3, V_4, V_5, respectively. The notation *u* and *d* are used to indicate upward and downward as discussed in Chapter 3.

base/lateral *pnp* collector must be considered carefully when reconciling the meaning of current gain (beta) as used for CAD with the physical model discussed earlier in this section. The *n* and *p* subscripts indicate the *npn* and *pnp* device parameters, respectively. Note that each output corresponds to a transport generator of value I_{sn} and has an associated set of base-current components I_{sn}/β_{u2} and one of I_{sn}/β_{d3}, I_{sn}/β_{d4}, or I_{sn}/β_{d5}. The notation of *u* and *d* follows that discussed in Chapter 3. The exponential junction voltages for each diode are multiplied by the appropriate current factor, for example, $(I_{sn}/\beta_{u2})e^{V_2/V_T}$. It is important to note that the β_{u2} given in Fig. 5.22*c* corresponds to the intrinsic value [β_{ui} given in Eq. (5.1a)]. The total base current is given by

$$I_B = \left(\frac{I_{sp}}{\beta_{p2}} + \frac{NI_{sn}}{\beta_{u2}} \right) e^{V_2/V_T}$$

where N is the number of collectors. In practice, the first term dominates and the base current is independent of the number of collectors. For each collector the current is

$$I_C = I_{sn} e^{V_2/V_T}.$$

Hence the extrinsic (measured) current gain is

$$\beta_{\text{meas}} = \frac{I_C}{I_B} \approx \left(\frac{I_{sn}}{I_{sp}} \right) \beta_{p2}$$

and by adding extra collectors, all connected together, the measured current gain for N collectors in parallel is

$$\beta_{\text{meas}}^{(N)} = \frac{I_C}{I_B} = \left(\frac{NI_{sn}}{I_{sp}} \right) \beta_{p2}$$

In terms of specifying model parameters for CAD one should confirm the dependence of β_{meas} on N and, based on the values for I_{sn} and I_{sp}, confirm that β_{u2} is large and therefore can be neglected in contributing to I_B compared to the I_{sp}/β_{p2} term.

Now consider a cascaded series of these gates generalized to include N outputs per gate instead of just three. A portion of this cascade is shown in Fig. 5.23*a*. Two aspects of this chain should be emphasized. First, the series of gates represents a data path, which carries a signal from input (path in) to output (path out). Second, at the instant in time when gate D is on and Q_D is saturated, the local considerations involve all N_D collectors as shown in Fig. 5.23*b*. The V_I node is at a logic high and the V_D node is at a logic low, thus isolating the Q_I and Q_L devices from dc considerations. In essence, for dc consideration the D device interacts only with its input current device $I_{\text{inj}}(D)$ and N_D output current loads; $I_{\text{inj}}(L)$.

To expand on the dc constraints for device Q_D, consider the partial circuit model in Fig. 5.24. Each *pnp* source driving a collector for device Q_D might be modeled as shown. The L subscripts are used to indicate that the injectors for the Q_L devices actually provide the current to the N_D collectors of Q_D. V_{sat} represents the collector of Q_D. If V_{sat} is small, then the circuit model for each collector load is simplified as shown in Fig. 5.24*a* assuming that the contributions due to junction voltage V_{2L} are negligible (i.e., $e^{V_{2L}/V_T} \ll e^{V_{1L}/V_T}$). The current I_0 is the maximum current that can flow to each collector of device D,

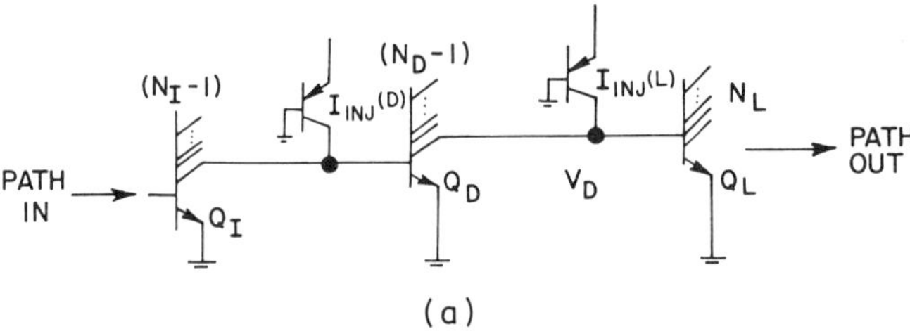

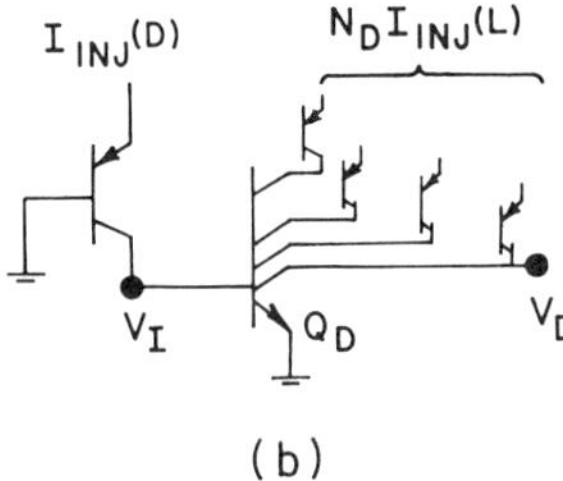

Figure 5.23 (*a*) The basic cascade of I²L gates with N_I, N_D, and N_L collectors. (*b*) The details associated with Q_D.

assuming V_{sat} is small. Thus I_0 is given by

$$I_0 = I_{sp} e^{V_1/V_T}$$

as can be seen in Fig. 5.24*a*. In addition to the fact that I_0 is the maximum load current, it can be readily shown that *any pnp* device with the same applied V_1 can be characterized using I_0 provided that I_{sp} values are the same. As a result of these circuit simplifications, Fig. 5.24*b* shows the gate with respective load currents. We can make one further simplification. For typical I²L gates the contributions of the diodes I_{sn}/β_{d3}, I_{sn}/β_{d4}, and I_{sn}/β_{d5} can be neglected since $\beta_d \gg 1$. As a result, the expanded circuit representation from Fig. 5.24*b* is shown in Fig. 5.24*c*. Figure 5.24*c* defines the essential parameters that control the dc operation of an I²L gate.

Using Fig. 5.24*c*, a minimum condition can be defined to guarantee proper dc operation of the gate. The maximum current each collector must "sink" is I_0. If the following condition is not met,

$$I_{sn} e^{V_2/V_T} \geqslant I_0 \tag{5.6}$$

then gate Q_D is not saturated. Hence, we must ensure that Eq. (5.6) is satisfied to achieve proper gate operation. To determine the necessary conditions, we can write the node equation at V_2:

$$I_{sp}\left(e^{V_1/V_T} - e^{V_2/V_T}\right) - \frac{I_{sp}}{\beta_{p2}} e^{V_2/V_T} - \frac{N_D I_{sn}}{\beta_{u2}} e^{V_2/V_T} = 0 \tag{5.7}$$

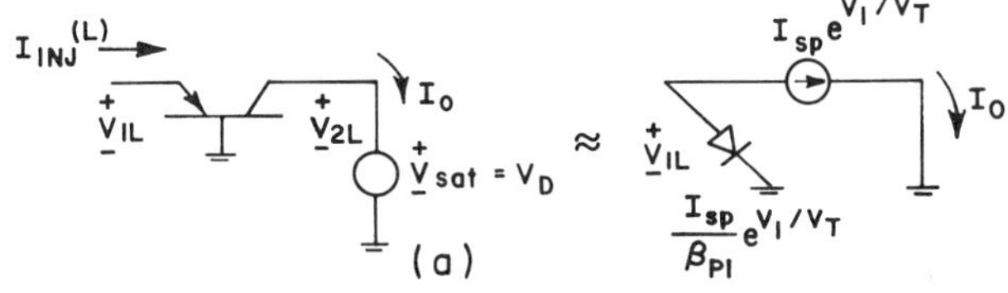

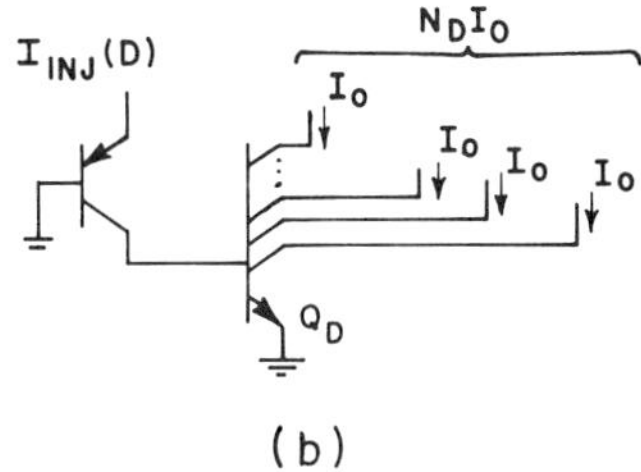

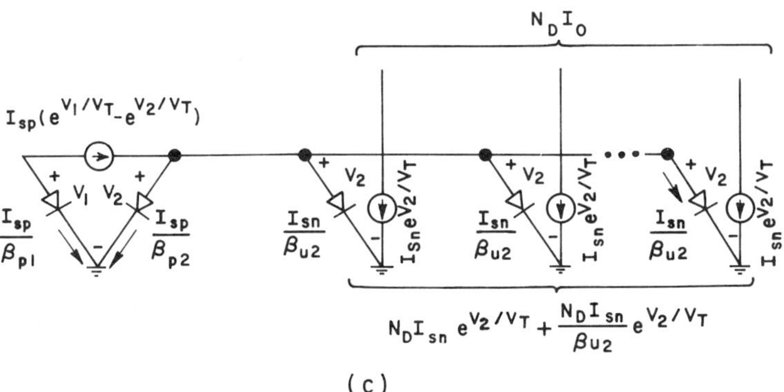

Figure 5.24 (*a*) Simplified equivalent load circuit, (*b*) gate Q_D with maximum output load current, and (*c*) expanded gate circuit diagram.

Replacing $I_{sp}e^{V_1/V_T}$ with I_0 and solving for e^{V_2/V_T}, one obtains

$$e^{V_2/V_T} = \frac{I_0}{I_{sn}N_D/\beta_{u2} + I_{sp}/\beta_{p2} + I_{sp}} \tag{5.8}$$

Hence to satisfy Eq. (5.6):

$$\frac{I_C}{I_0} - \frac{I_{sn}e^{V_2/V_T}}{I_0} = \frac{1}{N_D/\beta_{u2} + I_{sp}/I_{sn} + I_{sp}/\beta_{p2}I_{sn}} \geq 1 \tag{5.9}$$

Equation (5.9) implies that the following should be satisfied: $\beta_{u2}/N_D > 1$, $I_{sn}/I_{sp} > 1$, and $\beta_{pn}I_{sn}/I_{sp} > 1$. The first condition is a *fan-out* limitation and the last two imply intrinsic device design criteria. Although the dc fan-out constraint is significant and certainly cannot be violated, the transient effects place additional limits on circuit performance.

5.3.2 Transient Analysis of I²L

The transient behavior of I²L gives an additional set of constraints to those discussed above. Two limiting-case solutions will be considered. First, effects of only junction capacitance will be modeled. This corresponds to low-current operation or the "extrinsic" region, which implies that parasitic effects dominate. Second, the devices will be modeled considering only minority-charge storage. This corresponds to high-current operation or the "intrinsic" region, which implies that internal device minority-charge storage dominates. Although the combined effects indeed control performance, the two "limiting cases" define useful conceptual models for understanding I²L transient performance.

Consider first the "extrinsic" model shown in Fig. 5.25*a*. The current sources are assumed ideal with value I_0 and all capacitances have been averaged over the logic swing to give correct charge storage and transfer. $\overline{C}_{BE}$ is the average base–emitter capacitance per collector. Figure 5.25*b* shows the waveforms at nodes *I* and *D*.

Now let us calculate the fall time t_f of the waveform V_D. At the moment when Q_D becomes active (Q_I turned off), $\overline{C}_{BC}$ is charged via I_0/N_D (the fractional base current since there are N_D collectors) and the load at node V_D is discharged. Over the switching event the capacitance $\overline{C}_{BC}$ of Q_D swings by V_l whereas both sides of $\overline{C}_{BC}$ of Q_L change by V_l so the total charge transferred is given by $2V_l\overline{C}_{BC}$. In addition to the charge transfer across $\overline{C}_{BC}$ at Q_L, node V_D must be discharged. Although both events occur simultaneously, for purposes of estimating switching speeds they will be added as if they were sequential events. This is in some sense a worst-case estimate. The dc current that flows has already been determined for establishing fan-out so that Eq. (5.9) gives the collector current of each of N_D collectors. That is, the collector current discharging the capacitive load at node V_D is given by $I_{sn}e^{V_2/V_T}$. The charge transferred at node V_D is $N_L(V_l\overline{C}_{BE} + 2V_l\overline{C}_{BC})$. To a first approximation, t_f, the total fall time at node *D*, is given as

$$t_f = \frac{V_l\overline{C}_{BC}}{\left(\dfrac{I_0}{N_D}\right)} + \frac{N_L\left(V_l\overline{C}_{BE} + 2V_l\overline{C}_{BC}\right)}{\left(\dfrac{I_0}{N_D/\beta_{u2} + I_{sp}/I_{sn} + I_{sp}/\beta_{p2}I_{sn}}\right) - I_0} \tag{5.10}$$

comparing the denominators of the two terms, in particular if the terms involving β_{u2} and β_{p2} are neglected, one can show that while the first term is discharged by I_0/N_D, the second term is discharged by I_0I_{sn}/I_{sp}, which is much greater than I_0. The total node loading at *D* is obtained assuming the

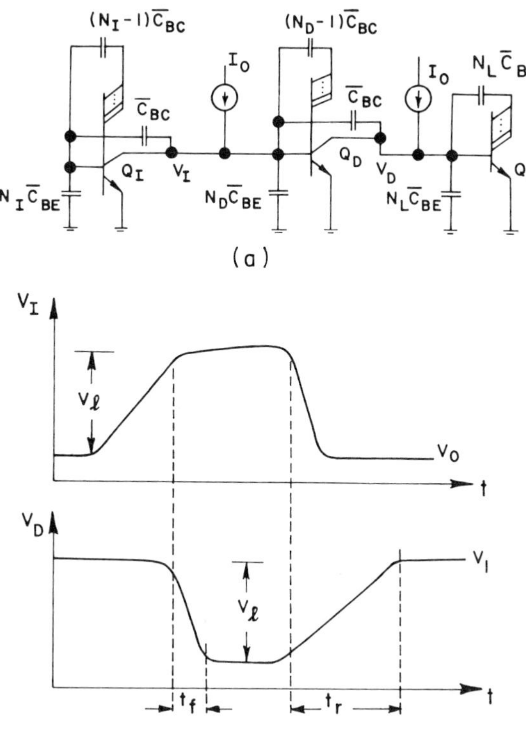

Figure 5.25 Extrinsic I^2L switching: (*a*) node capacitances and current sources and (*b*) switching waveforms at nodes *I* and *D*.

collectors of gate *L* provide a virtual ground. The series capacitance loading at *L* reduces the load contribution of gate *D*. In terms of waveform propagation, it must be noted that the fall time does not totally control signal propagation. Only a small change of the total V_l will switch off the Q_L transistor and the signal will be propagated. The comparison of calculated and simulated t_f provides an important cross-check of the transient modeling. However, the signal rise time is much more important in the overall signal delay of waveforms and it is considered next.

The rising waveform edge at *D* (Q_I is saturated) experiences the full V_l before gate *L* propagates the signal. Thus t_r can be written

$$t_r \approx \frac{2V_l\overline{C}_{BC}}{I_0} + \frac{V_l N_L(\overline{C}_{BE} + \overline{C}_{BC})}{I_0} \tag{5.11}$$

Note that $2V_l$ is due to the fact that the charge supplied by I_0 at the output requires contributions in voltage changes at both *I* and *D*. Also note that I_0 must charge the node.

Consider now the limiting case, when minority-charge storage dominates, Fig. 5.26*a*. All junction capacitances are neglected with respect to minority-charge storage. The base charging and discharging currents for gate Q_D are shown in Fig. 5.26*b*. Using a simplified model for dynamic charge-storage effects, two conditions for stored base charge are shown in Fig. 5.26*b*. In order to turn Q_D on, a total charge of q_F must be stored in the gate. Each collector contributes $\tau_{F2} I_0$ and the total base charging current is I_0, where τ_{F2} is the corresponding collector time constant. Hence the total q_F is

$$q_F = N_D I_0 \tau_{F2} \tag{5.12}$$

The turn-on delay to achieve the necessary q_F is hence

$$t_{\text{on}} \approx \frac{q_F}{I_0} = N_D \tau_{F2} \tag{5.13}$$

t_{on} is the time required to just store q_F across the base–emitter junction of gate D assuming N_D collectors. As in the extrinsic case, the time to turn Q_D "on," which can be considered analogous to t_f, is a small fraction of the propagation delay since only a small voltage swing at V_D is required to switch Q_L.

The saturation base charge is also shown in Fig. 5.26*b*. For each collector this charge is defined in terms of its excess base saturation current (I_{BS}).

$$I_{BS} = \frac{I_0}{N_D} - \frac{I_0}{\beta_{u2}} \tag{5.14}$$

where the first term is the actual base current and the second term represents the amount of current needed to just saturate the device. The saturated charge per collector is $\tau_s I_{BS}$, where τ_s is the saturation time constant and since there

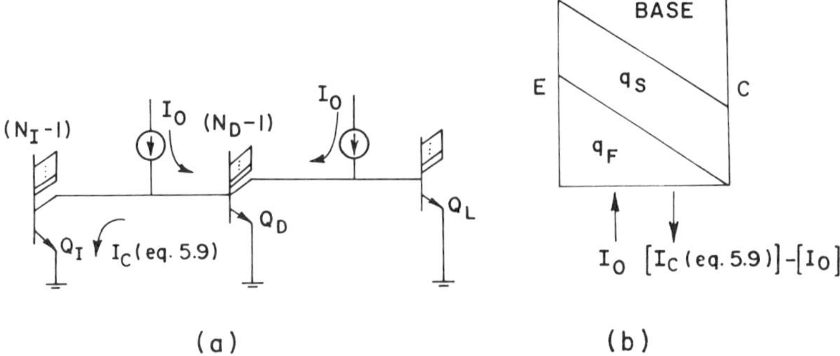

Figure 5.26 Limiting case for saturated I²L switching: (*a*) circuit schematic and (*b*) stored charge in the base of Q_D and base drive currents during transient: I_0 during turn-on and $\{I_c$ [Eq. (5.9)] $-$ $[I_0]\}$ during turn-off.

are N_D collectors q_S is given by

$$q_S = N_D \tau_s I_{BS}$$

$$= \tau_s \left(I_0 - \frac{I_0 N_D}{\beta_{u2}} \right) \tag{5.15}$$

Once this charge is stored, the previous gate must remove it before gate Q_D can come out of saturation. To calculate a switching time the discharge current can be estimated (using Eq. (5.9) to estimate the current drive at one collector of Q_I):

$$I_{B2} = \frac{I_0}{N_I/\beta_{u2} + I_{sp}/I_{sn} + I_{sp}/\beta_{p2} I_{sn}} - I_0 \tag{5.16a}$$

and again making the assumption that the terms involving β can be neglected with respect to the I_{sp}/I_{sn} term, then

$$I_{B2} \approx I_0 \left(\frac{I_{sn}}{I_{sn}} \right) - I_0 \tag{5.16b}$$

If it is assumed that the saturation delay is linear with $1/I_{BS}$, then the delay time is given by

$$t_{sd} = \frac{\tau_s (I_0 - I_0 N_D/\beta_{u2})}{\left[I_0 (I_{sn}/I_{sp}) - I_0 \right]} \tag{5.17}$$

Now for devices with reasonable β_{u2}, $\beta_{u2}/N_D \gg 1$, and $I_{sn}/I_{sp} \gg 1$, t_{sd} reduces to

$$t_{sd} \approx \tau_s \left(\frac{I_{sp}}{I_{sn}} \right)$$

Thus the turn-off is controlled by τ_s and the input drive removing q_S, which is proportional to I_{sn}/I_{sp}.

Several conclusions can be identified for the intrinsic switching waveforms. First, intrinsic device time constants dominate (as assumed). Second, the times are determined both by N_D and N_I, the fan-out of the gate and its input driver.

The intrinsic switching expressions given previously treat the saturation phenomena exclusively in terms of the *npn* τ_s term. It is useful to examine the device structure more closely and see exactly the technological factors that affect both τ_s for the *npn* as well as other charge-storage effects in the gate.

Figure 5.27*a* shows a section view of a single-collector I^{2}L gate with the base, epitaxial layer, and lateral charge components shown. The lateral terms include the injector stored charge Q_{IL} as well as contributions associated with the n^+ collar Q_{l1} and adjacent bases Q_{l2}. The terms with lower case subscripts will be neglected here for convenience. This assumption would also be reasonable if oxide isolation were used. The base charges can be considered as two subsets. The first subset involves terms due to electron injection from the epitaxial n layer which result in charge storage below the metal regions, denoted by Q_{BM}, and under the oxide, denoted by Q_{BO}. The second subset involves the charge that communicates with the n^+ collector, which is denoted as Q_{Bi}. Owing to the heavily doped n^+ region, we will ignore storage in this region. Also shown on the figure are critical dimensions W_e, W_{pL}, and W_b and W_p. A detailed analysis of charge effects is given in References 8 and 10. For simplicity of discussion the relative controlling effects for each charge component will be mentioned and then a simplified representation will be constructed. Begin by considering the charge storage time constants. To first order the following relationships are valid: Time constants for the injector region, under metal and under oxide:

$$\tau_{IM}, \tau_{IO} \propto \frac{W_p^2}{D_n} \tag{5.18a}$$

Time constant for intrinsic base region:

$$\tau_{Bi} \propto \frac{W_b^2}{D_n} \tag{5.18b}$$

Time constant for the lateral injector base region:

$$\tau_{IL} \propto \frac{W_{pL}^2}{D_p} \tag{5.18c}$$

Time constant for the epi emitter region:

$$\tau_E \propto \frac{W_e^2}{D_p} \tag{5.18d}$$

To compute the charge components one uses the expression $Q = \tau I$, where I is the relevant stored transit current. For diodes this is simply the terminal current, whereas for transistors the current is typically the pertinent junction contribution to collector current. For the special case of saturation an effective τ is used. In this case the relative contributions of both emitter and collector are collapsed into an equivalent diode model with I_{BS} and τ_s controlling the switching. For the I^{2}L gate shown in Fig. 5.27*a* the contributions Q_{BM} and Q_{BO} are predominantly diodelike. The terms Q_{IL} and Q_{Bi} are clearly transistor effects and Q_E has elements of both diode and transistor. Two assumptions

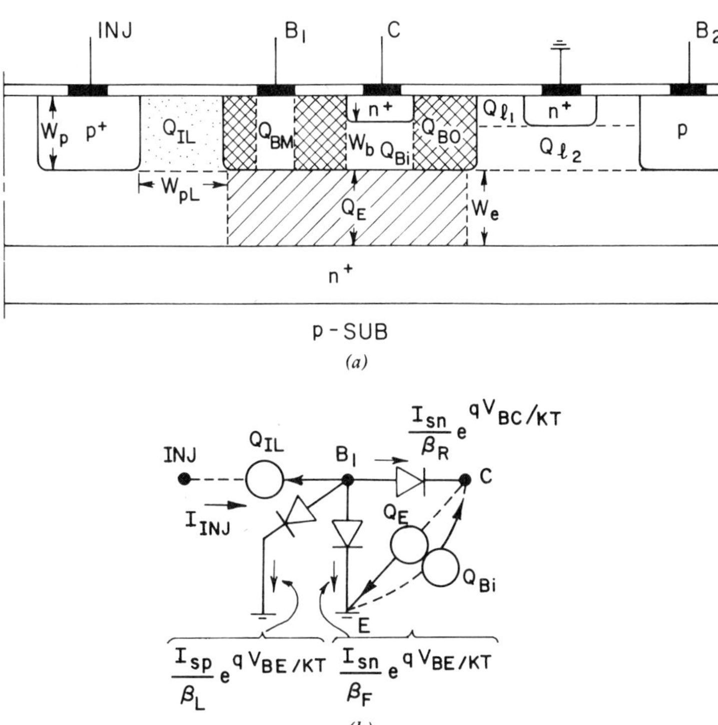

Figure 5.27 Charge storage in I^2L gate: (*a*) Physical and layout-related identification and (*b*) relationship of charge to dc equivalent model (i.e., Fig. 5.22*c*).

will allow us to drastically simplify the model of Fig. 5.27*a* and go on to derive a relationship pertinent for determining saturated switching limitations. First, assume that Q_{BM} and Q_{BO} can be equivalently included in Q_E. Since all these charge components are coupled to the *BE* junction, this seems reasonable. Moreover, since typically $\tau_E > \tau_{IM}$, τ_{BO} one might expect the Q_E contribution to dominate. Second, assume that Q_E can be considered as a transistor-controlled charge. This assumption is somewhat more difficult to swallow since the areas shown in Fig. 5.27*a* suggest the collector is a relatively small fraction of total surface area associated with Q_E. Nonetheless, we will proceed with this limitation, and the reader can rethink the problem in a more-complete manner if so desired [10].

Figure 5.27*b* shows the simplified equivalent circuit representation, which includes only the charge components Q_{Bi}, Q_E, and Q_{IL}. The corresponding junction diodes are shown based on their transport currents and current gains (note the I_{sn} and I_{sp} subscripts). Also note that in essence the charge storage has been associated with the pertinent contribution to the current generator of each device model. Hence, Q_{Bi} is associated with the *BC* current contribution. Since the primary objective is to determine how these components affect saturated switching, we will go directly to the problem of defining a τ_s(eff).

Assume that the total current entering node B_1 is I_{inj}, which in turn establishes a voltage V_{BE}. Node C is at V_{sat}, which to first order, neglecting series resistance, is given by

$$V_{\text{sat}} \approx \frac{kT}{q} \ln \frac{\beta_R + 1}{\beta_R} \tag{5.19}$$

Now since $V_{BC} = V_{BE} - V_{\text{sat}}$, the BC diode current can be rewritten as

$$I_{BC} = \frac{I_{sn}}{\beta_R} \exp\left\{ \frac{q}{kT} \left[V_{BE} - \frac{kT}{q} \ln\left(\frac{\beta_R + 1}{\beta_R} \right) \right] \right\} = \frac{I_{sn}}{\beta_R + 1} e^{qV_{BE}/kT} \tag{5.20}$$

Now all diodes are referenced to V_{BE}. By a simple current divider calculation the fraction of I_{inj} flowing through each diode can easily be determined. In turn charge contributions are computed by multiplying by the pertinent $\beta\tau$, which results in

$$Q_E = \frac{(I_{sn}/\beta_F) I_{\text{inj}}}{I_{sp}/\beta_L + I_{sn}/\beta_F + I_{sn}/(1 + \beta_R)} \tau_E \beta_F \tag{5.21a}$$

$$Q_{Bi} = \frac{[I_{sn}/(1 + \beta_R)] I_{\text{inj}}}{I_{sp}/\beta_L + I_{sn}/\beta_F + I_{sn}/(1 + \beta_R)} \tau_{Bi} \beta_R \tag{5.21b}$$

$$Q_{IL} = \frac{(I_{sp}/\beta_L) I_{\text{inj}}}{I_{sp}/\beta_L + I_{sn}/\beta_F + I_{sn}/(1 + \beta_R)} \tau_{IL} \beta_L \tag{5.21c}$$

To compute the $\tau_s(\text{eff})$ one simply sums the charges and divides by I_{inj} which gives:

$$\tau_s(\text{eff}) = \frac{\beta_F (I_{sn}/I_{sp}) \beta_L [(1 + \beta_R)\tau_E + \beta_R \tau_{Bi}] + \beta_F \beta_L (1 + \beta_R) \tau_{IL}}{(I_{sn}/I_{sp}) \beta_L [1 + \beta_F + \beta_R] + \beta_F (1 + \beta_R)} \tag{5.22}$$

Two major observations can be made relative to Eq. (5.22). First, if $I_{sp} \to 0$ the second terms in both numerator and denominator are neglected and

$$\tau_s(\text{eff}) \approx \frac{\beta_F (1 + \beta_R) \tau_E + \beta_F \beta_R \tau_{Bi}}{1 + \beta_F + \beta_R} \tag{5.23}$$

which is the simple result. The second observation involves the fact that although τ_{IL} does contribute to $\tau_s(\text{eff})$, the first term is proportionally larger by a factor I_{sn}/I_{sp}, which will be shown in Section 5.4 to be at least greater than the fan-out. Hence the increase of $\tau_s(\text{eff})$ due to the presence of the *npn* injector is not a dominant effect.

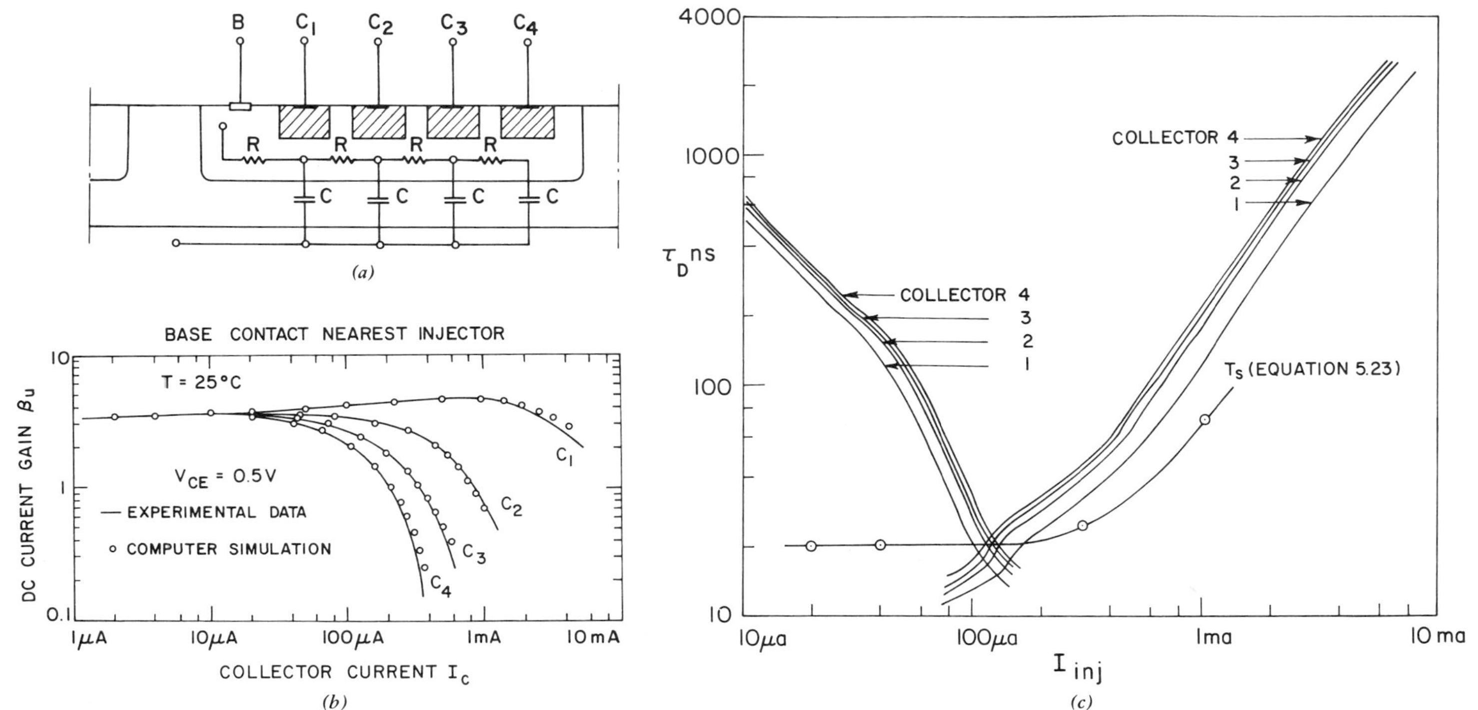

Figure 5.28 (*a*) The distributed base resistance model, (*b*) dc current gain β_u vs I_C with collector position as a parameter, and (*c*) τ_D vs I_{inj} for a four-collector I^2L.

5.3.3 I²L Device Effects

At high dissipation levels, the series resistance of the base prevents the fast injection or removal of the active charge [10]. The effect can be taken into account approximately by analyzing a distributed *RC* network [11, 13]. Figure 5.28*a* shows the modeled *RC* circuit under investigation for a four-collector structure. The nonlinear capacitance C is mainly the diffusion capacitance associated with each stage. The variation of C with current density J_n can be made available by using a dc model [9] to obtain the changes in Q_E with the junction voltage. The resistance R is the effective lateral base resistance between two consecutive *npn* stages [12]. Figure 5.28*b* shows the dc effects on current gain, and Fig. 5.28*c* shows the delays of the different collectors as a function of the injector current I_{inj}. One can observe that beyond certain current level, the total delay of the *npn* structure should be calculated using the distributed *RC* network.

5.4 SECOND-GENERATION I²L AND RELATED FAMILIES

Following the introduction of I^2L, research efforts have been directed mainly toward reducing the minimum delay per gate and thus increasing the area of applications where I^2L can be used. These efforts can be grouped into (1) improving the technology and (2) using new I^2L circuit structures. In the first group, dielectric isolation has been used [3, 14, 15], resulting in a delay on the order of 5 ns at a speed-power product of 0.7 pJ. Also, ion implantation has been used with similar results [16]. Using a second substrate as an injector has led to a higher packing density and a lower speed-power product [16]. In the second group, Schottky diodes have been used to reduce the logic swing and hence increase the speed of operation. A power delay improvement factor of 5 and a speed limit improvement factor of 2 have been reported at a reduced logic swing of 150 mV [16–19]. Also, means of controlling the saturation of the *npn* transistor have been proposed. The inherent saturation control of I^2L structures can be used. This is done by controlling the back injection of the common base region toward the injection [20]. This results in a reduction of β_u. Another method of controlling the saturation of the *npn* transistors is based on adding an extra "dummy collector" to the multicollector *npn* transistor and folding this back to the base [21]. This circuit technique is discussed in the following section and is similar to the transistor-clamped T^2L technique given in Section 4.7.8.

5.4.1 Folded-Collector I²L

The *npn* transistor in an I^2L gate is working in the saturation region when it is on. The degree of saturation can be controlled by controlling the back injection of the common base region toward the injection [20].

Another method of controlling the saturation of the *npn* transistor is based on adding an extra "dummy collector" to the multicollector *npn* transistor and folding it back to the base as was described in Reference 21. The technique not only improves the transient performance of the structure, but also provides a flexible circuit to be used in other digital applications where a current scaling factor could be used in realizing a logic function, for example, threshold logic, multivalued logic, and D/A, A/D converters.

Figure 5.29*a* shows three I^2L inverters. The transistor Q_2 is heavily saturated in its conduction state because of the excessive base current supplied to its base. In this case, $I_B = I_C = I_0$. The excess stored charge contributes to the minimum delay of the inverter. To improve this delay, the saturation of Q_2 has to be controlled. This can be achieved by adding an extra dummy collector to the *npn* transistor and folding it back to the input base, as shown in Fig. 5.29*b*. It was assumed that the dummy collector is of minimum dimensions (area A_d) and the area of the output collector A_C is x times A_d.

The performance of folded-collector structures has been compared experimentally to that of conventional structures. This was done using two seven-stage ring oscillators. The measured delay per stage is shown in Fig. 5.29*c*, which were also compared to a computer simulation. The calculated results are also plotted for the folded IC structures, taking the performance of the I^2L structures as a reference (dashed line). It is shown (Fig. 5.29*c*) that the folded structures improve the minimum delay per gate. However, the delay time of the low-current levels increases due to larger devices. The multicollector structure

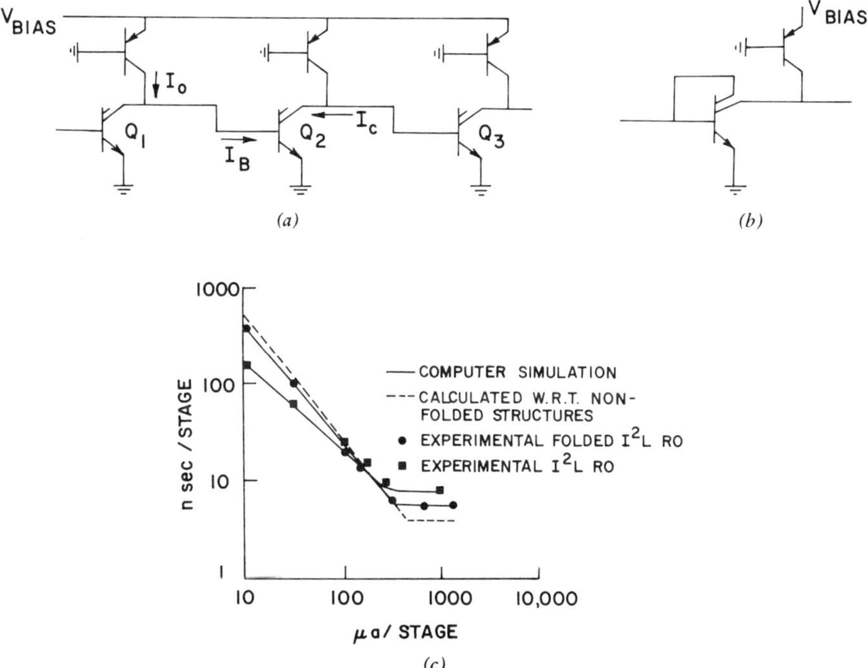

Figure 5.29 (*a*) Three I^2L inverters. (*b*) Folded-collector I^2L. (*c*) Delay per stage for seven-stage ring oscillators [21].

can be optimized for an operating current range by controlling the size and the location of the dummy collector.

The folded-collector I^2L structures offer convenient circuit elements to be used in applications where a current threshold is required. In the next two sections we shall discuss the applications of folded-collector I^2L in threshold and multivalued logic.

I^2L Threshold Logic Gates. The folded-collector I^2L circuit can be used to realize threshold gates. Threshold gates can realize a class of logic functions more efficiently than the conventional realizations using NAND or NOR gates. The class of logic functions that can be realized by a single threshold gate is called threshold functions, linearly separable function, 1-realizable function, linear-input functions, etc.

A logic function $F(X)$ of n binary variables X is a threshold function if the following conditions are satisfied:

$$F(X) = 1 \quad \text{if } f_A(X) = \sum_{i=1}^{n} a_i x_i \geqslant T$$

and

$$F(X) = 0 \quad \text{if } f_A(X) = \sum_{i=1}^{n} a_i x_i < T$$

where x_i = a binary variable, $i = 1, \ldots, n$
$X = (x_1, x_2, \ldots, x_n)$
a_i = weight of x_i
$A = (a_1, a_2, \ldots, a_n)$
T = the threshold value
$F(X)$ = a Boolean function of X
$f_A(X)$ = an algebraic function of X

A threshold logic gate has the inputs $x_1, \ldots, x_n$ and the output F. The gate computes f_A; the weighted sum of the input and compares it to the threshold value T. In I^2L, the folded-collector circuit configuration is used to compute f_A. The weighting is obtained by varying the relative size of the output collectors to the folded collector [21, 31]. Figure 5.30 shows the realization of four threshold gates: the inverter, OR (or NOR), AND (or NAND), and the 2 out of 3 majority gate. The ratio of the output collector(s) to the folded collector (x) in the case of the inverter is taken to be x_1. The injector current (I_{inj}) is taken to be I_0. The injector currents of all the other threshold gates and the value of x is referenced to that of the inverter case, namely, I_0 and x_1. It should be noted that the conventional realization of the 2 out of 3 majority gate requires three gates, thus a factor of 3 saving in gate count results by using threshold gates. The reader is referred to textbooks on threshold logic for

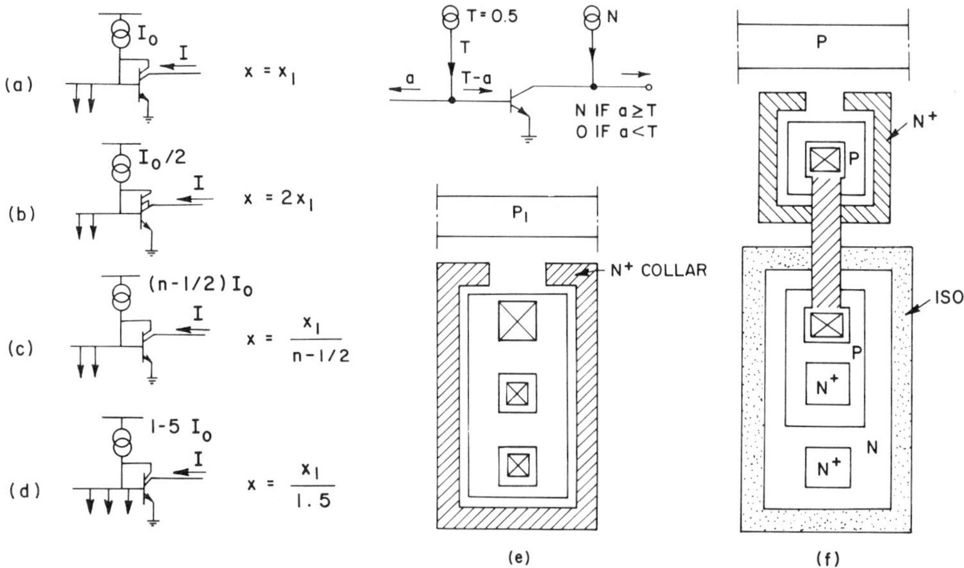

Figure 5.30 Threshold gate realization using folded-collector I^2L [21]: (*a*) Inverter, threshold level I_0. (*b*) OR [NOR], threshold level $I_0/2$. (*c*) AND [NAND], threshold level $(n - \frac{1}{2})I_0$. (*d*) 2 out of 3, threshold level $1.5I_0$. (*e*) Threshold detector [31] circuit diagram and layout, threshold level is controlled by the size of the opening in the n^+ collar. (*f*) Threshold detector layout [31] where the threshold level is controlled by the size of opening in the n^+ collar but the *npn* is isolated.

further discussion of the advantages of using threshold logic gates (see, for example, Reference 22).

To compare the weighted sum f_A to a threshold value T, an I^2L current differencing amplifier is used [31]. An injector (Fig. 5.30*e*) injects current into the base of the *npn* transistor. For a threshold value of T, T units of current are injected. This is controlled by the opening of the n^+ collar between the injector and the base region of the *npn* transistor as shown in the layout of Fig. 5.30*e*. Better control of the threshold could be obtained, as shown in Fig. 5.30*f*, by using an isolated multicollector *pnp* transistor where the sizes of the collectors are scaled to obtain the required value of T.

I^2L Multivalued Logic Gates. In multivalued logic, nonbinary representation of information is used. Thus three or more values of voltages (or currents in the case of I^2L) are used to represent the logic signals. Multivalued logic realizations of logic functions reduces the complexity of the interconnection paths. Thus it offers an efficient method of obtaining high-density VLSI chips. One problem associated with multivalued logic is the realization of a simple logic gate. The folded-collector I^2L circuit can realize such a gate [21, 31].

Figure 5.31 shows three ternary logic gates where the three logic levels (0, 1, 2) are represented by current levels of 0, I_0, and $\geqslant 2I_0$. The different

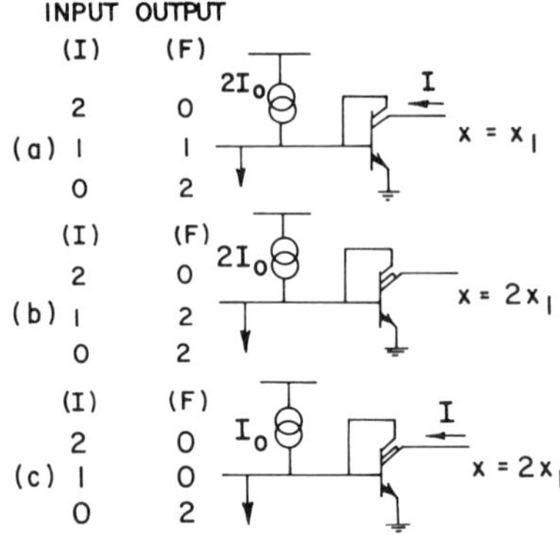

Figure 5.31 Ternary gate realization using folded-collector I^2L [21] with logic levels "0," "1," "2" represented by current levels 0, I_0, and $\geq 2I_0$: (*a*) simple ternary inverter, (*b*) positive ternary inverter, (*c*) negative inverter.

gates are obtained from the basic folded-collector configuration simply by changing both the injector current and the output collector area. The reader is referred to Reference 23 for a discussion of the advantages of multivalued logic gate realizations.

5.4.2 Stacked I^2L

One of the most important feature of I^2L is that it can be used on bipolar chips where analog circuits are integrated [24]. In these LSI circuit environments, the available power supply is greater than the minimum power supply required by I^2L, which is on the order of 800 mV. By stacking I^2L structures [25, 26], that is, blocks of logic gates are connected such that the same supply current flows through them, I^2Ls inherent high level of power efficiency is restored.

There are different approaches for stacking I^2L gates. One is suitable for the design of regular (logic or memory) arrays. It is based on stacking groups of I^2L structures (logic gates or memory cells) such that the *total* injector current of all the levels are the same. Figure 5.32 shows such a stack [25]. It is noted that each level of the stack has its own injector. The total emitter current of a level (except for the top one) feeds the injector of the lower level. The maximum number of levels is determined by the available V_{CC}. Level-shifting cells are required to couple between the levels [25]. In memory arrays, the coupling between levels is not required because each level contains a block of independent memory cells. In regular logic arrays, coupling between levels may be required and the efficiency of this approach is reduced.

Another I^2L stacking approach [26] is suitable for the design of LSI random logic blocks and is explained by referring to Fig. 5.33. Each stack consists of a base-current-driven top level gate followed by a number (two in Fig. 5.33) of stacked gates. The input currents to Q_1 and Q_2 are logically controlled, thus the logical outputs F_1 to F_3 are functions of the wired-OR inputs A to C. As

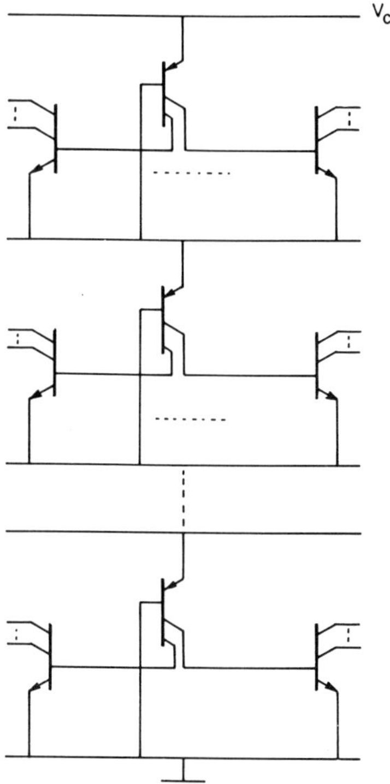

Figure 5.32 Stacked I^2L structures for regular arrays ($S1$).

shown later, level shifting stages are eliminated by using dummy stacked levels. The use of this stack in logic design has to follow certain logic constraints. This is because the possibility of faulty logic operations due to current sharing. These constraints are taken into account in developing a CAD program, STACK [27], which offers an optimum set of stacks for a given set of logic functions.

In order to study the circuit performance of stacked I^2L, let us consider Fig. 5.34 where a three-level stack is shown and the emitter–substrate

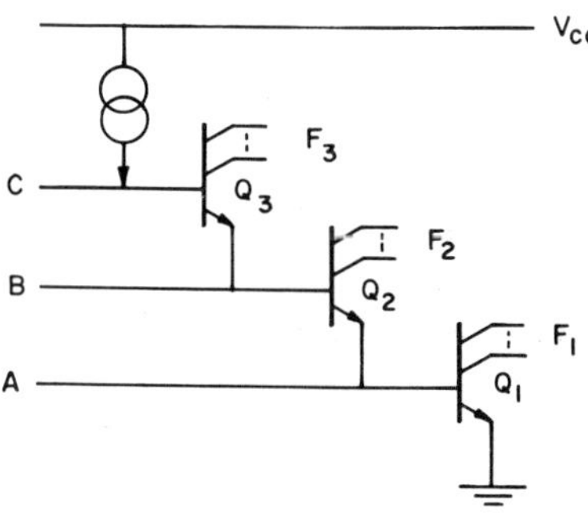

Figure 5.33 Stacked I^2L structures for random logic arrays ($S2$).

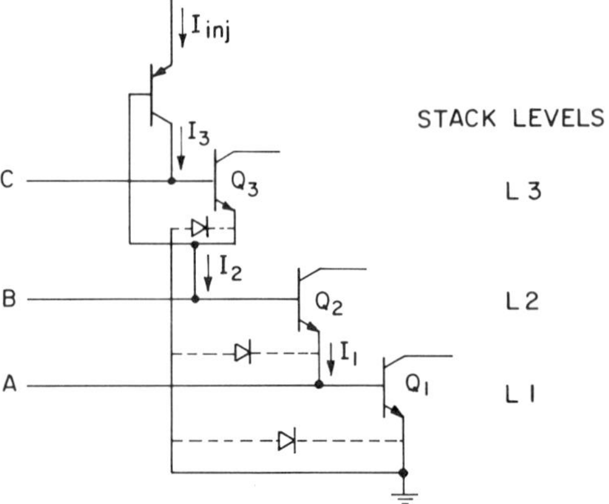

Figure 5.34 Circuit diagram of three-level *S*2 stack.

(eptiaxial–substrate) diodes are also shown. Single-collector structures have been shown for simplification.

The stack is operating with an injector current, I_{inj}. The logic levels at the input *A* are (the same as in conventional I^2L) V_j and V_{sc}, where V_j is the V_{on} voltage of the *BE* junction of the bottom level at an operating current I_1 and V_{sc} is the saturation voltage of the bottom *npn* transistor at the same operating current. The logic levels at input *B* are $(V_j@I_2 + V_j@I_1)$ and $(V_{sc}@I_2 + V_j@I_1)$. Similarly the logic levels at the input *C* are $(V_j@I_3 + V_j@I_1 + V_j@I_2)$ and $(V_{sc}@I_3 + V_j@I_1 + V_j@I_2)$. As a result the logic swing at each input is approximately the same and in the order of V_j.

In general, the operating currents for the different levels are not equal. Moreover, the node capacitances at the input are not equal. Hence, the speed of operating each level differs. In order to demonstrate this effect, each level, in different stacks, has been connected as a ring oscillator. The unused collectors are folded-back to their respective bases. In this case

$$I_3 = \lambda I_{\text{inj}}, \; I_1 = I_2 = I_{\text{inj}}$$

where λ is the injection-region efficiency. Because the lower value of the operating current I_3, the top-level (*L*3) ring oscillator is expected to have higher delays. This is shown in Fig. 5.35. The difference between the delays of the middle-level (*L*2) and the bottom-level (*L*1) ring oscillators is due to the difference between the node capacitances associated with the inputs *A* and *B*. The difference in the delays of the different levels is smaller at high-current levels where the epi storage time dominates.

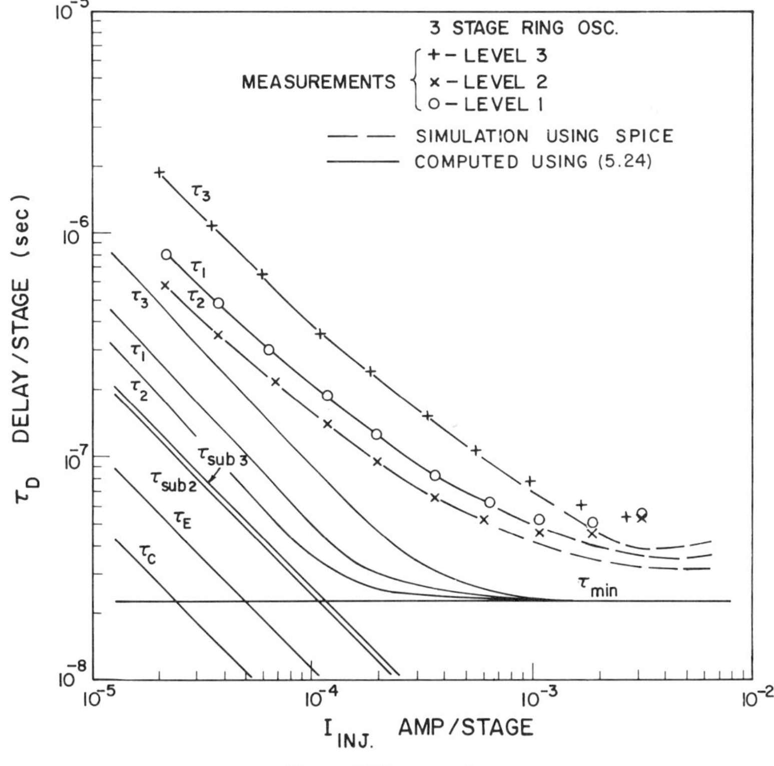

Figure 5.35 τ_D vs I_{inj}.

The total delay for a given level in the stack at a given I_{inj} can be calculated approximately by adding the corresponding delays [26]:

$$\tau_1 = \tau_E + \tau_c + \tau_{\text{min}} + \tau_{\text{Sub2}} + \tau_{\text{Sub3}}$$

$$\tau_2 = \tau_E + \tau_c + \tau_{\text{min}} + \tau_{\text{Sub3}}$$

$$\tau_3 = \left(\tau_E + \tau_c + \tau_{\text{min}}\right)\frac{1}{\lambda} \tag{5.24}$$

It is clear that $\tau_1 > \tau_2$ and τ_3 will be function of the injector region efficiency λ. If the current provided to the top level of the stack ($L3$) is obtained from a more ideal current source than the *pnp* transistor (e.g., a resistor), then $\lambda \to 1$ and L_3 will be relatively fast. On the other hand, as $\lambda \to 0$, then $L3$ will be the slowest level.

In summary, although the logic swing at each level of stacked I^{2}L structures is the same and is on the order of a V_j, the delay time through each level is different. If an injector is used to provide current to the top level, this level operates from a lower current than the other levels, hence it has a larger delay.

The delay times of the other levels in the stack depends on the logic levels at that particular level as well as the areas associated with each junction.

Figures 5.32 and 5.33 show two basic approaches to stacking I^2L structures. These two approaches will be referred to as S_1 and S_2, respectively. The two approaches offer a number of features to circuit designers, and they are summarized in the following:

1. $S1$ requires a common *pnp* injector at each level, hence the operating current of each level is λI_{inj}. $S2$ requires only an injector at the top level. In order to eliminate any faulty logic operation because of the $(1 - \lambda)I_{inj}$ base current, which will be fed to the lower levels, the injectors for $S2$ should be resistors or isolated *pnp* transistors.
2. In general $S1$ requires less isolation are than $S2$. This is because in $S1$ the grouping of different logic gates at each level is possible.
3. In $S2$ the emitters of each level are used as logic inputs. Thus the number of inputs (collectors of previous stages) required to realize a given logic function using $S2$ is less than that required using $S1$. This is an important feature of $S2$, which results in a saving of overall silicon area and an increase in yield.
4. The level-shifting stages required using S_1 or S_2 can be, in general, minimized by careful manipulation of the logic functions and their inputs. However, the logic realization using $S2$ can follow a formal method using STACK. In $S2$, the required level shifting, if any, can easily be implemented using dummy stacked levels. In $S1$, the required shifting stages are generally complex.
5. In $S1$, each level must contain a minimum number of transistors to allow the current to be distributed uniformly in the other levels. There exists a minimum operating current for each level because of current sharing between levels (even though no level-shifting stages are used). In $S2$, this is not the case since each level contains only one *npn* transistor.
6. For both $S1$ and $S2$ a lower-level gate can directly feed any upper level gate without level-shifting stages.

The above features are explained using three examples: control logic function, a full-adder, and a seven-segment display decoder.

Figure 5.36 shows the realization of the control logic functions:

$$F_1 = P_1 + P_2 + P_3 = \overline{A}B + BC\overline{D} + A\overline{B}\,\overline{C}\,\overline{D}$$

$$F_2 = P_4 + P_5 + P_6 = A\overline{B} + \overline{A}B\overline{C} + \overline{A}BC\overline{D}$$

$$F_3 = P_7 + P_8 + P_9 = C\overline{D} + A\overline{B}\,\overline{C} + \overline{A}B\overline{C}\,\overline{D}$$

and Table 5.2 shows the comparison between the realizations using $S1$, $S2$, and conventional I^2L gates.

(a)

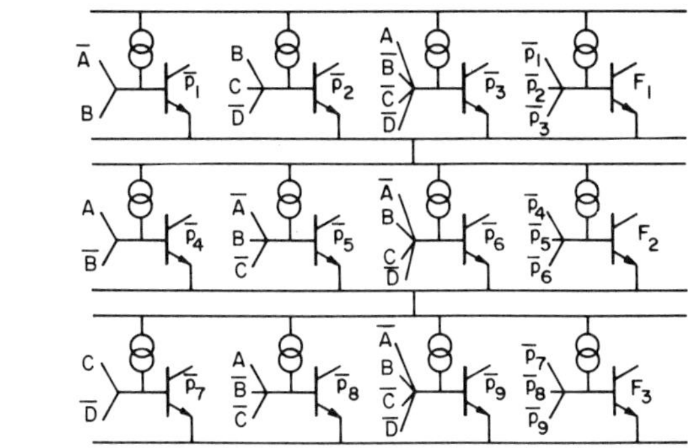

(b)

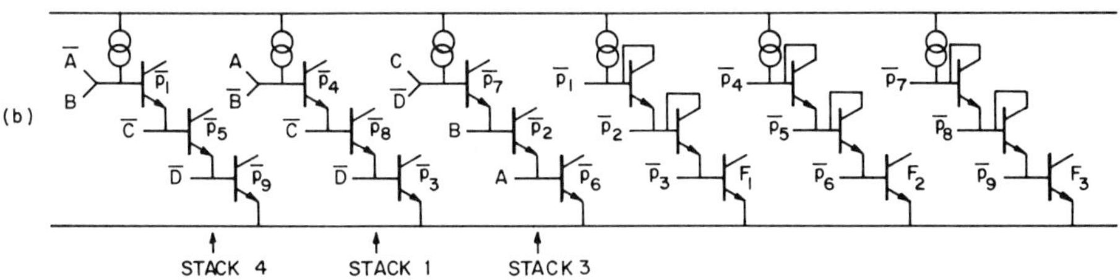

Figure 5.36 Control logic function: (*a*) using *S*1 and (*b*) using *S*2.

Table 5.2 Control Logic

Control Logic	Conventional	*S*1	*S*2
Total current	$12I_{\text{inj}}$	$4I_{\text{inj}}$	$6I_{\text{inj}}$
Max. fan-in	4	4	2
Max. fan-out	1	1	1
Number of injectors (*pnp*s or resistors)	12	12	6
Number of *npn*s	12	12	18
Min. current/gate	λI_{inj}	λI_{inj}	I_{inj}
Number of inputs (collectors)	27	27	12
Number of levels	1	3	3
Gates/level	12	4	6
Power used/total power	(V_j/V_{CC})	$(3V_j/V_{CC})$	$(3V_j/V_{CC})$
Relative area	1	1.1	0.9
Relative max. delay	1	1	0.7

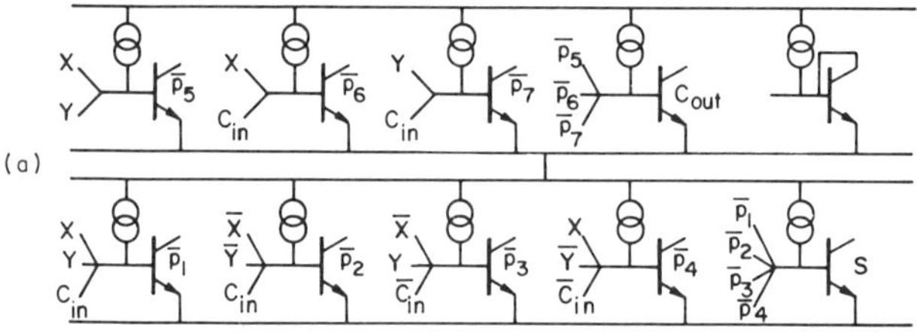

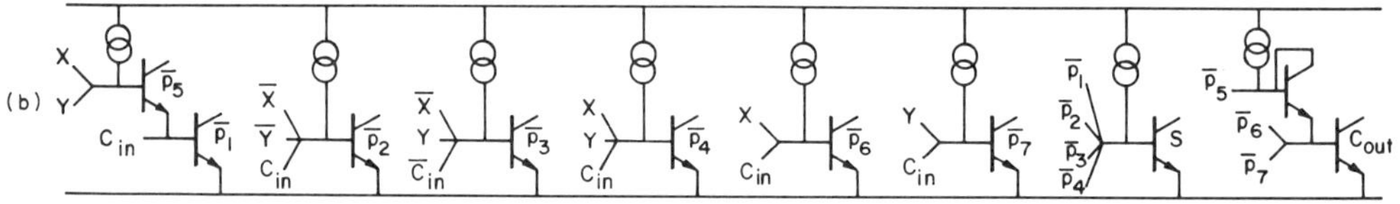

Figure 5.37 Full-adder: (*a*) using *S*1 and (*b*) using *S*2.

Figure 5.37 shows the realization of the full-adder functions:

$$S = P_1 + P_2 + P_3 + P_4 = XYC_{in} + \overline{X}\,\overline{Y}C_{in} + \overline{X}Y\overline{C}_{in} + X\overline{Y}\,\overline{C}_{in}$$

$$C_{out} = P_5 + P_6 + P_7 = XY + XC_{in} + YC_{in}$$

and Table 5.3 shows the comparison between the realizations using *S*1, *S*2, and conventional I²L gates. In Fig. 5.37*a* the top level includes a dummy I²L gate to allow the same injector current to be fed to the lower levels.

Table 5.3 Full-Adder

Full-Adder	Conventional	*S*1	*S*2
Total current	$9I_{inj}$	$5I_{inj}$	$8I_{inj}$
Max. fan-in	4	4	4
Max. fan-out	1	1	1
Number of injectors (*pnp*s or resistors)	10	10	8
Number of *npn*s	9	10	10
Min. current/gate	λI_{inj}	λI_{inj}	I_{inj}
Number of inputs (collectors)	18	18	16
Number of levels	1	2	1–2
Gates/level	9	5	2–8
Power used/total power	(V_j/V_{CC})	$(2V_j/V_{CC})$	$(1.25V_j/V_{CC})$
Relative area	1	1.1	0.9
Relative max. delay	1	1	0.7

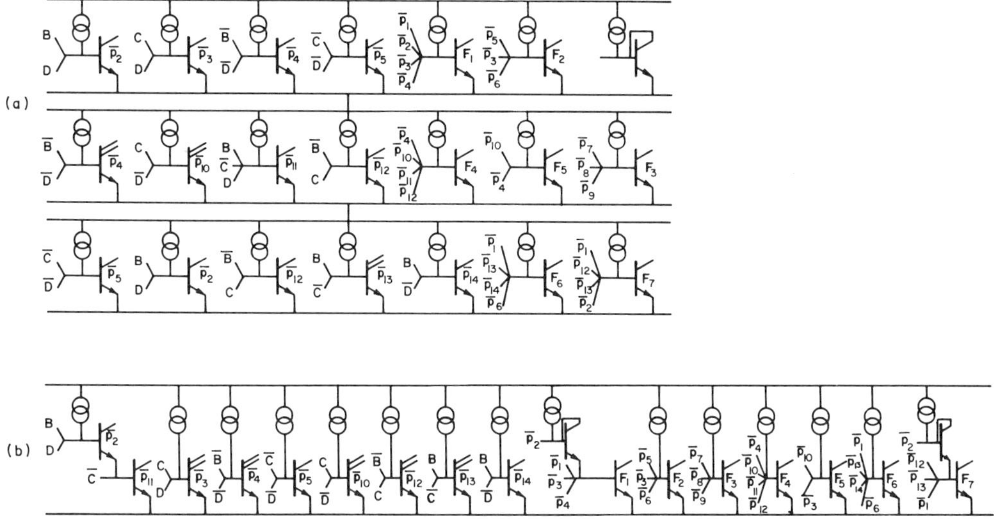

Figure 5.38 Seven-segment display decoder: (*a*) using *S*1 and (*b*) using *S*2.

Figure 5.38 shows the realization of seven-segment display decoder functions:

$$F_1 = P_1 + P_2 + P_3 + P_4 = A + BD + CD + \overline{B}\,\overline{D}$$

$$F_2 = P_5 + P_3 + P_6 = \overline{C}\,\overline{D} + CD + \overline{B}$$

$$F_3 = P_7 + P_8 + P_9 = B + \overline{C} + D$$

$$F_4 = P_{10} + P_{11} + P_{12} + P_4 = C\overline{D} + B\overline{C}D + \overline{B}C + \overline{B}\,\overline{D}$$

$$F_5 = P_4 + P_{10} = \overline{B}\,\overline{D} + C\overline{D}$$

$$F_6 = P_5 + P_{13} + P_1 + P_{14} = \overline{C}\,\overline{D} + B\overline{C} + A + B\overline{D}$$

$$F_7 = P_2 + P_{13} + P_{12} + P_1 = BD + B\overline{C} + \overline{B}C + A$$

and Table 5.4 shows the relevant comparison. In Fig. 5.38 a dummy I^2L gate is added to the top level for the same reason explained in relation with Fig. 5.37*a*. Moreover, the products P_2, P_4, and P_5 are generated at two levels to eliminate level shifting, for example, P_2 is generated at $L1$ and $L3$.

5.4.3 I^2L-Related Logic Families: SI^2L, STL, ISL

Schottky I^2L (SI^2L). In I^2L because a logic swing of about 700 mV is still larger than required by the transconductance of the switching transistor

Table 5.4 Seven-Segment Display Decoders

SSD Decoder	Conventional	$S1$	$S2$
Total current	$16I_{inj}$	$7I_{inj}$	$15I_{inj}$
Max. fan-in	4	4	4
Max. fan-out	2	2	2
Number of injectors (*pnp*s or resistors)	16	21	16
Number of *npn*s	16	21	18
Min. current/gate	λI_{inj}	λI_{inj}	I_{inj}
Number of inputs (collectors)	28	27	17
Number of levels	1	3	1–2
Gates/level	16	7	3–15
Power used/total power	(V_j/V_{CC})	$(3V_j/V_{CC})$	$(1.2V_j/V_{CC})$
Relative area	1	1.2	1
Relative max. delay	1	1	0.7

($\geqslant$ 150 mV), power-delay improvement is expected if the logic swing is reduced. This is achieved by using Schottky diodes for the output coupling as shown in Fig. 5.39*a*. This reduces the logic swing from about 700 mV to about 200–300 mV resulting in a power-delay improvement by a factor of 5 [18].

Schottky Transistor Logic (STL). If the I^2L collectors are replaced by Schottky collectors as shown schematically in Fig. 5.39*b*, a similar improvement in power-delay performance is expected because of the reduction in the logic swing and the reduction in the total stored charge [17]. STL also employs a Schottky clamp on the *npn* transistor. The clamp and collector Schottky diodes are made with different barrier-metal systems so that the difference in their forward voltages at any current level is about 200 mV. This low logic swing contributes to the high speed of the family. Moreover, the clamp Schottky diode, which is made of Pt-Si, keeps the *npn* transistor out of saturation and hence reduces the stored charge. Figure 5.40 shows a cross section in STL with dielectric isolation and two-layer of metalization are used.

Integrated Schottky Logic (ISL). In SI^2L and STL the *npn* transistor is operating in the upward mode of operation. In ISL the *npn* transistor operates in the downward mode of operation. The basic gates are shown in Fig. 5.39*c*, which consists of a *npn* transistor and a set of Schottky diodes used as outputs. To control the saturation of the *npn*, a merged lateral *pnp* transistor is added. The base-current source I is realized using an integrated resistor or a non-saturated *pnp* transistor [28]. We note here that this *pnp* transistor cannot be merged into the *npn* transistor as in I^2L because of the fact that the *npn* transistor is operating in the downward mode. This requires an extra isolated *n*

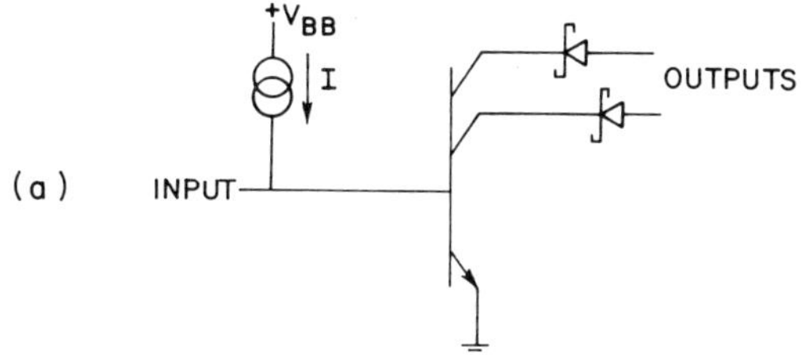

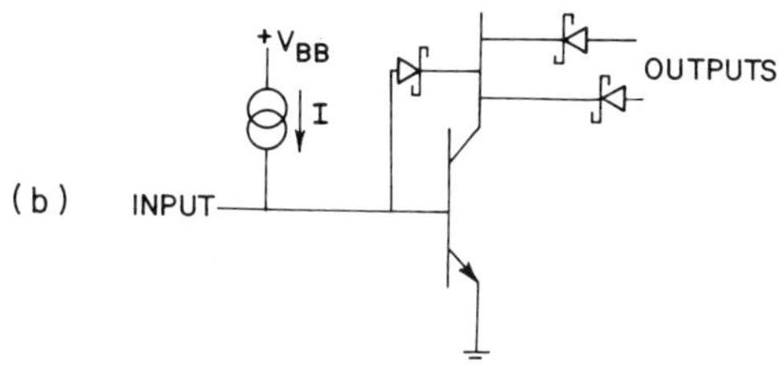

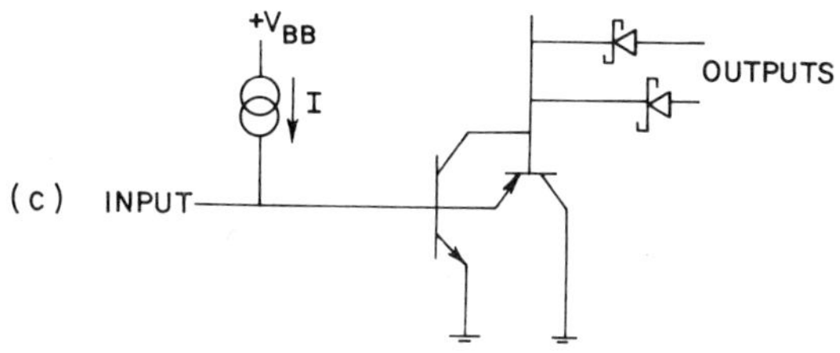

Figure 5.39 (*a*) SI^2L, (*b*) STL, and (*c*) ISL gates.

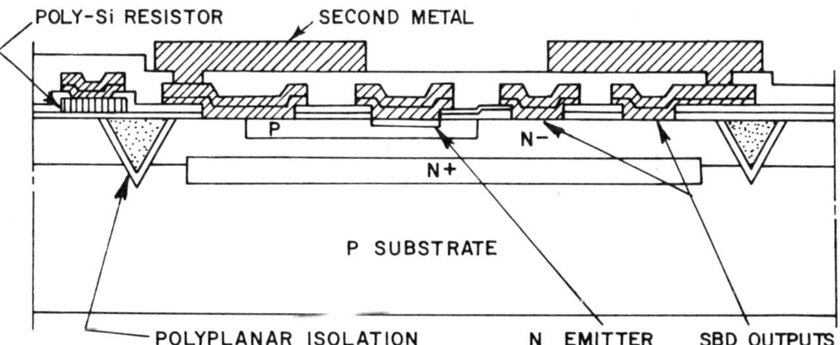

Figure 5.40 STL cross section. A poly resistor is used as a current source.

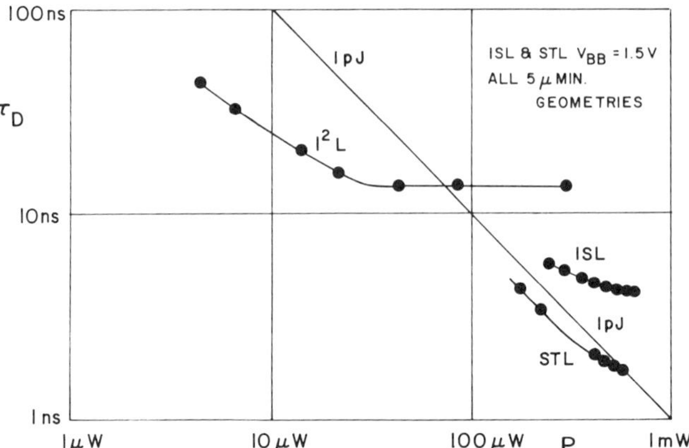

Figure 5.41 Power-speed comparison of I^2L, ISL, and STL.

island for the current source where a *pnp* or a resistor is integrated. This results in lowering the packing density compared to I^2L. However, mainly because the logic swing of the family is low (about 200 mV) and controlling the saturation of the *npn*, the family speed of operation is higher than conventional I^2L.

Figure 5.41 shows a power-speed comparison of I^2L, ISL, and STL using the same layout design rules [32]. I^2L offers a lower speed-power product, while ISL and STL offers lower delay. For the same basic dielectric-isolated double-metal ion-implanted bipolar process, I^2L requires 9 masks, ISL 10 masks, and STL 12 masks. Thus, in general, I^2L is the simplest technology, STL offers higher-speed performance, and ISL is between both in terms of complexity and speed.

5.5 I^2L PERFORMANCE EVALUATION

The previous sections have been given considerable detail concerning technology, device, and circuit implementation factors, which are pertinent for I^2L. The objective of this section is to return to the outside I/O world which I^2L sees and put our earlier discussions of T^2L in perspective. A few general comments can be stated. The primary advantages of I^2L are high packing density and linear bipolar circuit compatibility. Although this second point is beyond the scope of the present volume, these considerations are definitely relevant for all interface A/D and D/A circuitry. The major disadvantages of I^2L are its poor speed relative to Schottky T^2L or ECL/EFL. The speed limitations will continue to persist, and it is expected that for this reason the other bipolar technologies will continue to hold the edge for speed. On the other hand, the novel device and circuit merging techniques, which are the hallmark of I^2L, can be expected to continue. Possibly with these technology, topology, and circuit changes major speed improvements can be achieved as well.

Now let us briefly summarize the advantages of I^{2}L. As mentioned before, the logic swing V_l or I^{2}L is on the order of a junction voltage V_j. Also the required voltage to operate the structure is equal to V_j (the injector-epi voltage). Given these data, the delay-power product of the I^{2}L structure can be estimated.

The delay time can be expressed as

$$\tau = \frac{CV_l}{I_0} = \frac{CV_j}{I_0}$$

where C is the output node capacitance and I_0 is the injector current.

The power dissipation of the structure is given by

$$p = V_j I_0$$

Hence the delay-power product is given by

$$\tau p = CV_j^2$$

A typical node capacitance of minimum collector area is $\approx$ 0.1 pf, hence the delay-power product could be as low as 0.05 pJ, which is much less than that of T^{2}L. Table 5.5 compares the performance of I^{2}L and T^{2}L [11].

The following summaries the advantages offered by the I^{2}L logic family.

1. Wide Range of Supply Requirements: The family can operate from a wide range of supply as low as 0.7 V because of the minimum requirement of the base–emitter voltage of 0.7 V for the multicollector *npn* transistor (gate transistor) and as high as 15 V.
2. Wide Range of Frequency of Operation and Power Consumption: As discussed earlier in this chapter I^{2}L gives a wide range of current operation and so gives a wide range of frequency of operation. I^{2}L can

Table 5.5 Comparison of Typical I^{2}L and T^{2}L Properties

	I^{2}L	T^{2}L
Packing density (7-μm mask details)	120–200 gates/mm^2	20 gates/mm^2
Speed-power product	0.1–0.7 pJ/gate	100 pJ/gate
Minimum delay	10 ns	5 ns
Supply voltage	1–15 V	3–73 V
Logic voltage swing	0.6 V	5 V
Current range (per gate)	1 nA–1 mA	2 mA

operate between a clock frequency of a few hertz to a frequency up to 10 MHz (with improves technology, it is possible to increase this frequency up to 100 MHz). The power dissipation will change according to the frequency of operation and the trade-off existing between power dissipation and frequency of operation. As discussed earlier, the delay-power product being proportional to CV^2 is independent of frequency of operation, which is in contrast with other logic families, for example, CMOS, where the speed-power product increases with an increase in the operating frequency. A very-low-speed-power product obtained from this family is of the order of 0.1–0.7 pJ, as shown in Table 5.5.

3. High Yield and Low Cost: I^2L structures are easy to integrate. Two diffusion steps and four masks are required if *n*-type substrates are used (compared to four diffusion steps and six-masks in normal bipolar LSI). This simplicity results in a higher production yield and hence lower cost.
4. Interface to Other Logic Families: I^2L structures can interface with other logic families, with minimum requirements of interface circuits. This makes it easy to use the I^2L family in existing design of communications and digital systems.
5. Linear and Digital Circuits on the Same Chip: The family allows the integration of linear and digital circuits, on the same chip [23, 28, 29]. This will save on the cost of interconnection between packages, reduce the number of packages used in a system, increase system reliability, and reduce total cost. However, a bipolar technology with a *p*-type substrate is required.
6. High Packing Density: The basic structure of the I^2L family consumes an area of a single transistor. As a result up to 1500 gates can be integrated on the same chip.

REFERENCES

1. H. H. Berger and S. K. Wiedmann, Merged Transistor Logic (MTL)—A Low Cost Bipolar Logic Concept, *IEEE JSSC* **SC-7**, 340–346 (1972).
2. K. Hart and A. Slob, Integrated Injection Logic: A New Approach to LSI, *IEEE JSSC* **SC-7**, 346–351, (1972).
3. R. A. Allen and K. K. Schnegraf, Oxide-Isolated I^2L, *Dig. Tech. Papers*, IEEE ISSCC, **21**, Feb. 1974, Lewis Winner, New York, pp. 16–17.
4. H. H. Berger, Overview of I^2L Memory Structures, *Dig. Tech. Papers*, IEEE ISSCC, **21**, Feb. 1974, Lewis Winner, New York, pp. 28–29.
5. R. A. Heald, A Four-Device Bipolar Memory Cell (ISSCC Paper THAM 9.3, San Francisco, CA, February 15–17, 1978), *Tech. Dig.* **25**, 102–103 (1978).
6. F. J. Hill and G. R. Peterson, *Digital Systems: Hardware Organization and Design*, Wiley, New York, 1973.
7. T. R. Blakeslee, *Digital Design with Standard MSI and LSI*, Wiley, New York, 1975.

8. H. H. Berger, The Injection Model: A Structure-Oriented Model for MTL, *IEEE JSSC* **SC-9**, 218–227 (1974).

9. S. S. Rofail, M. I. Elmasry, and E. L. Heasell, Functional Modeling of I^2L—DC Analysis, *IEEE Trans. ED* **ED-24** (3), 234–241 (1977).

10. S. S. Rofail, M. I. Elmasry, and E. L. Heasell, Functional Modeling of I^2L—Transient Analysis, *IEEE Trans ED* **ED-25**, (9), 1120–1125 (1978).

11. N. C. De Troye, I^2L—Present and Future, *IEEE JSSC* **SC-9**, 206–211 (1974).

12. D. V. Kerns, The Effect of Base Contact Position on the Relative Propagation Delays of the Multiple Outputs of an I^2L Gate, *IEEE JSSC* **SC-11**, 712–717 (1976).

13. S. S. Rofail and M. I. Elmasry, Effect of Collector Location on β_u of I^2L Structures, *IEEE Trans. ED*, **24**, 289–291 (1977).

14. C. Mulder and H. E. J. Wulms, High Speed I^2L, in the Proceedings of the First European Solid State Circuits Conf., Sept. 1975, pp. 28–29, IEE Conf. Publ. 130.

15. R. Muller and J. Graul, CHIL and I^2L with Passive Isolation, *IEEE JSSC* **SC-11**, 30–31 (1976).

16. V. Blatt, P. S. Walsh, and L. W. Kennedy, Substrate Fed Logic, *IEEE JSSC* **SC-10**, 336–342 (1975).

17. H. H. Berger and S. K. Wiedmann, Schottky Transistor Logic, *Dig. Tech. Papers*, IEEE Int. Solid-State Circuits Conf., **22**, Feb. 1975, Lewis Winner, New York, pp. 172–173.

18. F. W. Hewlett, Jr., Schottky I^2L, *IEEE J Solid-State Circuits* **SC-10**, 343–348 (1975).

19. P. S. Walsh, V. Blatt, and L. W. Kennedy, Substrate Fed Logic—A Novel Form of Injection Logic, in *Proceedings of the First European Solid-State Circuits Conf.*, Sept. 1975, IEEE Conf. Publ. 130, pp. 122–123.

20. H. H. Berger, The Injection Model—A Structure-Oriented Model for Merged Transistor Logic (MTL), *IEEE J. Solid-State Circuits* **SC-9**, 218–227 (1974).

21. M. I. Elmasry, Folded-Collector I^2L, *IEEE JSSC* **SC-11** (5), 644–647 (1976).

22. C. L. Sheng, Threshold Logic, Academic, New York, 1969, Chap. 2.

23. Special Issue on Multiple-Valued Logic, *IEEE Computer*, Sept. 1974.

24. M. I. Elmasry *et al.*, I^2L for a Linear/Digital LSI Environment, *IEEE Trans. on ED*, **ED-25** (3), 351–357 (1978).

25. K. Kaneko, T. Okabe, and M. Nagata, Stacked I^2L Circuit, *IEEE JSSC*, **SC-12**, 210–212 (1977).

26. W. F. Petrie and M. I. Elmasry, Stacked I^2L Structures, *IEEE JSSC*, **SC-15**, 240–244 (1980).

27. W. F. Petrie and M. I. Elmasry, Computer-Aided-Design of Stacked I^2L Structures, *22nd Midwest Symposium on Circuits and Systems*, June 1979, Western Periodical, New York, pp. 611–615.

28. J. Lohstroh, ISL, A Fast and Dense Low-Power Logic, Mode in a Standard Schottky Process, *IEEE JSSC* **SC-14**, 585–590 (1979).

29. J. L. Saltich, W. L. George, and J. G. Soderberg, Processing Technology and AC/DC Characteristics of Linear Compatible I^2L, *IEEE JSSC*, **SC-11** (4), 478–485 (1976).

30. D. J. Allstot *et al.*, A New High-Voltage Analog-Compatiable I^2L Process, *IEEE JSSC*, **SC-13** (4), 479–483 (1978).

31. T. T. Dao, Threshold I^2L and Its Applications to Binary Symmetric Functions and Multivalued Logic, *IEEE JSSC*, **SC-12** (5), 463–472 (1977).

32. T. J. Sanders and B. R. Doyle, Bipolar Gate Array Technology: I^2L versus ISL, *Custom Integrated Circuit Conference*, May 1980, IEEE Press, New York, pp. 87–91.

Chapter 6

Logic Families for High-Speed LSI: ECL and EFL

6.1 INTRODUCTION

The previous chapters have considered the basic circuit-design factors involved with bipolar digital logic. Moreover, the saturated-mode logic circuits of T^2L and I^2L have been studied in some detail as examples. In addition, the use of Schottky clamping to prevent hard saturation was explored. The purpose of this chapter is to move totally into the realm of nonsaturating logic. In essence this is achieved by limiting current and voltage swings to ensure only active-mode device operation. The benefits are major speed improvements due to decreased stored minority charge and circuit topologies that result in less capacitance—hence more-efficient switching. One question which you may keep in mind as you study the emitter-coupled logic (ECL) and emitter-function logic (EFL) circuits presented here is the following: Is it possible to achieve all the density of I^2L and still get the performance of nonsaturating logic? The historic answer has been that those seeking speed did not necessarily look for density and low power as well. However, the new era of VLSI will push hard on technologies like I^2L to increase speed and, on the other side, technologies like EFL will be molded to optimize power and packing density. In short, Chapters 4, 5, and 6 represent illustrative examples of digital IC technologies and the future evolution will tend to trade-off and hopefully obliterate the present differences.

6.2 EMITTER-COUPLED LOGIC (ECL)

As mentioned in Section 4.4, in order to increase the frequency of operation of a bipolar logic family it is necessary to prevent the transistor from operating in the saturation region. One method to do this is to make sure that the operating V_{CE} is greater than the soft saturation voltage and that the base current is just enough to maintain the transistor operating in the active region. Because it is difficult to control the base current at a level corresponding to the active

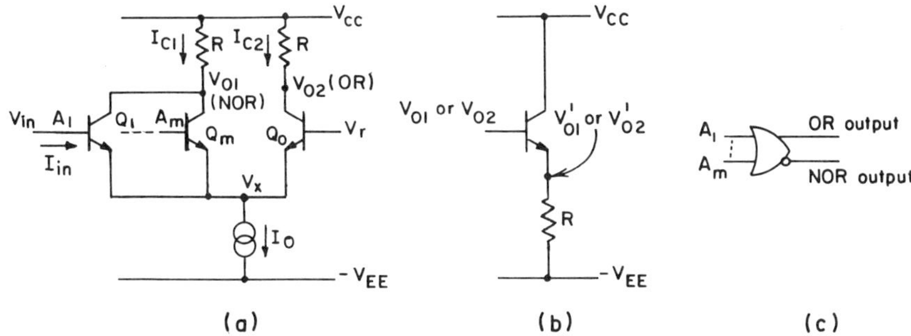

Figure 6.1 (*a*) ECL basic gate. (*b*) Emitter–follower stage. (*c*) Logic symbol.

region of operation ($I_B = I_C/\beta$) due to the wide tolerance in β, it is more practical to control the emitter current (or the collector current for that matter) at a constant value that is determined by the bias circuit.

This leads us to the basic emitter-coupled logic gate of Fig. 6.1. It consists of a fixed bias transistor Q_0 whose base is connected to a reference voltage V_r and a number of input transistors connected in parallel (Q_1 to Q_m).

The input transistors are able to steer the current I_0 between Q_0 and (Q_1 to Q_m). As the current is steered, the logic output voltage at V_{01} and V_{02} also will change giving a NOR and an OR logical output, respectively. Because the fact that the steering of the current (Fig. 6.2) changes the output voltage, the name current-mode logic (CML) is sometimes used. The current in Fig. 6.2 is normalized with respect to $\alpha_0 I_0$, which is the maximum collector current since I_0 is the emitter current and $\alpha \triangleq I_C/I_E$, the emitter efficiency. The abscissa represents the voltage excursion needed to steer current between Q_0 and the input transistors. Since the collector currents are exponentially dependent on their base-emitter voltages, it only requires a change of mkT/q (m is about 2–3) in the input voltage with respect to V_r to switch the currents. This transition region is shown in Fig. 6.2. In the examples used in this chapter, V_r is set at an odd multiple of $\frac{1}{2}V_{on}$ (assume $V_{on} = 0.7$ V). Because it is desirable to have symmetric noise margins at the logic high and low levels, a total logic swing of V_{on} is typical and it is higher than mkT/q. Referring to Fig. 6.1, the

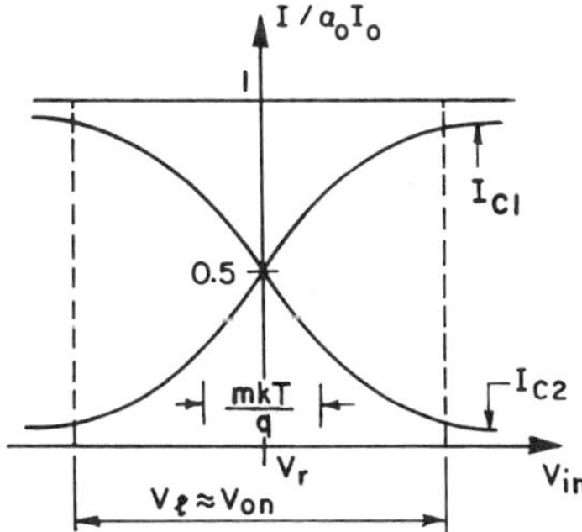

Figure 6.2 Collector currents I_{C1} and I_{C2} vs V_{in}. I_{C1} and I_{C2} are normalized with respect to the maximum collector current αI_0.

highest potential for V_{01} will be V_{CC} (typically ground) and the lowest potential will be $V_{CC} - V_{on}$, which is $\frac{1}{2}V_{on}$ below V_r.

The threshold voltage of the ECL gate is V_r (see Chapter 4 for the definition of the threshold voltage of a logic gate). If any of the input logic signals (A_i) is sufficiently higher than V_r, then Q_i is ON and Q_0 is OFF. If all the inputs logic signals are sufficiently lower than V_r, then Q_0 is ON and Q_i ($i = 1, \ldots, m$) are OFF. Allowing for a 0.7 V drop across the current source to avoid saturation, V_r has to be 1.4 V, i.e. two diode drops above ($-V_{EE}$). If Q_0 is allowed to operate at V_{CE} of 350 mV, then its collector will be at 1.05 V above ($-V_{EE}$). If the logic swing across R_2 is a diode drop of 0.7 V, then the required *minimum* V_{CC} is 1.75 V above ($-V_{EE}$). However, in order to allow cascoding (stacking structures sharing the same current source) of ECL structures, V_{CC} is usually taken to be 5.2 V above ($-V_{EE}$). This allows three ECL structures to be cascoded. If cascoding is not required, V_{CC} could be 1.75 V above ($-V_{EE}$), assuming the current-source operating voltage and the logic swing are both 700 mV and the emitter-coupled pair operates with a minimum V_{CE} of 350 mV when they are on.

In ECL circuits the logic output is usually referred to a true ground potential, that is, V_{CC} is taken to be ground and the circuit will operate from a negative supply voltage. This provides a less noisy signal at the output. This can be explained as follows: (1) If the "current source" is a resistor R_E, $R_E \gg R_L$, then noise at the V_{CC} node introduces a signal $= \Delta V_n R_E/(R_E + R_L)$, whereas noise introduced at the V_{EE} node creates a signal $= \Delta V_n R_L/(R_E + R_L)$. The voltage divider ratio suggests that noise at the V_{CC} node is larger. (2) If "current mirrors" are used as discussed next, changes at the V_{EE} terminal affect the bias level and hence the mirror current by the $\Delta V_n/R_{\text{bias}}$, whereas noise at the V_{CC} terminal appears directly at the output assuming the current sources have infinite output impedance.

In order to increase the driving capability of the basic ECL gate, an emitter–follower stage is added as shown in Fig. 6.1*b*. The addition of emitter followers lowers the logic levels at the outputs by a V_{on}. Thus the reference V_r must also be lowered to a value of $1\frac{1}{2}V_{on}$. Figure 6.3 shows the input current over the input logic swing about V_r. Normalized kT/q units ($V_T = kT/q$) are used, and it is observed that only a few V_T are required to go from off to on states.

From Figs. 6.1*a* and 6.2 it is clear that if the input voltage changes by a diode drop V_{on}, the collector current will change by approximately I_0. The logic swing of the circuit is $I_0 R$ (where $R = R_1$ or R_2 depending on the output node). Usually the output logic swing is limited to a diode drop and in many ECL circuits this is obtained by shunting the load resistance with a diode.

The current source is usually realized using a transistor working in the active region, as we shall discuss later.

In the circuit of Fig. 6.1*a* if we consider $-V_{EE}$ to be a reference voltage, the active current source needs a V_{CE} higher than the soft saturation voltage ($\simeq$ 300 mV). Let this voltage be a diode drop of 700 mV. As a result, when any of the transistors (Q_0–Q_m) is ON, it is working in the active region ($V_{CE} \geqslant$ the soft saturation voltage of the transistor). Figure 6.2 shows the relationship between the output currents I_{c1} and I_{c2} and the input voltage V_{in} assuming identical transistor characteristics and an ideal constant current source I_0. I_{c1} and I_{c2} is related to the input voltage by an exponential characteristic [1]:

$$I_{c1} = I_s e^{(V_{in} - V_x)/V_T}$$

$$I_{c2} = I_s e^{(V_r - V_x)/V_T}$$

where V_x is the common emitter node voltage shown in Fig. 6.1 and I_s is the bipolar transistor transport reference current. Hence, the ratio of the two collector currents is given by

$$\frac{I_{c1}}{I_{c2}} = e^{(V_{in} - V_r)/V_T}$$

Using the equation $I_{c1} + I_{c2} = \alpha_0 I_0$, then

$$\frac{I_{c2}}{\alpha_0 I_0} = \frac{1}{1 + e^{(V_{in} - V_r)/V_T}} \tag{6.1}$$

where α_0 is the common-base short-circuit current gain and $V_T = kT/q$ is the thermal voltage. It follows that the input-current–input-voltage characteristic [2] (Fig. 6.3) is given by

$$\frac{I_{in}}{(1 - \alpha_0) I_0} = \frac{1}{1 + e^{(V_r - V_{in})/V_T}} \tag{6.2}$$

Figure 6.4a shows a complete realization of the classical two-input ECL OR/NOR gate with typical element values [4]. Transistors Q_{1A} and Q_{1B} provide the standard NOR gate and transistors Q_3 and Q_4 provide emitter–

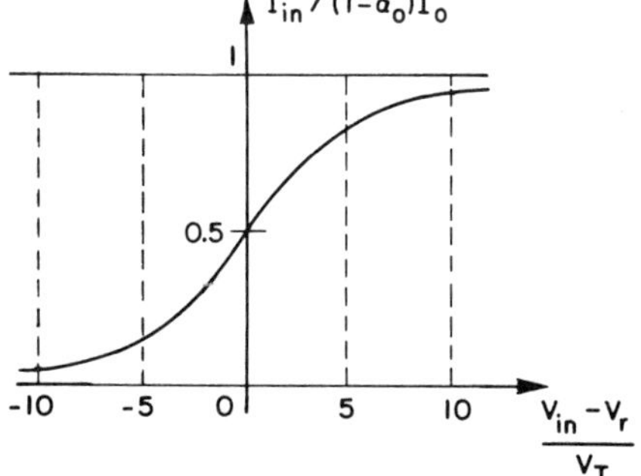

Figure 6.3 Input current I_{in} vs $(V_{in} - V_r)$. I_{in} is normalized with respect to the maximum input current $(I_0 - \alpha_0 I_0)$, while $(V_{in} - V_r)$ is normalized with respect to the thermal voltage V_T.

(a)

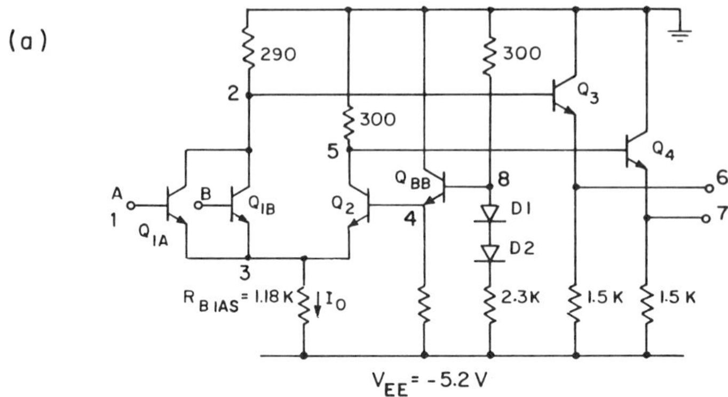

(b)

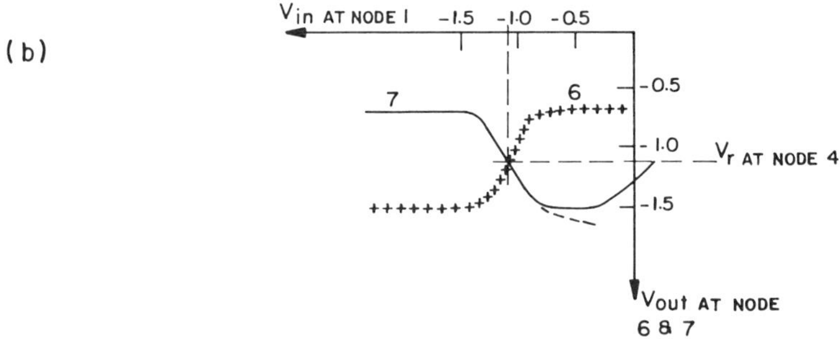

Figure 6.4 (a) A typical ECL NOR/OR gate with a biasing chain for V_r and source followers. (b) V_{out} vs V_{in} for Fig. 6.4a

follower outputs from nodes 2 and 5. The 1.18-kΩ resistor from node 3 to V_{EE} sets the circuit operating current [for this circuit $I_0 = (5.2V - 2.5V_{on})/1.18$ kΩ $= 2.85$ mA]. The voltages at nodes 3 and 4 are established at two and one V_{on} drops below the bias at node 8 which is at -0.44 V, a value about $\frac{1}{2}V_{on}$ below ground. The operation of the basic ECL gate dictates that I_0 controls the maximum swing at nodes 2 and 5, hence $V_{l2} = 290\ \Omega\ I_0 = 0.825$ V. The output voltages at nodes 6 and 7 are shifted downward by one V_{on} of Q_3 and Q_4. Hence the output levels of -0.7 V and 1.525 V shown in Fig. 6.4b are observed. One can see that the choice of the bias voltage at node 4 of $V_r = -1.14$ V is exactly consistent with the midpoint output voltage swing.

It is useful to define capabilities of ECL for more-complex logic structures and the critical timing constraints of the basic logic. Figure 6.5 shows more-complex ECL logic gates. The novel features are apparent as compared with Fig. 6.4. Diode clamps have been used as load devices (Fig. 6.5b) and a stack logic realization is used to generate two levels of functional complexity per bias current (Fig. 6.5a). The diode clamp is useful in limiting the logic swing. The stacked circuit functions is an efficient use of both power and device area. The

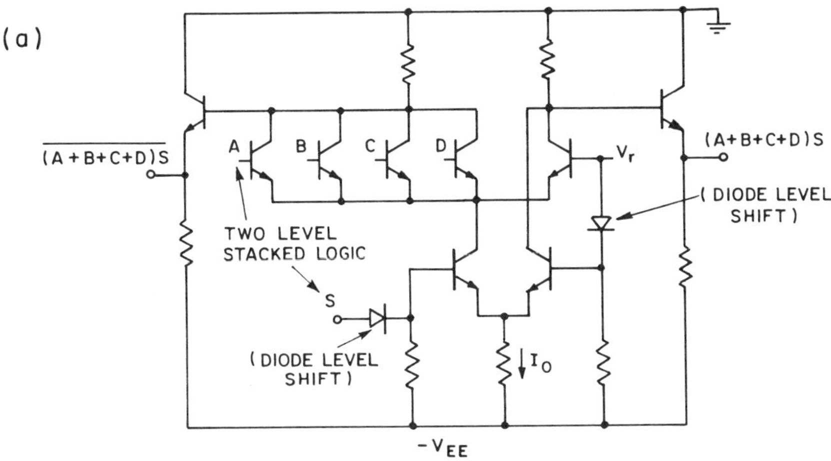

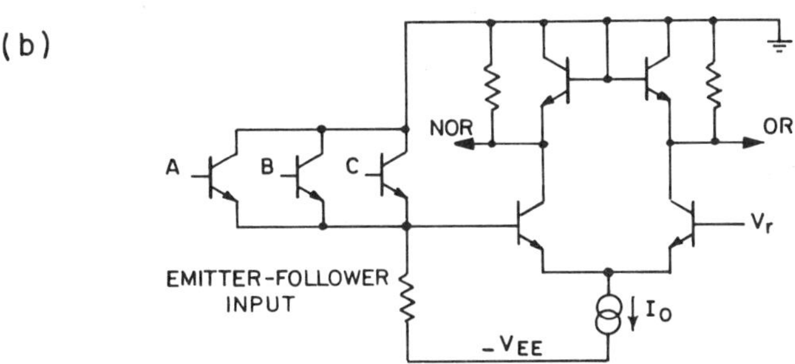

Figure 6.5 Modified ECL gates: (*a*) two-level stacked gates; (*b*) a high-speed realization with diode clamps and emitter–follower inputs.

one penalty is that two bias voltage levels are required although this is a minor internal problem as can be noted. This stacking feature is explained in Section 6.4 in association with EFL.

6.2.1 ECL Current Sources

Figure 6.6 shows a simple biasing circuit to obtain n active current sources for ECL gates. The circuit consists of a biasing chain, a resistor R, and diode-connected transistor. For identical transistor characteristics

$$I_0 = \frac{I_{\text{ref}}}{1 + (n+1)/\beta} \tag{6.3}$$

If $(n+1)/\beta \ll 1$, then $I_0 \approx I_{\text{ref}} = (V_{CC} - V_{\text{on}})/R$.

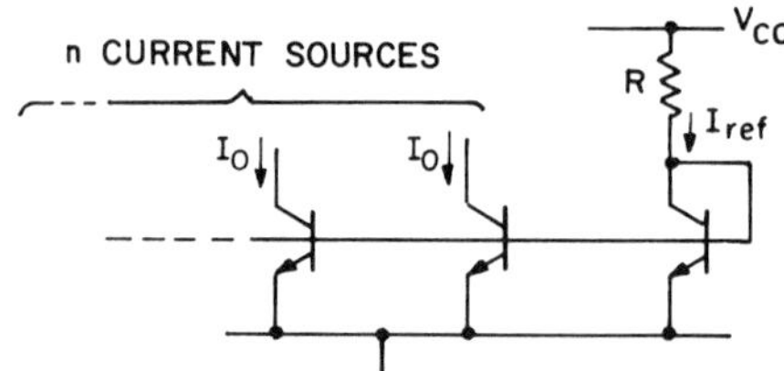

Figure 6.6 Biasing circuit to obtain I_0.

The simple circuit shown in Fig. 6.6 can be modified to accommodate different requirements. For example, Fig. 6.7 shows a biasing circuit that provides different values of current sources. This may be required where different ECL gates are operating at different speeds. In Fig. 6.7

$$I_1 = I_{\text{ref}} \frac{R_0}{R_1}, \qquad I_2 = I_{\text{ref}} \frac{R_0}{R_2}$$

$$I_{\text{ref}} \simeq \frac{V_{CC} - V_{\text{on}}}{R + R_0} \tag{6.4}$$

It should be mentioned here that extensive work has been done on current sources for operational amplifiers [3], and it is applicable to the design of current sources for ECL.

6.2.2 ECL Reference Voltages

Figure 6.8*a* shows a simple biasing circuit to provide a reference voltage V_r to ECL gates. It consists of a diode-resistor chain followed by an emitter–follower stage. The emitter–follower stage provides a low output resistance, hence V_r can feed more than one ECL gate. Neglecting the base current of the emitter–follower transistor,

$$V_r = V_{CC} \frac{R_2}{R_1 + R_2} - V_{\text{on}} \frac{R_2 - R_1}{R_1 + R_2} \tag{6.5}$$

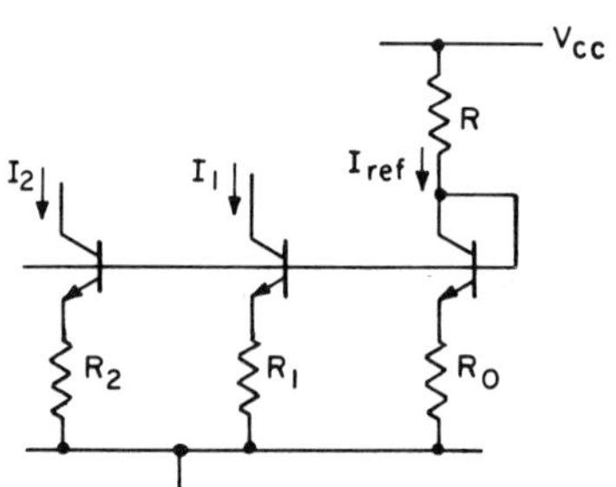

Figure 6.7 Biasing circuit to obtain different values of I_0.

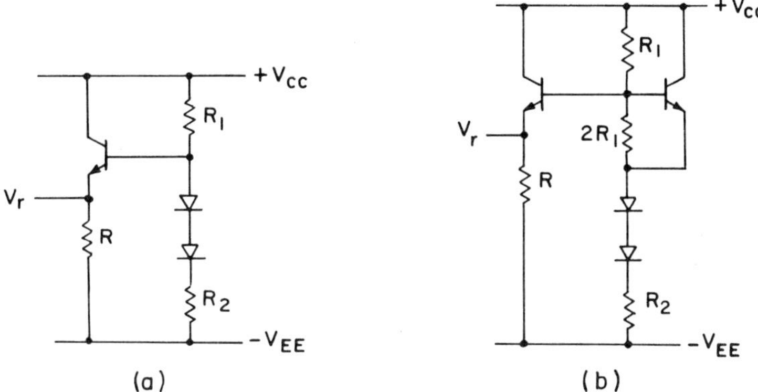

Figure 6.8 (*a*) and (*b*) Biasing circuits to obtain the reference voltage V_r.

The purpose of the resistor R is to bias the emitter–follower stage in the active region. The circuit can be modified to meet different requirements [4].

As will be shown in Section 6.6, it is often desirable to set V_r at $1.5V_{\text{on}}$. The circuit shown in Fig. 6.8*a* can be designed to achieve the desired bias level by choosing R_1 and the bias current to set the reference voltage. A more-exact method to set V_r is to replace R_1 with the circuit shown in Fig. 6.8*b*. In this case the resistive divider of R_1 and $2R_1$ gives $0.5V_{\text{on}}$ across R_1, which is then applied at the base of the emitter follower. The output V_r is then $1.5V_{\text{on}}$.

6.3 ECL MASTER-SLICE GATE ARRAY

System designing with LSI can often use a predesigned universal gate array chip, where the chip can be personalized to a specific function by customizing higher-level masks—for example, the final metal mask. This approach has the advantage of quick turnaround time: the time taken from the final stage of logic design to the fabricated chip. The approach has the disadvantage that in some cases chip area can be underutilized. However, the approach has been widely accepted, and an example using ECL [14] is considered in this section.

The basic logic gate of the array is an ECL gate with Schottky-diode-clamped load as shown in Fig. 6.9. Both the OR and NOR function outputs are available. The gate operates from $V_c = 1.7$ V provided from an off-the-chip power supply. The gate can operate on two levels of power dissipation. This is made possible by providing the gate resistors with center taps, as shown in the layout of Fig. 6.10. The particular design under consideration allows an operation of 0.85 mW at 1.5 ns per gate or 1.7 mW at 0.8 ns per gate.

The logic "1" level of the gate is $V_C(= 1.7 \text{ V})$ while the logic "0" level is $V_C - V_{SD}(= 1.7 - 0.4 = 1.3 \text{ V})$, where V_{SD} is the turn-on voltage of the Schottky diodes. Thus the reference voltage V_r should be midway between V_1 and V_0: 1.5 V.

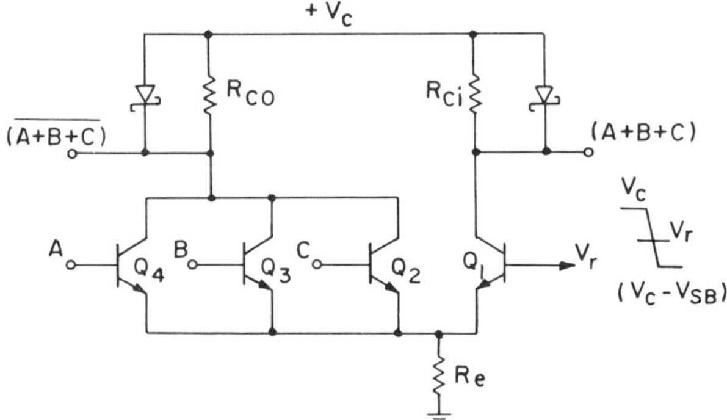

Figure 6.9 Schottky-diode-clamped ECL.

Figure 6.11 shows the on-chip reference voltage (V_r) generator circuit. It operates from a high-power supply V_{CC} of 5 V, which is provided off-the-chip. The reference voltage V_r is established by the biasing chain, which consists of the resistor R_1 (which is centered taped) and R_2, the Schottky diode D_1, and the diode-connected transistor Q_1. The established V_r is fed into one end of a differential pair. The other input of the pair samples the output of the circuit, and feedback is used to minimize the effect of the variations in V_{CC} on the value of V_r. A Darlington configuration output stage is used to minimize output impedence and maximize drive capability. Capacitors are provided for closed-loop stability.

In order to provide T^2L compatibility at the inputs and outputs the two circuits shown in Figs. 6.12 and 6.13 are used. Figure 6.12 shows an on-chip receiver level-shifting circuit, which converts T^2L logic levels to the internal logic levels. It operates from V_C and uses a Schottky-clamped transistor, a Schottky input diode, and two resistors. Figure 6.13 shows the off-chip driver level-shifting circuit. It operates from V_{CC} and uses a totem-pole output circuit.

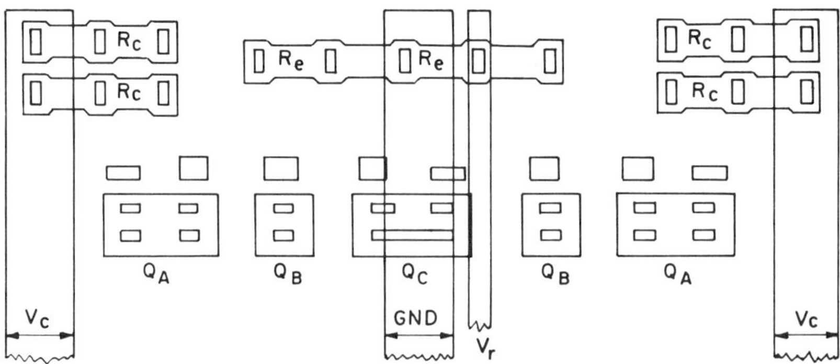

Figure 6.10 A typical ECL cell layout for a master gate array.

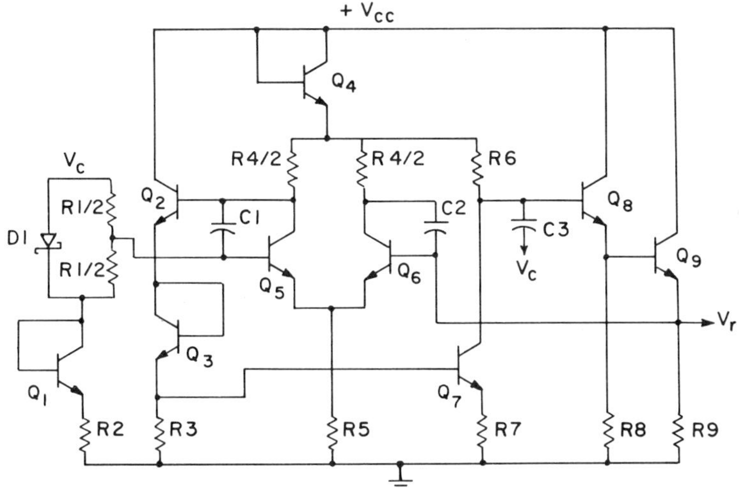

Figure 6.11 On-chip reference generator V_r circuit.

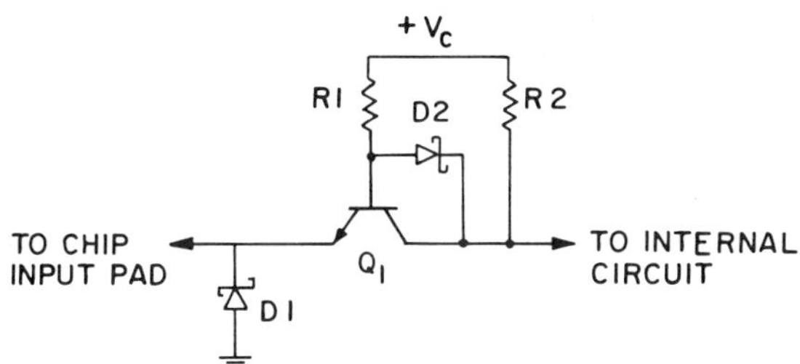

Figure 6.12 On-chip receiver level-shifting circuit.

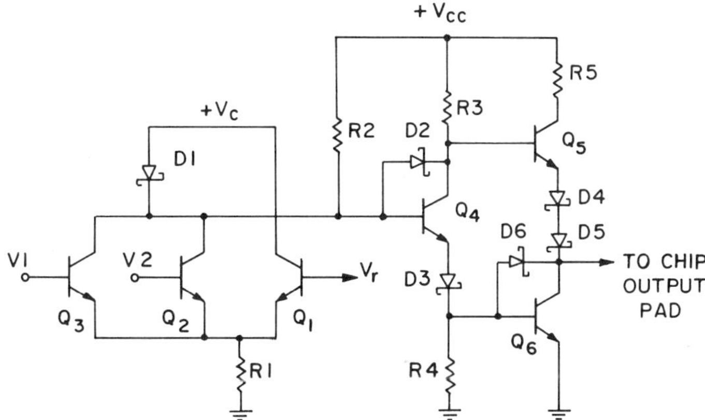

Figure 6.13 Off-chip driver level-shifting circuit.

The physical arrangement of the chip layout is shown in Fig. 6.14. The level-shifting drivers (64 of them), the reference generating circuits (2 of them), and the receiver circuits (88 of them) are along the periphery of the chip. The gate array is organized into 34 rows and 22 columns.

6.4 EMITTER FUNCTION LOGIC (EFL)

Emitter-function-logic (EFL) cells can be viewed as a repartitioned form of ECL [5–8]. The main feature of EFL is the increase in logic-function complexity at minimum increment in the device and circuit complexity. This is achieved without additional power dissipation in comparison to the single-function current switch gate. This increase in functional complexity results in significant improvements in relative density, power per gate, and speed-power product. It should also be noted that all of these gains are realized only if this high functional complexity can be fully utilized. Figure 6.15 shows the similarities with ECL. Figure 6.15*a* shows the conventional ECL gate with two collector loads, shown as boxes for generality. The time delay between the input and the OR output is less than that between the input and the NOR output. This is because of the Miller effect at the NOR output. At the NOR side the input and output voltages swing simultaneously in opposite directions. On the other hand, for the OR side the base is fixed so that only the collector-voltage swing affects speed. The time delay of the circuit would be that of the OR output if the NOR output is not used. Thus if only a single load, the one connected to the fixed-biased transistor, is left in the circuit; two ECL gates connected in cascade will appear as shown in Fig. 6.15*b*. Furthermore, if a simple repartition of the circuit is done, the input now is at the emitter of the fixed-biased transistor Q_{1N} and the output(s) is at the output transistor Q_{OUT}. This is shown in Fig. 6.15*c*. The structure now realizes a noninverting gate, that is, the logical output is the same at the logical input, which is of limited use in logic design. However, if Q_{1N} is made as a

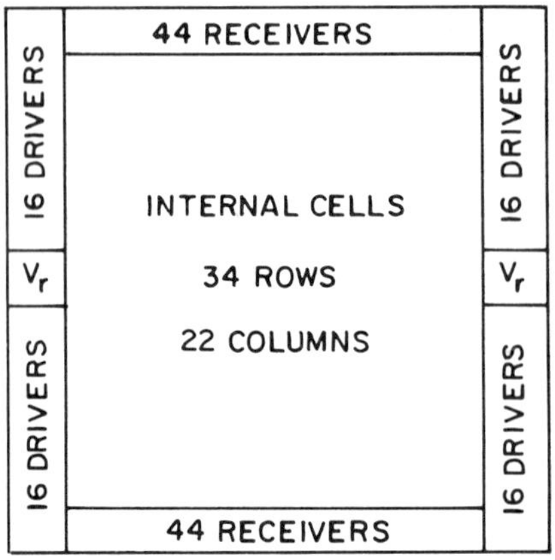

44 RECEIVERS
16 DRIVERS
V_r
16 DRIVERS
INTERNAL CELLS
34 ROWS
22 COLUMNS
16 DRIVERS
V_r
16 DRIVERS
44 RECEIVERS

Figure 6.14 General layout arrangement of a masterslice.

multiemitter transistor as shown in Fig. 6.15*d*, the structure now realizes an AND gate, in that a_1, a_2, and a_3 must be high for the outputs to be high. The input pulldowns can be implemented as (1) a resistor, (2) active current source, or (3) logically controlled switched current source. The logical capability of the structure is further enhanced if a two-level EFL cell, as shown in Fig. 6.15*e*, is used as a building block for logic design. Figure 6.15*e* shows Q_{1N} with two emitters (the number of emitter inputs can be more than two as we shall see later), and one current source is used which is controlled by a second level of current steering transistors Q_1 and Q_2, where Q_2 is a fixed bias transistor with reference voltage V_{r2}. As a result the output logic function F is given by

$$F = A_1 a_1 + \bar{A}_1 a_2 \tag{6.6}$$

where A_1, a_1, and a_2 are in general wired-OR functions. That is, multiple logic outputs can be directly attached (*OR*ed) to a_1, for example.

Equation (6.6) can be explained more physically as follows. If A_1 is logical "1," that is, its voltage level is greater than V_{r2}, the current I_0 is steered through Q_1 and, hence, the output logical level is controlled by a_1: $F = a_1$. Alternatively if A_1 is logical "0," that is, its voltage level is less than V_{r2}, the current I_0 is steered through Q_2 and, hence, the output logical level is controlled by a_2: $F = a_2$.

In order to allow the outputs to be directly coupled either to an input emitter or an input base(s), V_{r_1} must be one V_{on} higher than V_{r_2}. Hence the threshold voltage of the EFL gate is V_{r_2}. It should be mentioned that each output emitter can fan-out to *only one* input emitter. The reason for this is that when the current I_0 is steered through Q_1 or Q_2 to the emitter input it will be further steered through Q_{IN} or through Q_{OUT} *of the previous stage*. Allowing output emitters to fan-out to only one emitter input prevents current sharing problem between inputs. On the other hand, Q_{OUT} can fan-out to any number of bases since these inputs have a negligible current input requirement.

Figures 6.16*a* and 6.16*b* show an EFL cell with more than two emitter inputs. The cell has $(2n + 1)$ inputs, where $A_1, \ldots, A_n$ are the base inputs and are referred to as *control inputs* and $a_1, \ldots, a_{n+1}$ are the emitter inputs and referred to as *direct inputs*. The output function F is logical "1" if both A_i and a_i are logical "1," for any i. If all A_i are "0" and a_{n+1} is "1," the output is also logical "1." Normally only one control input (A_i) is allowed to be "1" at a given time and when more than one "A_i" is "1," the output must be a DON'T CARE case. If $F =$ "0" for the DON'T CARE cases, F can be expressed as

$$F = A_1 a_1 + A_2 a_2 + \cdots + A_n a_n + \left(\bar{A}_1 \bar{A}_2 \bar{A}_3 \ldots \bar{A}_n\right) a_{n+1}$$

Each A_i and a_j can be a first-order wired-OR function from the emitter outputs of previous stages.

In practice, the maximum wired fan-in numbers are limited because of chip-layout considerations or because of the required delay. A maximum

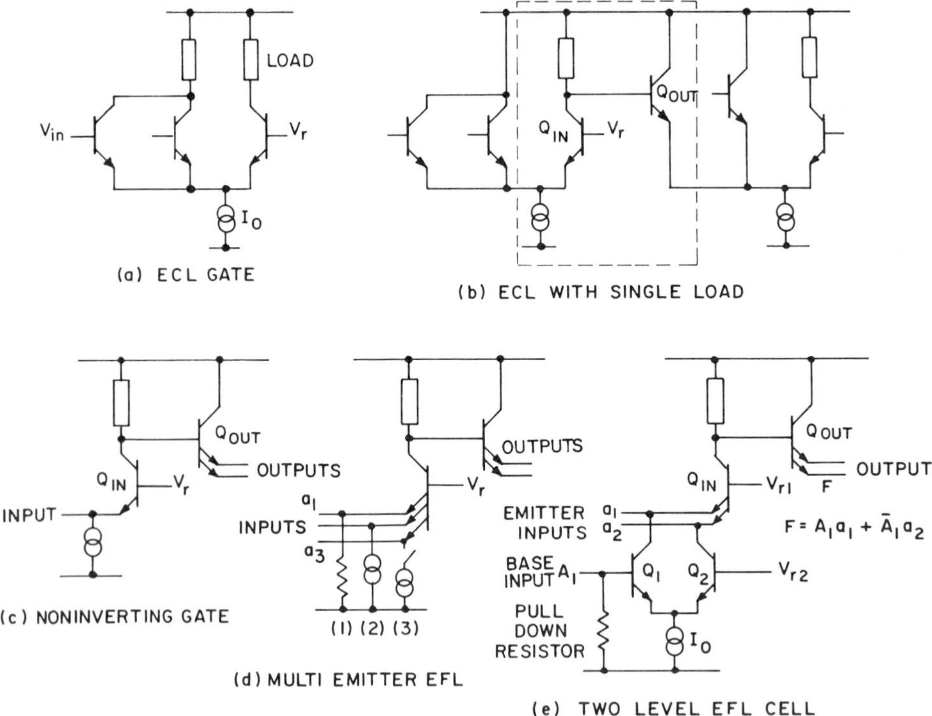

Figure 6.15 Developing EFL circuit structure from ECL: (*a*) Basic ECL gate with NOR/OR outputs. (*b*) Basic ECL gate with only OR output. (*c*) Repartitioning the gate of Fig. 6.15*b*: an emitter-function logic (EFL) gate. (*d*) Multiemitter EFL gate with (1) a resistance, (2) an ideal current source, or (3) a logically controlled switchable current source. (*e*) Two-level EFL cell.

number of four is considered feasible and it satisfies the requirements of most logic realizations.

In Figs. 6.16*a* and 6.16*b* only one control input is allowed to be "1" at a given time to prevent the current source I_0 from being shared between Q_{OUT} of the previous stages. Multiple outputs will result in a nonspecific variation in delay. Although this restriction of allowing only one control input to be "1" at a given time presents a limitation on logic designs, in practice it does not present a serious drawback.

In EFL, the collectors of the lower transistors can be tied together as shown in Fig. 6.16*c* and 6.16*d* to form a wired-OR function of a_1 and f_1. Where f_1 is the NOR of $A_1, \ldots, A_m$:

$$F = (f_1 + a_1)(\bar{f}_1 + a_{n+1})$$

where

$$f_1 = \overline{A_1 + A_2 + \cdots + A_m}$$

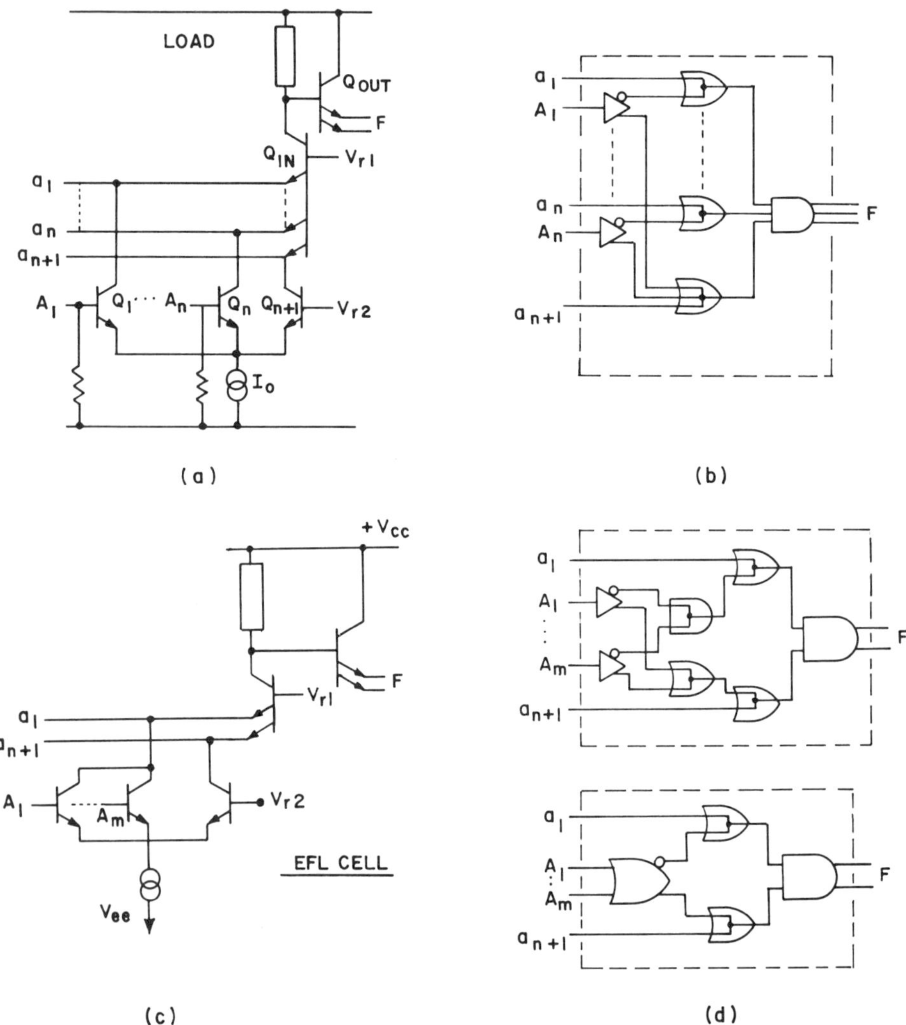

Figure 6.16 (*a*) Circuit diagram of a two-level EFL cell, with *n* emitters. (*b*) Logic gate equivalent. (*c*) Circuit diagram of a two-level EFL, with two emitters. (*d*) Logic gate equivalents.

In this case no restriction on control inputs applies and any number of control inputs can be high, since all the control currents are summed at only one direct input.

6.4.1 EFL Cell Implementation

Having discussed the operation of an EFL cell it is useful to illustrate how the technology is used to realize the basic block shown in Fig. 6.16. Figure 6.17 shows one possible layout of a four-input/four-output cell excluding the

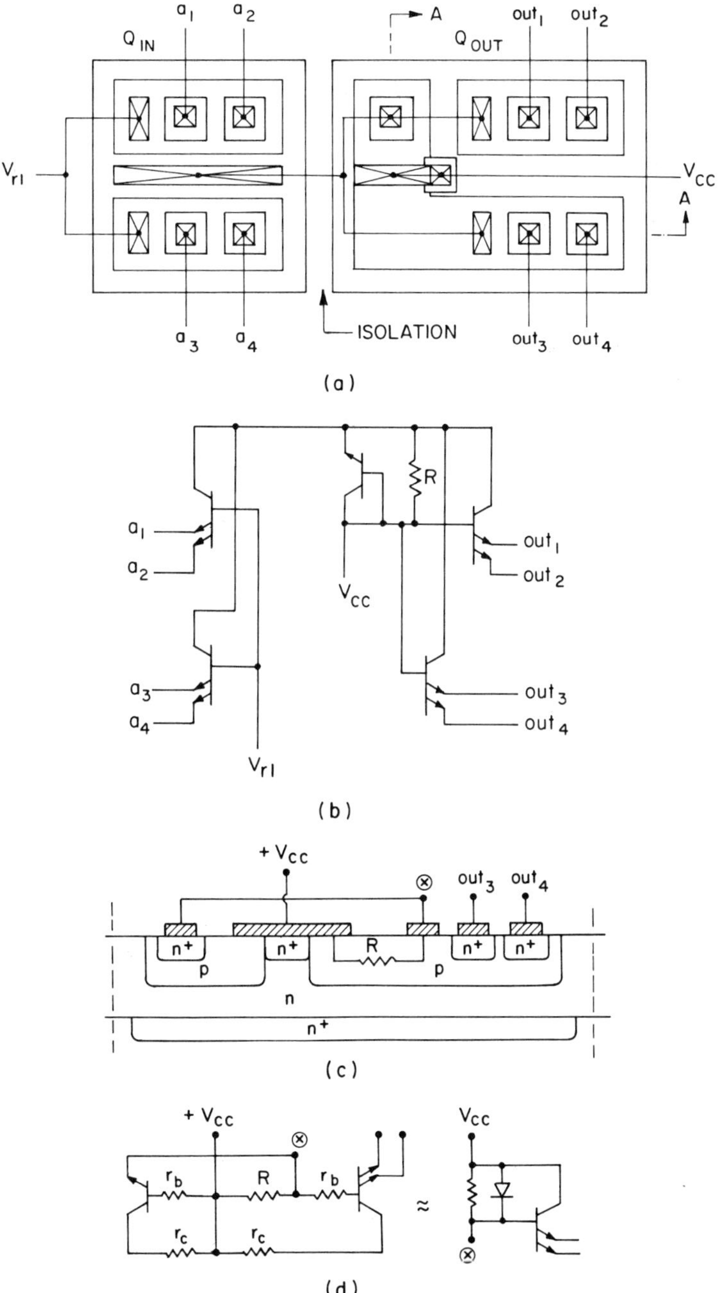

Figure 6.17 Basic $Q_{\rm IN}$, $Q_{\rm OUT}$ and the load for an EFL gate. (*a*) Layout. (*b*) Circuit diagram corresponding to the layout. (*c*) Cross section *AA*. (*d*) Equivalent circuit of cross section *AA* including parasitics.

differential stages used to create $Q_1, Q_2, \ldots, Q_{n+1}$ [6]. Two isolation regions are used. The load resistor R is created by extending the base diffusion in the "L" shape between the base contact of out_3 and out_4 to the pn^+ metal contact to the collector of Q_{OUT}. Figure 6.17*c* shows the section view of the devices as viewed along the L-shaped section line shown in Fig. 6.17*a*. In addition to the resistor and $\text{out}_3/\text{out}_4$ devices, a clamping-diode-connected transistor is shown. This diode appears in parallel with the resistor and guarantees a maximum logic swing of V_{on}. Note that the $+V_{CC}$ contact (usually ground potential) consists of ohmic contacts to *both* the p regions and the n^+ collector. A schematic circuit equivalent is shown in Fig. 6.17*d* and both the circuit element R as well as extrinsic device resistances are depicted. The input device(s) are nearly identical except for the resistor/diode combination. The design choice to use a distributed R is a "neat-trick" of IC technology and leads to higher packing density.

Unfortunately the superintegration of R also offers certain penalties—in particular the logic swing becomes sensitive to the control of sheet resistance and added parasitics must be considered during gate design. Moreover, sometimes well-isolated components are desirable, which can be controlled separately and even adjusted independently. The key advantage of removing R from the layout shown in Fig. 6.17*a* is that an alternate layout is possible that reduces r_c and r_b, which in turn increases the speed performance of the gate.

Figures 6.18*a*, 6.18*c*, and 6.18*e* show the $Q_{\text{IN}}/Q_{\text{OUT}}$ portion of another layout with separated R's and the clamping-diode-connected transistor. Figures 6.18*b*, 6.18*d*, and 6.18*f* show the current source/switch portion of a cell. Figure 6.18 shows a cell array approach to the layout in an oxide-isolated process. The layout cell is composed of two portions separated by an interconnect corridor. Figure 6.18*a* shows $Q_{\text{IN}}/Q_{\text{OUT}}$, two R for the load, and a clamping-diode-connected transistor. The number of inputs and outputs is programmable in modules of two for easy expansion. The symmetry of components enables the direction of inputs to outputs to be top to bottom, bottom to top, left to right, and right to left. The current-source switch portion enables realization of either individual current sources, as in Fig. 6.18*d*, or current switches with a variable number of inputs as in Fig. 6.18*b* and 6.18*f*. Each cell can implement up to four individual current sources or a 2 to 3 way current switch with a current source. Additional components for more complex cells are assembled, using neighboring cells. The arrows at the periphery of the drawings indicate the metal grid spacing. That is, center lines of metal can pass along any of these arrows. Obviously, for a single-level metal technology the orthogonal directions cannot cross. For the dual-layer metal case the perpendicular layers can cross without connecting. The X's shown in Fig. 6.18 indicate specific locations where the two metal layers do indeed connect by means of a "via."

The density of this spacing defines the actual cell size for given technology capability. Also the number of lines in vertical and horizontal interconnect corridors varies with chip size and complexity.

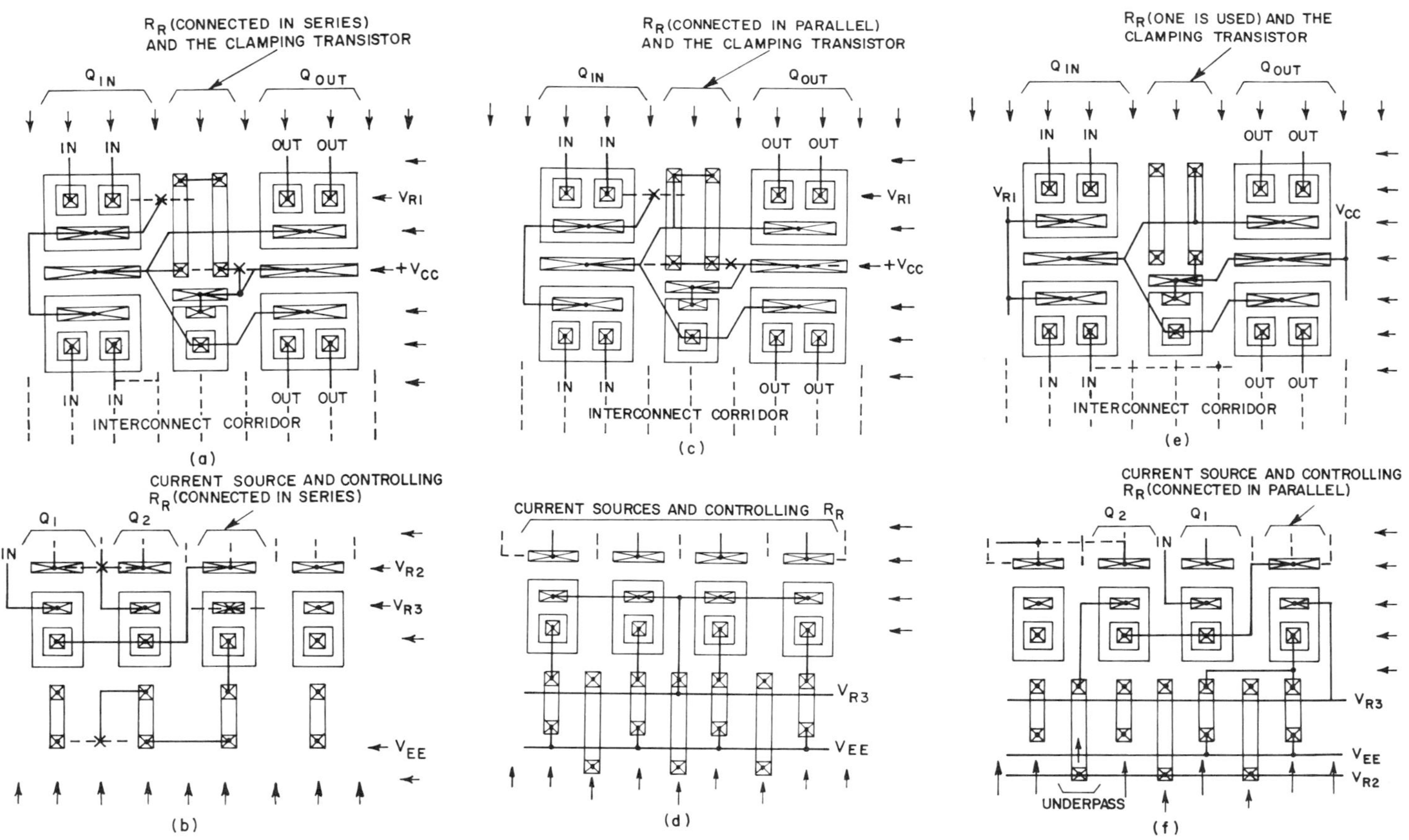

Figure 6.18 (a)–(f) EFL cell layouts for EFL cell array.

Figures 6.18*a* through 6.18*c* show dual-metal implementation of cells and Figs. 6.18*d* through 6.18*f* show single-metal implementation. Only a small variation can be seen in the device area with the two-layer metalization technology being a little more area efficient. Greater differences are in interconnect corridors and supply distribution. Necessary underpasses in single-layer implementation introduce unwanted parasitics and an increase in area is a requirement. High-performance implementations, therefore, require use of dual-metal technology. Various connections of resistors enable simple 1 to 4 trade-off in speed and power dissipation of cells. Figures 6.18*a* and 6.18*b* show series connections. Figures 6.18*d* and 6.18*e* show use of a single resistor, and Figs. 6.18*c* and 6.18*f* show parallel connections. This *power-level programming* enables the highest speed for critical path circuitry and reduction of power for slow logic.

The next few subsections will consider examples, of logic, memory, and register designs using EFL cells. In a manner similar to that used in Chapter 5 for I^2L, the pertinent details at both the circuit and layout level will be discussed.

6.4.2 EFL Logic Building Blocks: Combinatorial Logic Elements

Examples of simple combinational functions are given in Fig. 6.19. It should be realized that more than just single gate functions should be implemented in order to utilize fully the EFL cells. Figure 6.19*a* shows the basic AND-OR implementation. Note that the power dissipation is a function of the number of the wired-OR's, and no extra power is added due to the AND function. Two-input AND and INHIBIT are shown in Figs. 6.19*b* and 6.19*c*. The selector in Fig. 6.19*d* is the first function of sufficient complexity to utilize the EFL cell logic capability. This is, in fact, the most frequently used element in logic designs, as will be shown further in more-complex designs. Figures 6.19*e* through 6.19*g* show logic realizations that make more extensive use of the EFL gate structure. Figure 6.19*e* is the single-cell realization of the exclusive-OR function. Clearly this is a very efficient use of silicon area. The EFL realization uses five-base-diffused regions and a total of $7n^+$ emitters. A typical I^2L gate realization uses six base diffusions and eight n^+ collectors. Moreover, this EFL cell uses only one current source and the voltage references. The I^2L gate requires *pnp* injectors for each of the base diffusions. Although for these comparisons to be valid the full circuits and areas must be examined in detail, the bottom-line observation is clear that device counts per logic function are comparable, and these numbers are well below the equivalent *gate* count for MSI. (See Table 6.1.)

The coincidence (exclusive NOR) gate implementation in Fig. 6.19*f* uses one current source and requires two emitters for *A*, *B* inputs. Only the variables (not their complements) are used. The exclusive-OR realization shown in Fig. 6.19*g* has two current sources, and only direct inputs are used. The coincidence gate shown in Fig. 6.19*h* is a modified EFL cell circuit, which shows the use of

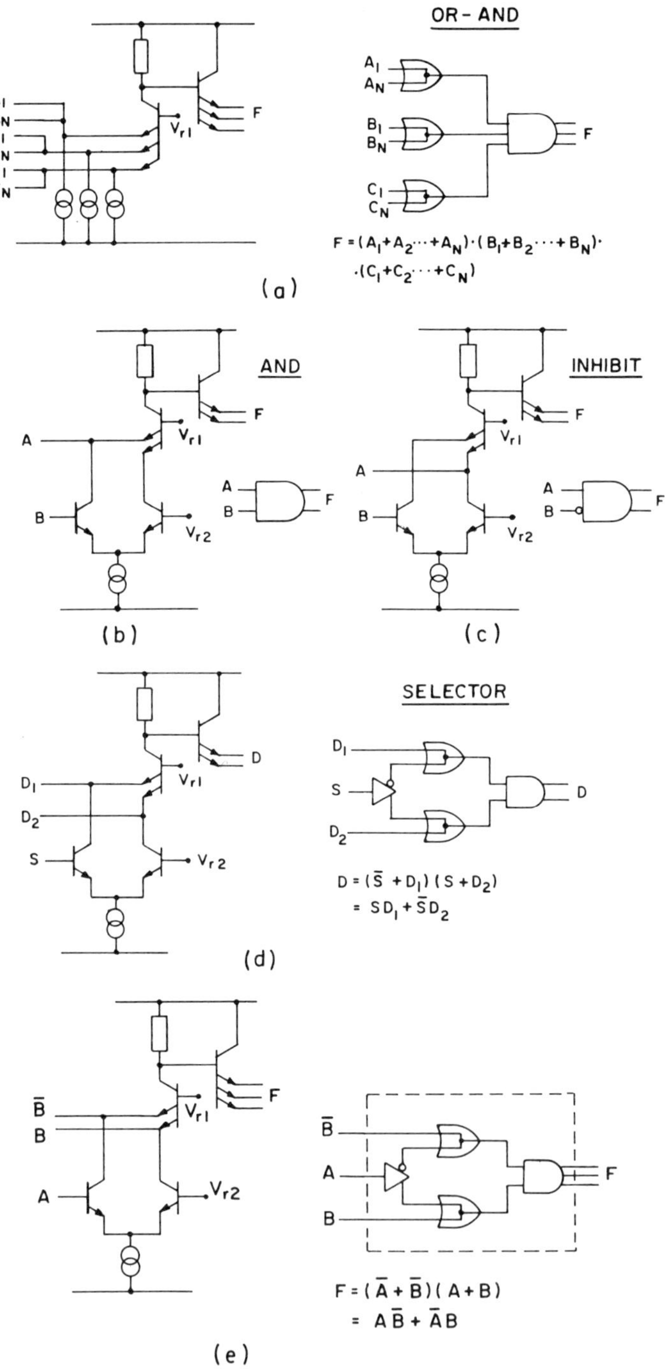

Figure 6.19 EFL combinatorial logic examples: (*a*) OR/AND, (*b*) two-input AND, (*c*) INHBIT, (*d*) selector, (*e*) exclusive-OR, (*f*) coincidence (exclusive-NOR), (*g*) exclusive-OR, (*h*) coincidence (exclusive-NOR), (*i*) 4 : 1 multiplexer, (*j*) 1 : 4 demultiplexer, and (*k*) full-adder.

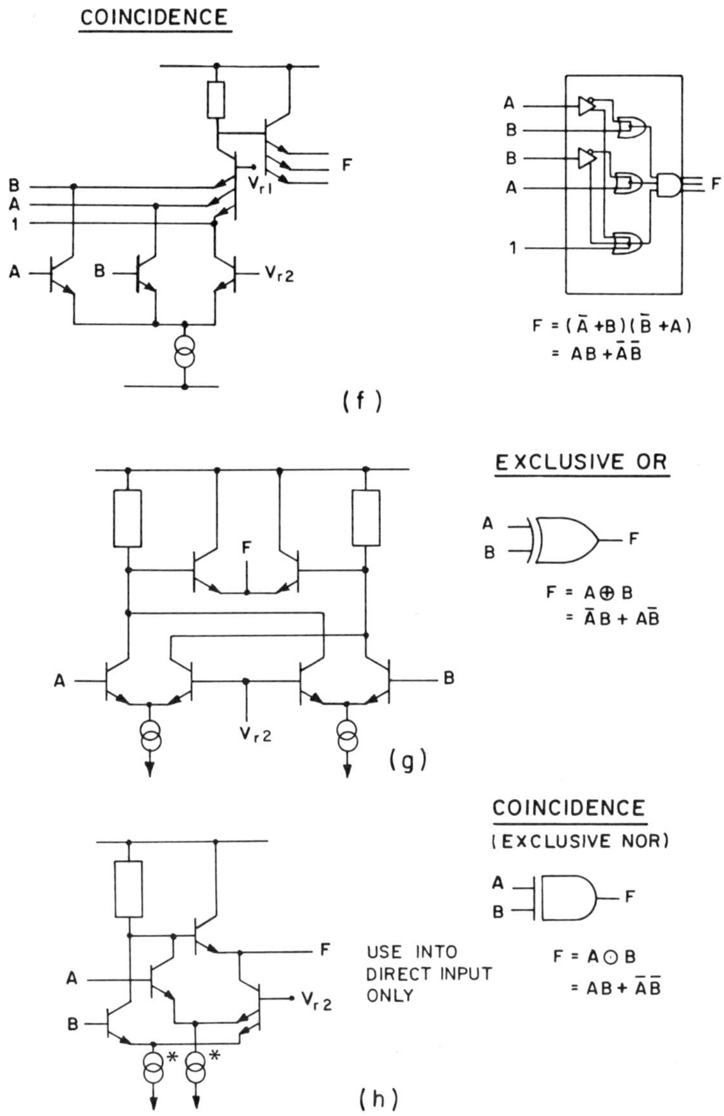

Figure 6.19 (*Continued*)

a multiemitter device at the lower switching level. Data multiplexing and demultiplexing are shown in Figs. 6.19*i* and 6.19*j*. Note the presence of only one noninverting buffer in the data path in order to achieve the high speed. For the demultiplexer circuit the current sources shown as I_0^* are driving the inputs of following cells and should be properly associated with inputs for the next stage. It should be clear that $\overline{D}_0, \ldots, \overline{D}_3$ may only drive the $a_1, \ldots, a_n$ level inputs since the current source I_0^* would make no logical sense if used on a base input on the $A_1, \ldots, A_n$ level.

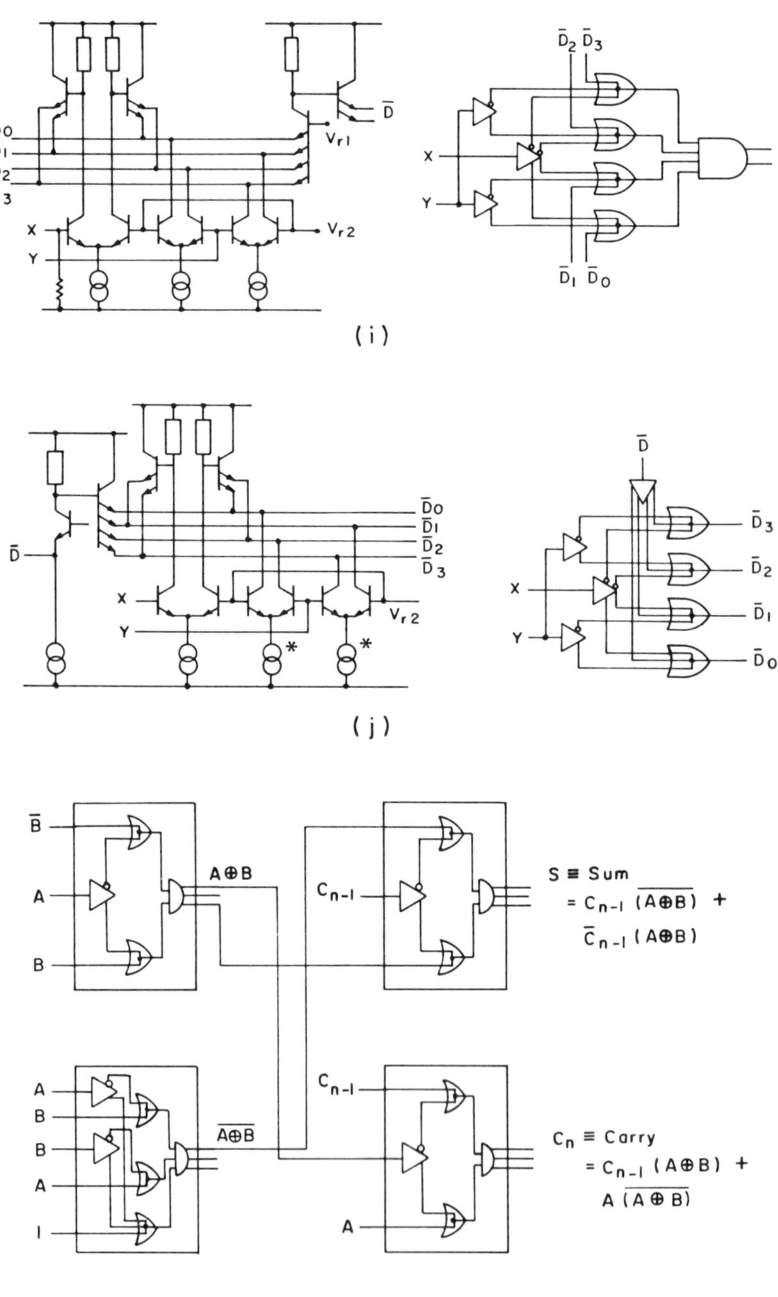

(k)

Figure 6.19 (*Continued*)

A full-adder function implemented with four EFL cells (mostly selectors) is shown in Fig. 6.19*k*.

6.4.3 EFL Logic Building Blocks: Sequential Logic Elements

The ability to hold information is an essential function in digital circuits. The basic cross-coupled *RS* latch was discussed in the context of I^2L gates. Figure 6.20*a* shows the simplest EFL realization of the *RS* latch constructed with a single bias reference. The operation of this circuit realization is as follows. One output and input are tied together with a bias resistor and provide the set (S) input. A second input is the reset (R). The set is normally low and the reset is high. For Q low the current through R_{EE} is provided by Q_{IN}. To set Q high a positive pulse must be applied to S which turns Q_{IN} off and forces Q_{OUT} to drive the current through R_{EE}. To reset the circuit a negative pulse is applied to R, which turns Q_{IN} back on again, moving base current from Q_{OUT} and thus returning to the original Q-low condition with bias to R_{EE} provided by Q_{IN}. This circuit provides the memory function.

The gated latch shown in Fig. 6.20*b* represents a circuit form more compatible with clocked logic and standard current-source biasing with two reference levels.

Starting from these basic EFL cells we can now move on to a more formal definition of several memory-element realizations. Two basic configurations are considered:

1. Configuration *A*, as shown in Fig. 6.21*a*, where one of the output emitters, representing the output *Q*, is connected to the control input *X*. The cell has similar characteristics to the *RS* latch, as can be seen from the state table and state diagram shown in Fig. 6.21*b*. This realization

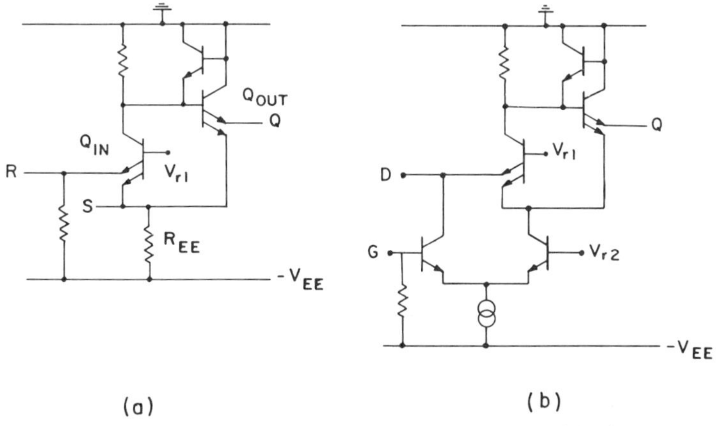

Figure 6.20 EFL latches: (*a*) simple *RS* and (*b*) gated latch.

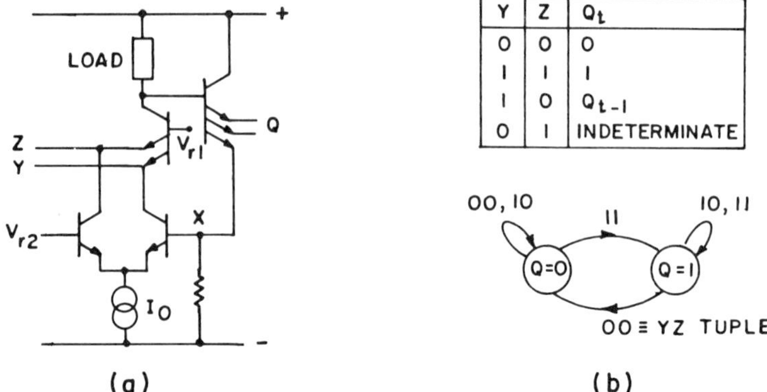

Figure 6.21 EFL latch, configuration *A* (analogous to *RS*-type latch): (*a*) circuit diagram and (*b*) state table and state diagram.

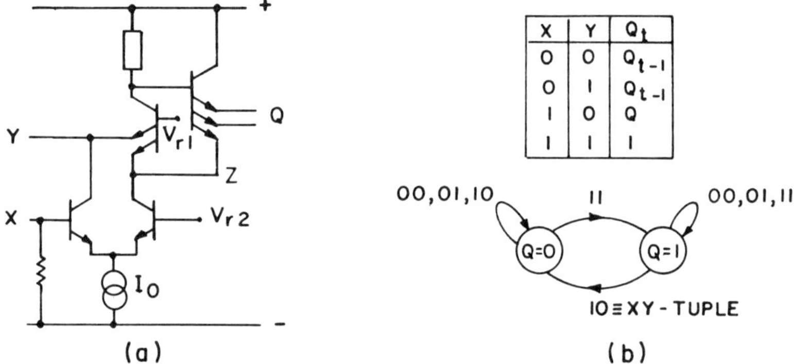

Figure 6.22 EFL latch, configuration *B* (analogous to *D*-type latch): (*a*) circuit diagram and (*b*) state table and state diagram.

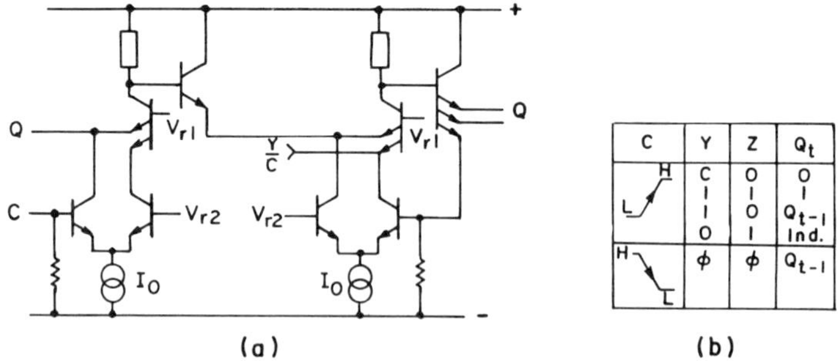

Figure 6.23 EFL gate *RS* flip-flop: (*a*) circuit diagram and (*b*) state table.

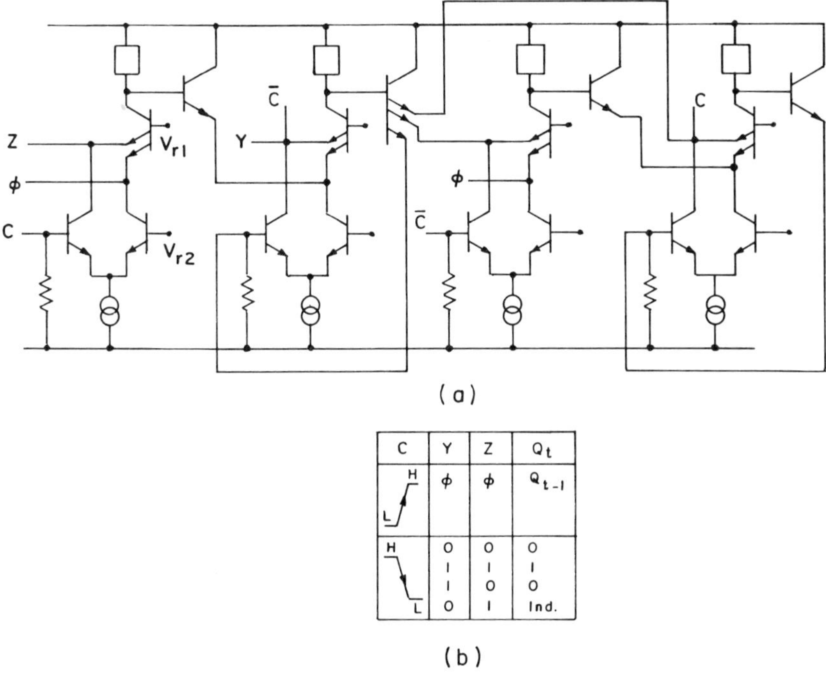

C	Y	Z	Q_t
L→H (rising)	φ	φ	Q_{t-1}
H→L (falling)	0	0	0
	1	1	1
	1	0	0
	0	1	Ind.

Figure 6.24 EFL master–slave *RS*-type flip-flop: (*a*) circuit diagram and (*b*) state table.

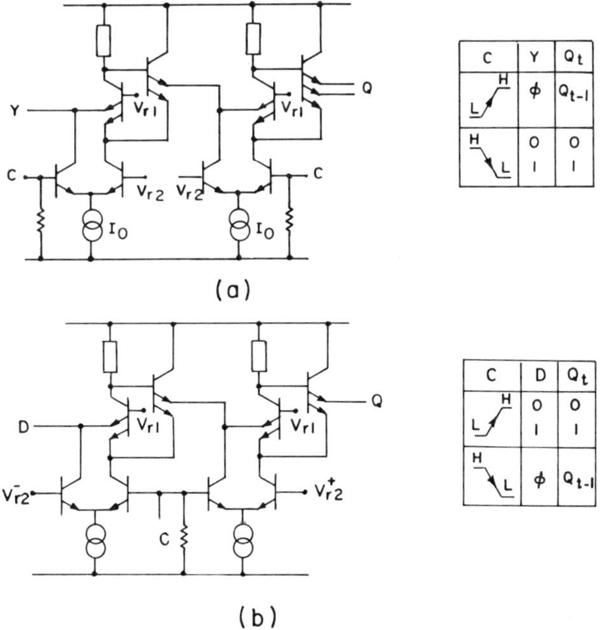

C	Y	Q_t
L→H (rising)	φ	Q_{t-1}
H→L (falling)	0	0
	1	1

C	D	Q_t
L→H (rising)	0	0
	1	1
H→L (falling)	φ	Q_{t-1}

Figure 6.25 EFL master–slave *D*-type flip-flop: (*a*) negative edge triggered and (*b*) positive edge triggered.

compares favorably with the conventional cross-coupled two-gate realization.

2. Configuration B, as shown in Fig. 6.22a, where one of the output emitters, representing the output Q, is connected to the direct input Z (or Y). As shown from Fig. 6.22, the structure has the characteristic of a data latch, where the input X is the gating input (Fig. 6.20). This EFL realization consumes less power and less silicon area than the conventional realization of four MSI gates.

The above two configurations can be used to design more-complex flip-flops. Two EFL cells are used to realize a gated RS flip-flop, as shown in Fig. 6.23, as compared with four gates used in conventional realizations. Figure 6.24 shows a master–slave realization of an RS flip-flop. Two EFL cells are used to realize a master–slave D flip-flop, as shown in Fig. 6.25 as compared with 10 gates used in conventional MSI realizations. The use of the same V_{r2} threshold for both master and slave latches is permitted only if minimum clock edge speed can be guaranteed. The use of two slightly offset thresholds, as in Fig.

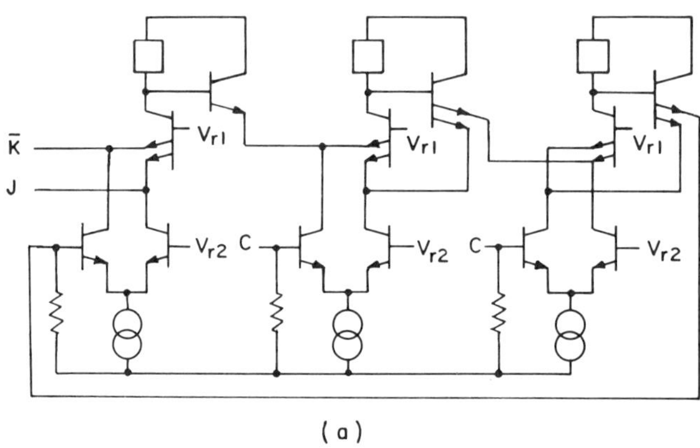

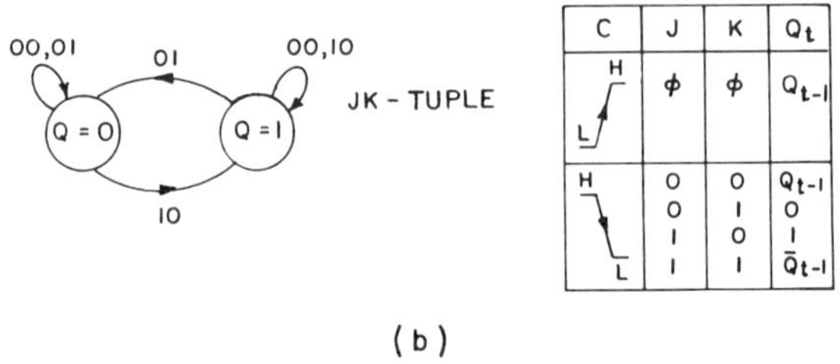

Figure 6.26 EFL master–slave JK flip-flop: (a) Circuit diagram and (b) state diagram and state table.

6.25*b*, guarantees hazard-free operation for even slow clock edges. The threshold difference of ± 100 mV will enable clock edge speed range from dc to subnanosecond. The JK master–slave circuit realization using EFL cells is shown in Fig. 6.26.

Although saving of silicon area and power consumption is possible for all commonly used flip-flop configurations, the most significant saving is for the master–slave *D*-type flip-flop. For this reason it is used in preference to the others, for the realization of functional blocks.

Table 6.1 compares the EFL to the NAND MSI realization of commonly used logic and storage functions, a factor of 2–5 improvement in terms of gate to cell counts results. However, because the functions, where the improvement factor is 2, are not frequently used in a fully developed design, it is fair to assume an average improvement of 4.

6.5 DC SPECIFICATIONS FOR A PRACTICAL EFL

There have been a variety of integrated realizations of EFL cells and some of these will be discussed. In the early MSI realization, Skokan [6] demonstrated that a complete flip-flop including 50-Ω line buffers could be fabricated in an area of 400 μm $\times$ 400 μm using conventional diffused transistors with f_T of 1.8 GHz. This realization gave a master–slave *D*-type flip-flop, which operates at a clock speed of over 500 MHz for a power dissipation of 16 mW. Other development yielded 4-mW circuit using 1-GHz transistors, which operated at a clock speed of 200 MHz. Figure 6.27 shows the voltage levels of EFL. This type of circuit is compatible in logic levels with several conventional on-chip emitter-coupled-logic (ECL) systems. The voltage levels in the system are dependent on the forward voltage of *pn* junctions and are shown for a junction voltage of 0.8 V.

In Fig. 6.27, the voltage at the collector of Q_{r1} can vary between 0 and -0.8 V, while the base is biased at $-\frac{1}{2}V_{\text{on}}$ or -0.4 V. Considering Q_{r2}, the collector potential can never fall below -1.2 V because it is connected to an emitter of

Table 6.1 EFL Cells versus NAND Realization of Logic Gates and Flip-Flops

Function	Number of EFL Cells	Number of NAND Gates
Exclusive-OR	1	3
Equality	1	3
Full-adder	4	9
RS flip-flop	1	2
D-type latch	1	4
Gated *RS* flip-flop	2	4
Master–slave *D*-type	2	10
Master–slave JK	3	9

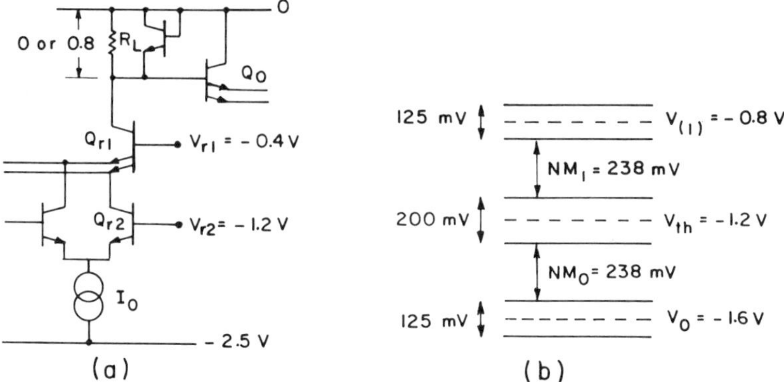

Figure 6.27 Logic levels and noise margins for an EFL cell.

Q_{r1} and its emitter cannot be more positive than 2 V while in conduction, so the minimum conducting value of V_{CE} is 0.8 V. For the lower-level transistors, with logic inputs high, the minimum value of V_{CE} is 0.4 V. For the output transistor Q_0, V_{CE} varies between 0.8 and 1.6 V. Since the worst-case soft saturation voltage (the V_{CE} at which f_T for the transistor falls below its specified value) is 0.2 V, there is a worst-case margin of 0.2 V for V_{CE} of all the transistors. This allows the circuit to be operated above 100°C, where this margin reduces to 0 V. A variety of realizations can be used for the current source I_0, and it is not a critical part of the circuit because there are no high-frequency requirements.

The logic level system shown in Fig. 6.27 provides a threshold tolerance of 200 mV, "0" and "1" level tolerances of 125 mV and noise margins of more than 200 mV.

6.6 EFL APPLICATIONS

To demonstrate the circuit potential of EFL and to provide the reader with a context for understanding more-complex digital IC designs, subsystem examples will be presented. To simplify the discussion the logic presentation of the EFL cells are used. For example, Fig. 6.28*a* shows how a master–slave *D* flip flop would be represented in this notation, and Fig. 6.28*b* shows a master–slave JK configuration.

In Chapter 7, the system and the LSI circuit-design aspects of digital filters are considered. In this section we shall consider the EFL realization of some of the digital filter building blocks.

Advantages of digital filtering, such as superior stability, higher accuracy, and flexibility of changing the filtering function are well known. However, the price for these advantages is the increase in hardware complexity, thus an increase in cost. This cost is reduced if the hardware can be shared by as many signal channels as the speed of the circuitry allows. Thus it is important to

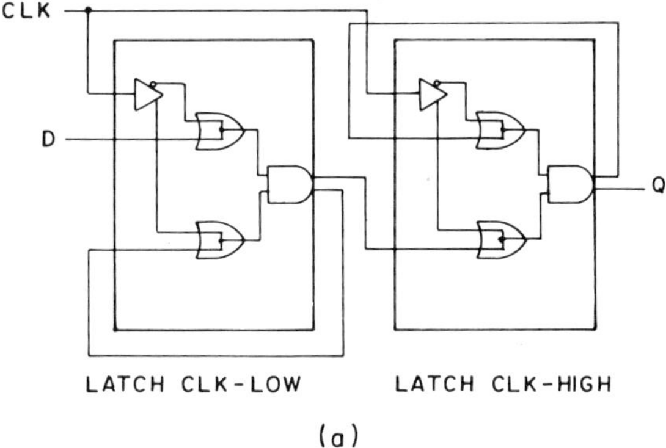

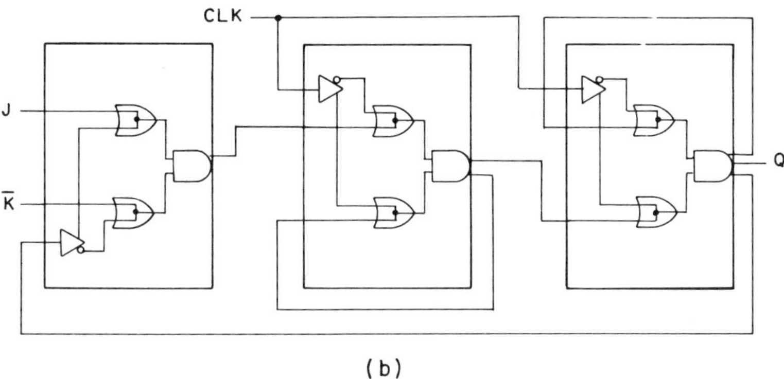

Figure 6.28 (*a*) Logic representation of a master–slave *D*-type EFL flip-flop. (*b*) Logic representation of a master–slave JK EFL flip-flop.

increase the speed at which circuit structures can handle the data in digital filtering. Moreover, these structures must have the following features:

1. The structure must be capable of performing simple storage function and a useful range of combinational logic, with minimum changes in interconnection patterns.
2. The structures must be economical in power consumed and silicon area and thus be suitable for large-scale integration (LSI).
3. To be able to adapt to various applications with as little as possible circuit change.
4. To be optimized to perform the basic filtering functions, for example, binary addition and unit delays, rather than the conventional NAND or NOR gates.

5. They can be fabricated using existing high-speed technology and also be easily adapted to advances in the technology.
6. They can be easily adapted to a highly modular pipeline organization (see Chapter 7), which offers a flexible system architecture with varying complexity.

These requirements can be met for high-speed digital-filtering applications using EFL cells [9]. In digital filtering, the effective processing rate at which the samples can be processed by the filter is an important parameter. A single high-frequency signal or many lower-frequency signals can be processed by the filter if this rate is high. To maximize the processing rate, the digital filter can be pipelined. Pipelining the circuit allows the logic to be partitioned into stages separated by registers to save the results of each stage. Storing the data in a register stage frees the previous logic to process new data. For better performance all stages should have approximately equal time delays since all stages operate at the rate of the slowest stage. When the partitioning is continued until only the smallest logic elements separate the registers, the system is fully pipelined. Fully pipelined system has the highest possible throughput but the longest input to output delay and requires the most logic compared to nonpipelined or partially pipelined design. In digital filtering, as in many other applications, the input to output delay is not critical compared to the throughput rate. Thus a fully pipelined digital filter yields the greatest performance for a given logic delay. In EFL realization the register stage consists of a master–slave D flip-flop. For uniform data flow the master and slave sections can be separated by a logic stage as shown in Fig. 6.29.

Since the data are processed in one clock cycle, the clock period must be greater than or equal to the sum of the propagation delays of (L_1), master, (L_2), and the slave. As an example Fig. 6.30 is an EFL fully pipelined full-adder that adds in one clock cycle. A second-order filter module shown in Fig. 6.31 requires four delay elements, five multiplier elements, and one five-input full-adder. An N-bit 2's complement serial binary arithmetic is chosen for the filter module (see Section 7.3). The 2's complement arithmetic reduces the random logic and simplifies pipeline construction by eliminating the sign manipulation necessary in signed magnitude computations. A realistic trade-off between speed, logic complexity, and chip interconnections is achieved by performing input/output, delay, addition, and multiplication in serial.

The filter module requires EFL realizations of delay, multiplier, and adder elements. The delay element is simply an N-bit shift register, realized using

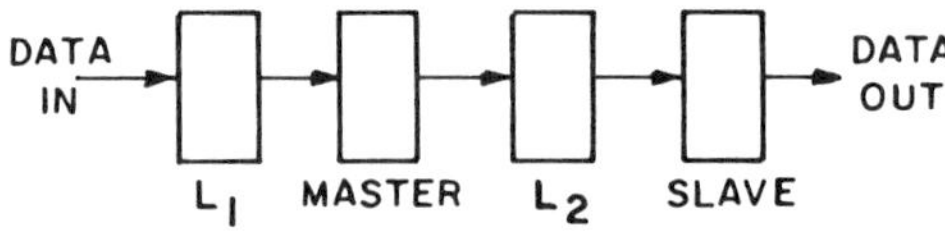

Figure 6.29 EFL pipeline system, master and slaves are separated by logic blocks L_1 and L_2.

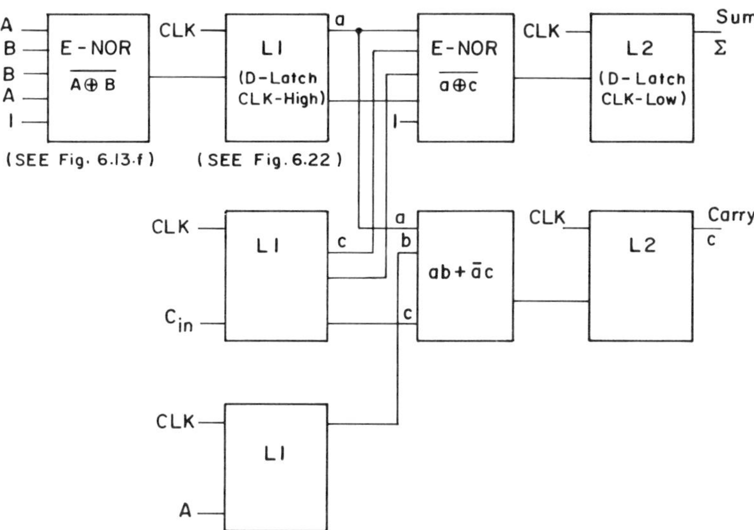

Figure 6.30 EFL fully pipelined full-adder.

EFL master–slave D-type flip-flops. The multiplier element forms a $2N$-bit product from the N-bit input data and N-bit coefficient. The coefficients are stored in shift registers, which can be serially loaded to change the filter parameters. Since the internal arithmetic is 2's complement, Booth's algorithm (see Section 7.4) is chosen for straightforward pipeline operation. The serial multiplier examines the multiplicand one bit at a time starting at the least significant bit (LSB) and adds or subtracts accordingly.

The addition is performed in parallel with carry save and an automatic shift right. For subtraction each bit of the multiplier (constant) is complemented, a carry is forced into the LSB of the adder and an addition operation performed. After N cycles the product is complete, but N more cycles are needed to shift out the result and add the carry–slave register.

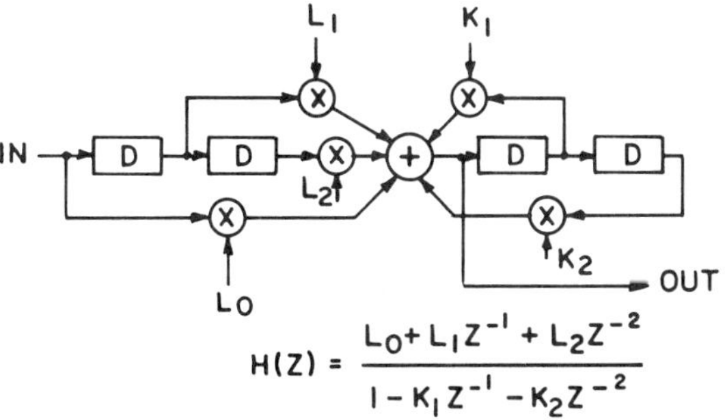

Figure 6.31 Block diagram of a second-order filter module, filter transfer function $H(z)$ is shown with the filter coefficients L_0, L_1, L_2, K_1, and K_2.

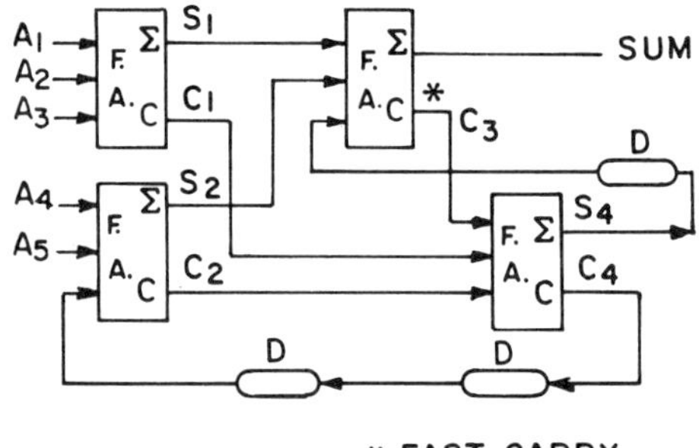

Figure 6.32 Block diagram of five-input serial adder.

Table 6.2 Number of EFL Structures for the Filter Blocks of Fig. 6.31

	Number of Structures/Element			Total Number
Number of Bits/Word	Five-Input Adder	Delay	Multiplier	of Structures
4	32	8	56	336
8	32	16	104	608
12	32	24	152	880
16	32	32	200	1152
20	32	40	248	1424
24	32	68	296	1696

The five-input adder is used so that all multiplier outputs can be summed simultaneously. A block diagram is given in Fig. 6.32. Note that a fast-carry circuit is employed to reduce the S_4, C_4 delays. The input-to-output delay of the adder is two clock cycles. The number of EFL structures for the individual filter blocks of Fig. 6.31 are tabulated for different word lengths in Table 6.2.

Other EFL circuit applications have been appeared in the literature, see, for example, References 10, 11, and 12.

REFERENCES

1. C. S. Meyer, D. K. Lynn, and D. J. Hamilton, *Analysis and Design of Integrated Circuits*, McGraw-Hill, New York, 1968.
2. M. I. Elmasry and P. M. Thompson, The Design of Load Structures for Current Mode Subnanosecond Logic, *IEEE JSSC* **SC-10** 72–75 (1975).
3. P. R. Gray and R. G. Meyer, *Analysis and Design of Analog Integrated Circuits*, Wiley, New York, 1977.
4. L. S. Garret, *Integrated-Circuit Digital Families—ECL*, *IEEE Spectrum* **7**, 30–42 (1970).
5. M. I. Elmasry and P. M. Thompson, Logic Partition for Multiemitter Two-Level Structures, *IEEE Transactions on Circuit Theory*, May 1974, IEEE Press, New York, pp. 354–359.
6. Z. E. Skokan, *Emitter Function Logic: Logic Family for LSI*, *IEEE JSSC* **SC–8**, 356–361 (1973).

7. M. I. Elmasry and P. M. Thompson, Two-Level EFL Structures for Logic-in-Memory Computers, *IEEE Transactions on Computers*, March 1975, IEEE Press, New York, pp. 250–258.

8. M. I. Elmasry, Logic Design Using Emitter-Function-Logic, *IEEE Transactions on Computers*, September 1976, IEEE Press, New York, pp. 952–956.

9. M. I. Elmasry, and R. C. Madter, Pipeline Digital Filtering at 250 MHz Bit Rate, Proceedings of the 1975 Midwest Symposium on Circuits and Systems, August 1975, Western Periodicals, New York, pp. 461–465.

10. J. Kane, A Low-Power Bipolar, Two's Complement Serial Pipeline Multiplier Chip, *IEEE JSSC* **SC–11** (5), 669 (1976).

11. G. L. Baldwin *et al.* A Modular, High Speed Serial Pipeline Multiplier for Digital Signal Processing, *IEEE JSSC* **SC–13** (3), 400–408 (1978).

12. R. J. Blumberg and S. Brenner, A 1500-Gate Random Logic LSI Masterslice, *IEEE ISSCC, 1979*, Lewis Winner, FL, pp. 60–61.

Chapter 7

Digital LSI / VLSI Subsystems

7.1 INTRODUCTION

Given a digital subsystem, should we realize it using SSI, MSI, or LSI/VLSI building units? The answer to this question depends on the overall system complexity under consideration, the number of subsystems which can benefit from LSI, and the state of the art of a given integrated technology at the time and its influence on the development and the production cost. It is clear that the answer does also involve an extrapolation of what LSI/VLSI can offer in the future. However, we can say, in general, that LSI/VLSI has the potential to offer improved performance and reliability at reduced cost. These gains are not only dependent on the LSI technology used, but also on the system architecture and its partition.

Digital design with *standard* SSI, MSI, and even standard LSI building blocks [1] follows certain constraints that are different from the constraints imposed by the LSI/VLSI *chip* environment. For example, component count is important in the first approach, while it is replaced with silicon area in the second approach. The basic building unit in the second approach is elastic, it can be considered as simple as a multiemitter transistor or as complex as an ALU. The optimization techniques along with the "subsystem-circuit tricks" used also differs in the two approaches. For example, a separate processing and memory units is a must for system partitioning in designing using standard components while it may be advantageous to consider merging the logic and memory functions in the LSI/VLSI chip approach.

The purpose of this chapter is to highlight some of the system consideration for the LSI chip involvement and to give some design examples.

7.2 SUBSYSTEM DESIGN FOR LSI/VLSI

The continual evaluation of digital integrated circuits toward higher degree of integration has been associated with an increase in performance. This has offered the circuit designer a challenging opportunity to integrate digital subsystems or even complete systems on a silicon chip. The applications of

VLSI in subsystem design can be grouped into the integration of:

1. Large capacity memory chips [2, 3].
2. Data processing subsystems [4, 5].
3. Signal processing subsystems [6–10].

The recent impact of VLSI, both in terms of performance and silicon area, has been in the area of signal processing. The demands of signal processing exceed that of data processing and memories. Signal processing requires real-time operations at high data rates. Moreover, data processing involves the integration of relatively small primitive operations (e.g., add, subtract, mask, compare), while signal processing involves the integration of larger primitives (e.g., multipliers). Owing to the nature of signal processing, there are few data-dependent jump operations. This influences the choice of the system architecture, for example, pipelining can be used to advantage. This will be explained in Section 7.3.

In data processing, VLSI impacts computer architecture especially at the microcomputer and microprocessor levels [11–14]. For example, combining arithmetic and memory functions on a single chip eliminates the need to pass data off-chip to access memory. This reduction in data-transfer requirements improves system performance and reduces the required I/O pins. Moreover, merging of arithmetic, logic, and memory functions in each cell of a data-processing array would offer a modular array suitable for VLSI. This will be explained in the next section.

7.3 LOGIC-IN-MEMORY ARRAYS

One approach to obtain high-speed processing is to combine logic and memory functions in basic building blocks. These blocks are arranged in regular arrays, commonly called logic-in-memory (LIM) arrays [15–21]. These arrays may be regarded either as a logically enhanced memory array or as a logic array whose cells can be programmed to realize a desired logical function. The main advantage of these arrays is their suitability for LSI and VLSI, because of the highly modular structure of the arrays and because digital machines can be built using few types of these arrays [22]. Moreover, these arrays offer the following features:

1. Function Flexibility.

Figure 7.1 shows a general form of a basic LIM array, which can be used in data-transfer or data-processing operations. It consist of N columns and M rows, and each cell is labeled, referring to the row and the column number, respectively. The cell has two sets of inputs: control inputs and data inputs. The control inputs determine the mode of operation of each cell and the data inputs carry the data to be processed. The control input set is partitioned into

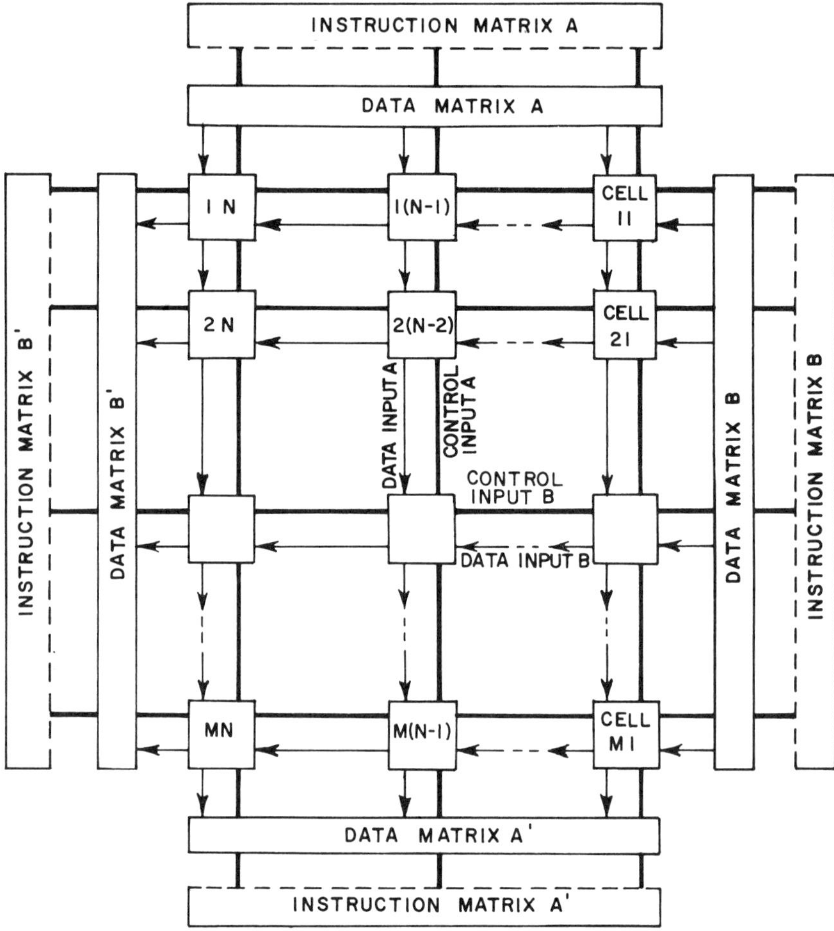

Figure 7.1 General form of logic-in-memory (LIM) array. The function of each LIM cell is changed using the contents of the instruction matrix A and B. The input data are contained in the data matrix A and B. The output data are contained in the data matrix A' and B'.

two sets, named A and B, similarly for the input and output information. The control inputs of the boundary cells are connected to an instruction matrix and the data inputs to a data matrix. These matrices are realized using read-only memories (ROMs). The function of each cell, and, hence, the function of the array, can be changed by changing the contents of the instruction matrix. Thus, a cell that behaves as a full-adder in one mode, may act in other modes as a counter stage or a shift-register storage.

2. Testability.

Because of their regular structure, LIM arrays should be substantially easier to test for faults and to diagnose than constructed networks having the same number of elements [24, 25].

3. Fault Accommodation.

Because of the flexibility explained in point 1, isolated faulty cells in an array may often be bypassed by programming the adjacent cells to avoid active connection to the faulty cell [23].

4. Subarray Interconnectability.

A cellular array may be realized as a macrocellular interconnection of cellular subarrays, each realized on an LSI chip. Since each cell has only a small number of connections to adjacent cells, the interconnection pattern on the chip is simple. Moreover, the number of external inputs–outputs that are connected to the edges of an array or subarray is relatively small.

5. Low Power and High Speed.

Because of the low fan-in and fan-out requirements, the cells can operate at low-power levels without sacrificing the high-speed operation offered by the technology used for cell realization.

6. Insensitivity to Improvement in Technology.

The future availibility of more-advanced technology allowing higher circuit density per chip, does not imply a complete redesign of an array-organized digital system, but requires only that a smaller number of subarrays (chips) be used in building up full arrays.

7. Simplified Design Effort.

Because each subarray consists of an iteration of a single, relatively simple cell, all design effort may be focused on the optimization of the circuitry of this cell with respect to speed, silicon area, tolerance, etc.

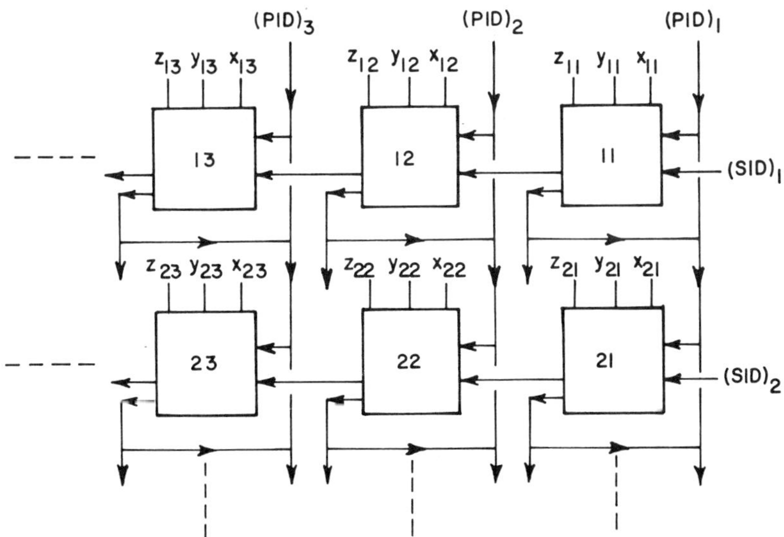

Figure 7.2 Information-transfer array for serial or parallel input and serial or parallel output.

The capability of EFL to realize logic and memory functions economically results in its being suitable for LSI LIM arrays. In the following sections, two specific examples are given: the EFL realization of information transfer arrays and sorting arrays [27].

7.3.1 Transfer Arrays

An information-transfer array can accept parallel or serial information and convert it from one form to another. Figure 7.2 shows a part of this array, and Fig. 7.3 shows a EFL realization of the basic cell. The cell has two data inputs —a serial input data (SID) and a parallel input data (PID)—and two data outputs—a serial output data (SOD) and a parallel output data (POD). All PIDs and PODs of a column are wired-OR to minimize the total number of inputs–outputs. Each cell has three control inputs, X, Y, and Z clocked with two phase-clocks—C_1 and C_2 to allow a master–slave operation. The EFL realization allows two logic operations to be performed at the master and one logic operation at the slave without increasing the total power dissipation and the silicon area.

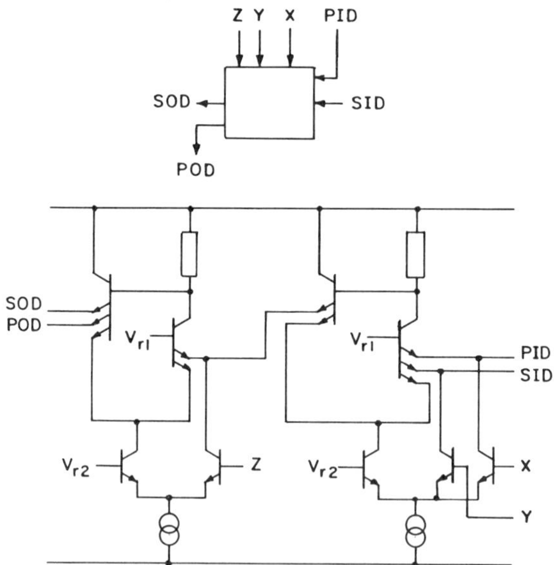

$X = C_1 \cdot$(PARALLEL CONTROL) $= C_1 \cdot PC$

$Y = C_1 \cdot$(SERIAL CONTROL) $= C_1 \cdot SC$

$Z = C_2 \cdot$(SERIAL CONTROL) $= C_2 \cdot SC$

Figure 7.3 EFL cell for information-transfer array.

7.3.2 Sorting Arrays

A sorting array is a multiple-word memory that keeps in sorted order all data words that are fed into it. Words that are read out are obtained in order of size, with the largest word first (or alternatively, with smallest first, as desired). The words are assumed to be of maximum length n. This array can be used as a functional unit within the central processor of a general-purpose computer or to a special-purpose computing system for which sorting capability is needed.

The sorting array described here is a two-dimensional iterative configuration of identical cells, each of which is a simple sequential logic circuit [27]. Each cell is connected only to its immediate neighbors and the cells around the edges of the array are connected to fixed signals or to registers. Figure 7.4 shows the array configuration along with an input–output register X and a word-selected register W. All cell input terminals on the right-hand side of the array are connected to a logical signal Z_0 (normally fixed at 1) and the outputs labeled $\hat{Z}$ from the left edge of the array serve as inputs to the stages of the W register. Each cell contains one flip-flop, whose content is designated Y, so that the set of n flip-flops in any one of the N rows of the array may be employed to store one n-digit word. This set of words is assumed to be encoded, for example, binary-coded decimal number system, with the most significant digits at the left ends of the words and one's or two's complement representation for negative numbers, with the minus sign encoded as a 0 at the left end of the word.

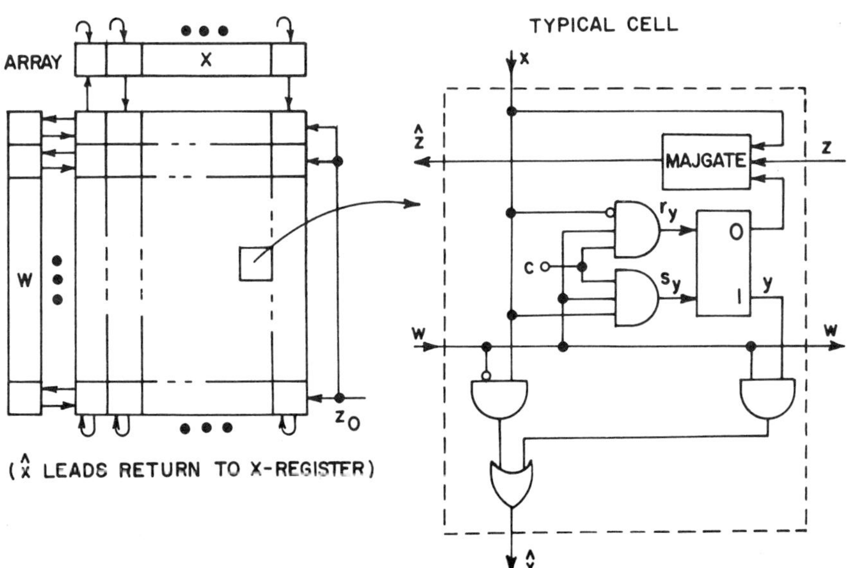

Figure 7.4 Cellular sorting array.

An entire word is handled as a unit during the input, output, and sorting operations. One cycle of operation of the array consists of two steps:

1. A comparison step, in which the word X in the X-register is simultaneously compared with N words stored in rows of the array; a "1" is injected into the W-register in those rows whose words (including blank words) are smaller than or equal to the word X, and a "0" is injected into the W-register in those rows whose words are larger than the word X.
2. An execution step, in which:

   (a) The set of all words that are stored in the subset or rows having a "1" in the W-register are moved downward one row within the set, while the word in the X-register is copied into the uppermost such row.

   (b) The lowermost such word is copied into the X-register. Words having a "0" in the W-register are not moved.

Sorting with this array is accomplished by maintaining a sorted file of previously entered words, with the largest at the top of the array and the smallest and any blank rows at the bottom. Each new word that is to be sorted is inserted into this file in a single operational cycle. To read out in-order the words in the file, largest to smallest, place a "1" in the uppermost stage of the W-register. Now carry out step 2 of the cycle repeatedly, shifting the "1" downward in the W-register, one row with each step. If the words are desired in the opposite order, a "1" may be started at the bottom of the W-register and shifted upward.

Figure 7.5 shows the array cell and its EFL realization; five EFL structures are used compared with 20 conventional gates. The majority gate at each cell forms part of a chain of n such gates. The inputs x and $\bar{y}$ of this gate allow it to act as a size comparator, so that the leftmost Z-output in each row takes on the value "1" when and only when the number represented on the set of x-lines entering this row is smaller than or equal to the number represented in the cascade of y-flip-flops in the same row. Normally, step 1 is carried out with the W-register initially empty, so that the W busses in all rows of the arrays carry the value "0." In this case the x-output in each cell carries the same value as the x-input. $\hat{x} = x$, that is, the contents of the X-register is bussed downward to all rows of the array. As a result, the comparisons in step 1 are made between the word X and every word Y stored in the array.

During step 2, a row for which the W-register contains a "0," we have $w =$ "0," so that each cell in this row behaves according to

$$\hat{x} = x, \qquad y = y_{t-1}$$

that is, the row is static and behaves as if it were not even present. A row for which the W-register contains a "1," we have $w =$ "1," so each cell behaves

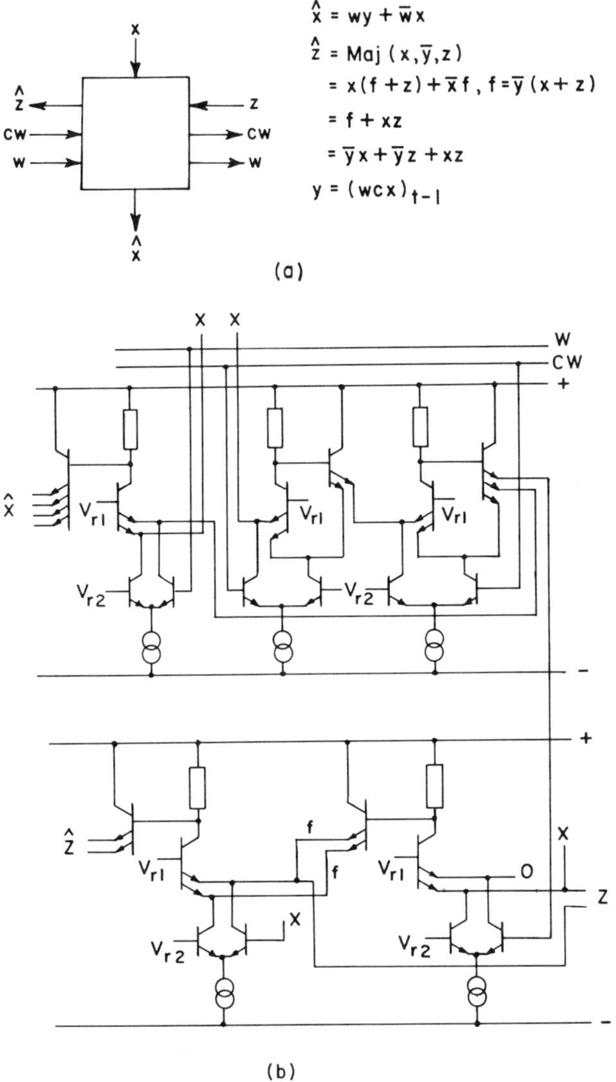

Figure 7.5 (*a*) Sorting array cell. (*b*) Its EFL realization.

according to

$$\hat{x} = y, \qquad y = (cx)_{t-1}$$

Thus the word stored in the flip-flops in this row is transferred onto the set of x-lines that pass downward from this row. For the array as a whole, all words in the subset of rows consisting of the X-register and those rows marked with a "1" in the W-register shift down cyclically one row position within the subset. The contents of the X-register fills the top position and the contents of the lowest-marked row passes back into the *X*-register.

The sorting array may also be used for the realization of an arbitrary logical function, as a pushdown memory (last-in, first-out) or a buffer memory (first-in, first-out) or as a content-addressed (associative) memory [28–31]. In these configurations, similar EFL realization could also prove economical.

7.4 DIGITAL FILTER MODULES

A major application of digital LSI/VLSI is in the area of signal processing [32, 33]. Until recently, most signal processors were implemented with analog devices because the complexity of equivalent digital processors was simply prohibitive. Recent advances in LSI/VLSI make the digital realization of these processors both feasible and economical. The digital approach offers many advantages, such as superior stability, higher accuracy, no need for calibration or adjustment, and the ability to be programmed to accommodate new or changing requirements. In this section we shall study the LSI realization of digital filter modules through an example.

Figure 7.6 shows a block diagram of the transmitter and the receiver sections of a coder and decoder system used in telephony. The coder converts an analog signal to a digital μ-law-decoded (or A-law-decoded) signal [34]. The μ-law is referred to the nonlinear decoding law used in the A/D converter to reduce the demand for large-sized words. The coder consists of a low-pass filter (LPF), A/D converter, and a digital processor (DP). The DP contains a

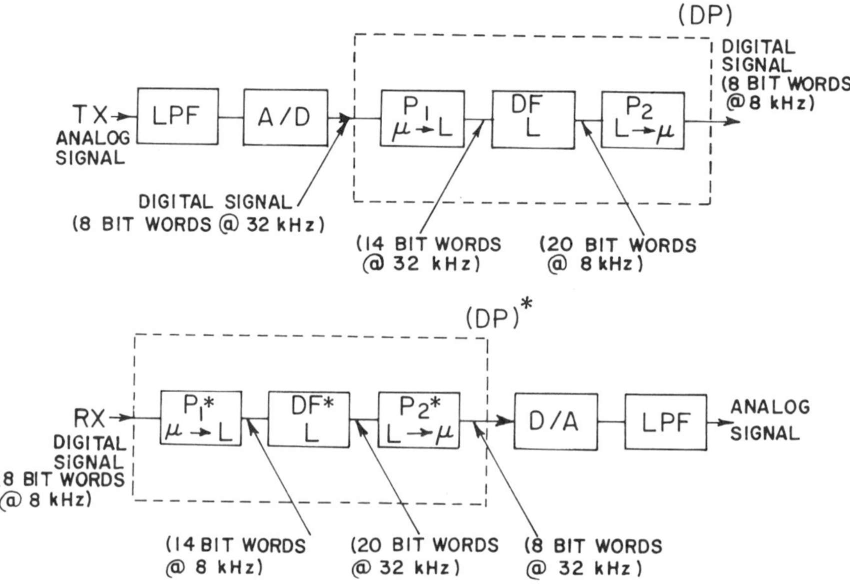

Figure 7.6 Coder and decoder block diagrams at the transmitter (TX) and at the receiver (RX) in a telephone system.

digital filter (DF) and two processors, P_1 (called expander) and P_2 (called compressor), to convert from μ-decoded signal to linearly decoded signal and vice versa. The conversion is necessary since the digital block is operating on linear codes. The block diagram of the receiver can be similarly explained. It is possible for the blocks DP and DP* to have identical LSI circuit implementation and only a change in the final metalization mask would differentiate between DP and DP*.

7.4.1 Subsystem Specifications

Now we will consider the LSI realization of DP and DP*. The following information is obtained from a general system considerations:

1. The sampling rate at the input of DP (and at the output of DP*) is 32 kHz. This is taken above the minimum sampling rate of 8 kHz (twice the bandwidth of the signal under consideration) to relax the design requirements on the prefilter (LPF). The sample size is 8 bits.
2. The sampling rate at the output of DP (and the input of DP*) is 8 kHz. The sample size is 8 bits.
3. It is assumed that a fourth-order DF is adequate for a range of applications. This range includes single telephone channels. From the LSI realization viewpoint and for stability reasons the fourth-order DF is realized as cascading of two second-order sections [9]. Moreover, in the LSI realization we time-share a single second-order section to do the job of the two second-order sections.
4. The overall transfer function of the fourth-order filter under consideration is given in the Z-domain as

$$T(z) = K_1\left(\frac{1 - a_{11}z^{-1} + a_{12}z^{-2}}{1 - b_{11}z^{-1} + b_{12}z^{-2}}\right)K_2\left(\frac{1 - a_{21}z^{-1} + a_{22}z^{-2}}{1 - b_{21}z^{-1} + b_{22}z^{-2}}\right)$$

   where K_1 and K_2 are scaling factors and a_{ij}'s are the filter coefficients [7]. The size of these words are chosen based on the maximum values. For example, for a case where

$$K_1 = \tfrac{3}{16},\ K_2 = \tfrac{11}{16},\ a_{11} = 1,\ a_{12} = 1,\ b_{11} = \tfrac{39}{32},$$

$$b_{12} = \tfrac{13}{32},\ a_{21} = \tfrac{21}{32},\ a_{22} = 1,\ b_{21} = \tfrac{23}{16},\ b_{22} = \tfrac{29}{32}$$

   K_1 and K_2 can be represented by a (4 + 1)-bit word and the filter coefficient can be represented by a (6 + 1)-bit word. The bit (+1) is a sign bit.
5. A data word size of (19 + 1) bits is assumed to be adequate for the DF arithmetic.

7.4.2 Subsystem Architecture

Now, having these system requirements let us proceed. One important aspect of LSI/VLSI designs is the choice of a certain *architecture*. We differentiate between three types:

1. Parallel-data system.
2. Serial-data system.
3. Pipeline-data system.

In parallel-data systems, the different bits of the data word appear on parallel inputs, and the time intervals associated with a bit and with a word are the same. In serial-data systems, the bits appear sequentially, usually with least significant bit first. It is obvious that the parallel system offers a high processing speed and highly complex hardware. Conversely, the serial system offers a slow processing speed and a lower hardware complexity.

A more-efficient compromise is a pipeline architecture where the serial logic blocks are separated by retiming flip-flops [35]. The approach offers modular building cells suitable for LSI. This modularity results in a saving of silicon area, reduces the LSI design effort in terms of layout, optimization, characterization, and testing. The pipeline architecture is to be used in the design presented.

The pipeline approach is explained as follows. In a serial system, combinational logic functions are realized in a circuit with serial multilevel logic gates. Since the total delay is the summation of the delays of each logic level in a signal path, large critical delays result. An alternative approach is to divide these multilevel logic gates into a number of smaller logic sections separated with retiming flip-flops (pipeline). Each section has a much shorter critical path delay than the single circuit, and the pipeline version will therefore operate at a higher clock rate. The logic function is thus performed in several shorter clock intervals rather than a single long interval. The overall time required to do the function may be longer than that of the serial system, but this increased latency is of no consequence on the system performance, the important improvement being the increased throughput. It should be noted that there is *no* increase of hardware required for the retiming flip-flops, since in digital-signal-processing systems and, in particular, in digital filters, the output of a given arithmetic operation is stored in a serial memory.

7.4.3 Subsystem Logic Implementation

The implementation of P_1 and P_2 [34] (and hence P_1^* and P_2^*) is simple and straightforward and is left as an exercise to the reader. The LSI implementation of DF is now considered.

As mentioned earlier, a time-multiplexed second-order filter section will be used to realize the fourth-order DF. Figure 7.7 shows the DF which consists of an input control section (I) and the second-order filter section (2D). The input section (I) accepts the input from P_1 or internally from the output of 2D. Thus, the input to 2D is a (19 + 1)-bit word at 64 kHz. Hence, a clock frequency of $\geqslant$ 1.28 MHz (64 kHz times 20 bits) is required for the processing of information in a pipeline architecture. We shall choose the clock frequency to be 2.048 MHz, since this frequency is usually available in a general system.

The second-order filter section (2D) can be implemented using a ROM realization [36] or a multiplier realization [37]. Figure 7.8 shows a multiplier realization of a second-order section. It consists of four pipeline multipliers, a five-input adder, four SR's, and a logic control section for truncating the output (T). The section shown dotted in Fig. 7.8 consists mainly of multipliers, hence the name M^+. This dotted section can be replaced by the dotted section shown in Fig. 7.9 where a read-only-memory (ROM) is used to store the partial product, an accumulator/subtractor is used to add up N partial products, where N is the size of the input word, and a parallel-in/serial-out shift register is used to clock the output data to the truncated logic control T. The dotted block of Fig. 7.9 is mainly ROM, hence the name ROM$^+$. The size of the ROM for a second-order section in this example is 32 $\times$ maximum size of a multiplier coefficient = 32 $\times$ 7 = 224 bits.

In order to time-multiplex the second-order filter hardware, the storage capacity of the SRs has to be increased as shown in Fig. 7.10. The storage SRs in the feedforward and feedbackward paths have to double in size, and an extra buffer shift-register SRO is required. The dotted block in Fig. 7.10 is realized using the ROM or the multiplier approach. In the case of the ROM approach, two ROMs are required, one for each second-order section (the size of each is 224 bits, that is, a 448-bit ROM is required). In the multiplier approach, the multiplier coefficients are changed according to the filter section. The switches shown in Fig. 7.10 represent the function of the input control unit (I) of Fig. 7.7.

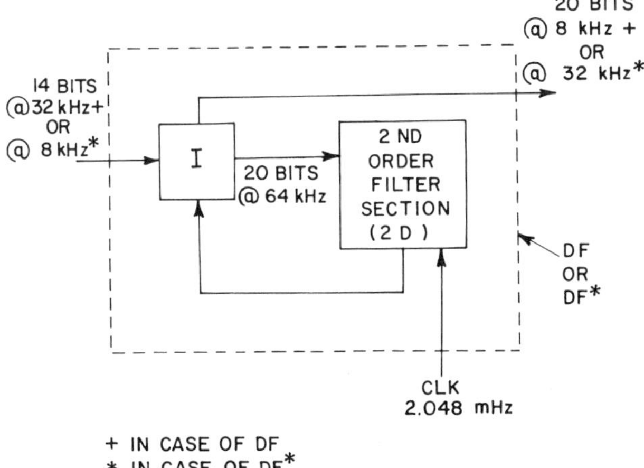

Figure 7.7 Time-multiplexed second-order digital filter to obtain a fourth-order filter section.

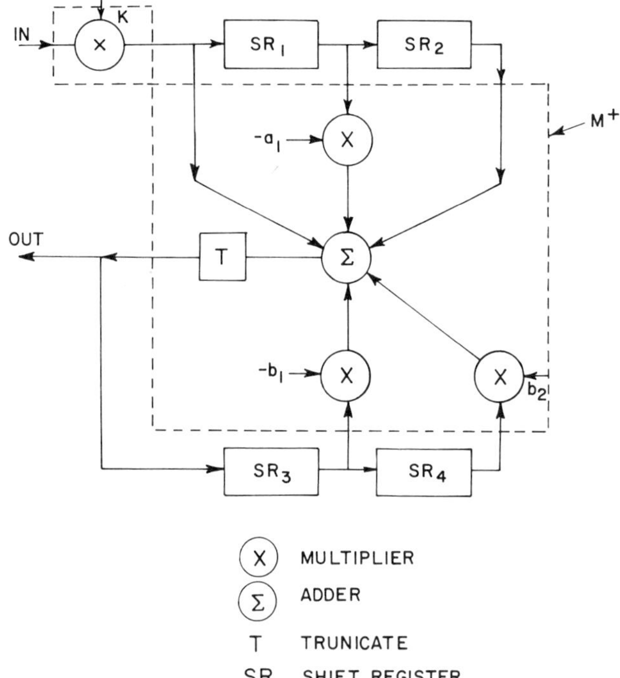

Figure 7.8 Multiplier realization of the second-order digital filter.

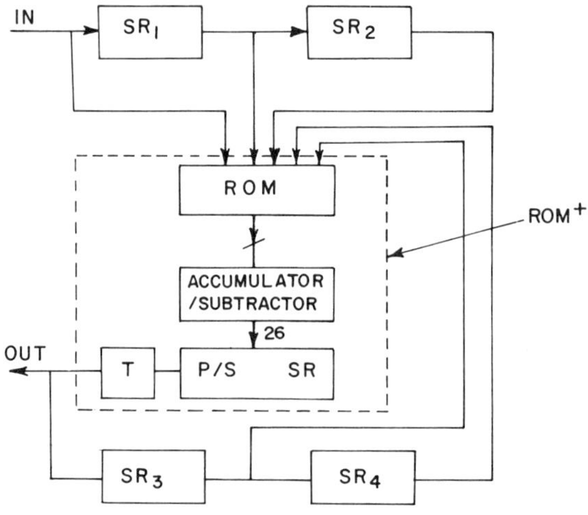

Figure 7.9 ROM realization of the second-order digital filter.

Although it is essential to use the filter configuration shown in Fig. 7.8 in association with the ROM approach, it is not necessarily the most efficient configuration for the multiplier approach. Thus, we consider the filter configuration shown in Fig. 7.11. For time-multiplexing, the filter section is shown in Fig. 7.12 where only five *SR*s are used versus nine *SR*s in Fig. 7.10.

Timing Considerations. As stated earlier, a pipeline architecture offers a compromise between hardware and speed of operation. Distributing the logic between masters and slaves reduces the required number of retiming flip-flops. The clock used to operate the pipeline digital filter is the 2.048 MHz. Assume, the maximum allowed delay between a master and slave is 194 ns (Fig. 7.13).

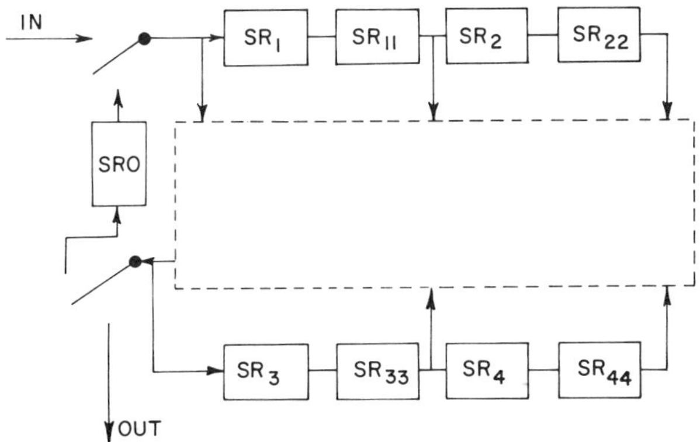

Figure 7.10 Time multiplexing a second-order digital filter for a fourth-order realization.

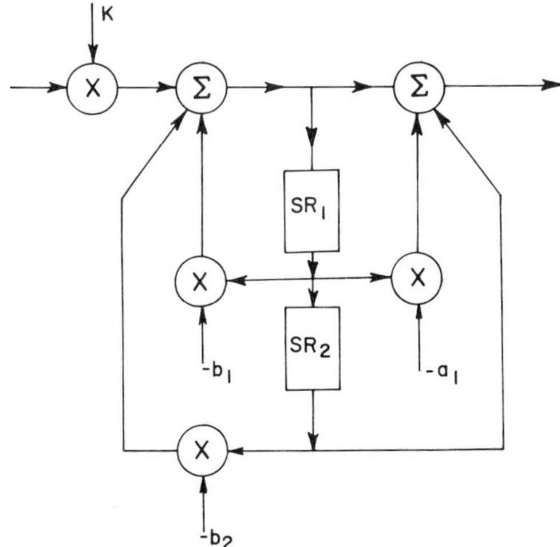

Figure 7.11 Another configuration for realizing a second-order digital filter.

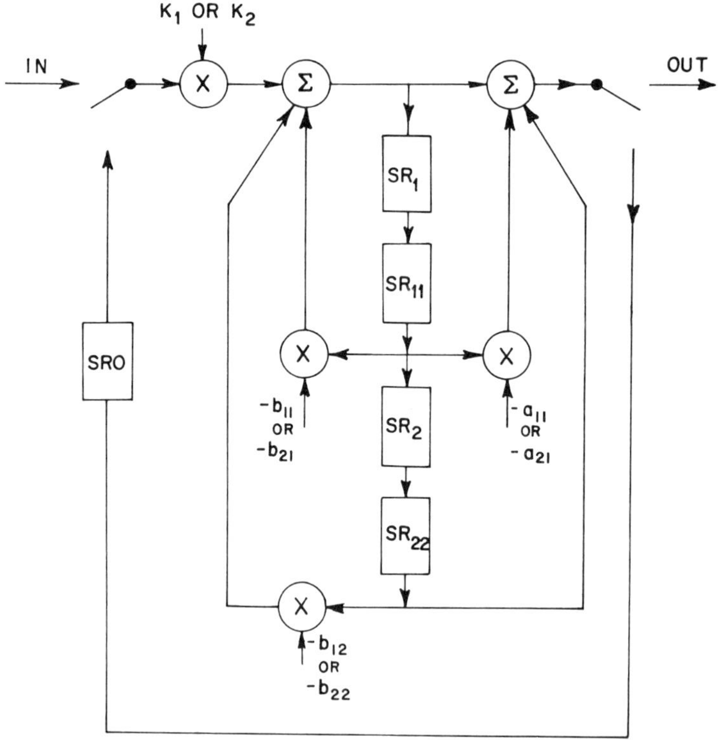

Figure 7.12 Time multiplexing of the configuration of Fig. 7.11 for a fourth-order realization.

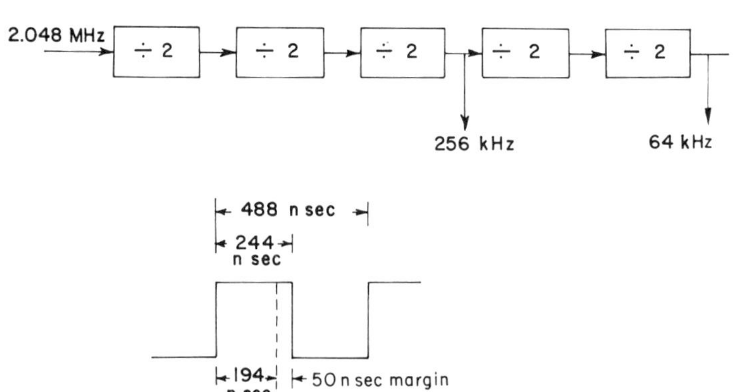

Figure 7.13 Clock generation using 2.048 MHz, a 50 nsec margin is allowed between the clock edge and the end of the logic operation.

Considering the ROM approach, the read time should be less than 194 ns. In the multiplier approach, a logic delay of less than 194 ns can be achieved easily since the approach has the flexibility of operating as fully pipeline system (one logic level between master and slave) or as a partial-pipeline system (more than one logic level between master and slave). The multiplier approach is chosen in the presented design example to pursue further.

Type of Arithmetic. Fixed-point arithmetic has been chosen for the filter realization rather than floating-point arithmetic. Although floating-point multipliers are simpler, floating-point adders are more complex [37]. Moreover, fixed-point arithmetic, when combined with two's complement representation and Booth multiplication algorithm [35], yields a highly modular multiplier cell with no special hardware to deal with the sign bit. Using two's complement representation in the filter section for both multipliers and adders avoids the conversion from one representation to another in the high-speed section of the system.

7.4.4 Pipeline Multipliers

The pipeline multiplier operates on the serial input data word (x), which is in the two's complement representation. The coefficients (Y_1 and Y_2) and their negative values ($-Y_1$ and $-Y_2$) are hardwired into the multiplier, where Y_1 and Y_2 are typical coefficients for the first and second sections of the DF. A Booth's algorithm is used to allow a highly modular multiplier cell, no ripple through carry, no special hardware for the sign bit, and the output is in two's complement. The product is $(N - 1 + M)$ bits where N is the size of the data word ($= 20$) and M is the size of the coefficient word (maximum $M = 7$). The data word is fed to the multiplier in serial with the LSB first and the sign bit tailing the word. The LSB of the product is delayed 1 bit from the LSB of the multiplicand (the data word $= X$). The adder operates directly on the multiplier outputs in two's complement.

Figure 7.14 shows the multiplier logic diagram. Figure 7.14*a* shows the 2-bit shift register which is used to generate the control word S_1S_2, that used in the multiplier algorithm, from the modified input data word x^* as shown next. Figure 7.14*b* shows the M-bit multiplier module, where M is the size of the coefficient word. Each cell has four inputs. For example, the ith cell of the multiplier has $(Y_i)^1$, the ith bit in the positive value of the coefficient word corresponding to the first section of the DF; $(-Y_i)^1$, the ith bit in the negative value of the coefficient word corresponding to the first section; $(Y_i)^2$, the ith bit in the positive value of the coefficient word corresponding to the second section; and $(-Y_i)^2$, the ith bit in the negative value of the coefficient word corresponding to the second section. The input S allows the multiplier cells to operate on $(Y_i)^1$ and $(-Y_i)^1$ or $(Y_i)^2$ or $(-Y_i)^2$. Figure 7.14*c* shows the pipeline multiplier cell logic in association with the input control section with inputs S, S_1, S_2, $(Y_i)^1$, $(-Y_i)^1$, $(Y_i)^2$, $(-Y_i)^2$ and output Y. The inputs for each cell are Y and the output of the previous cell.

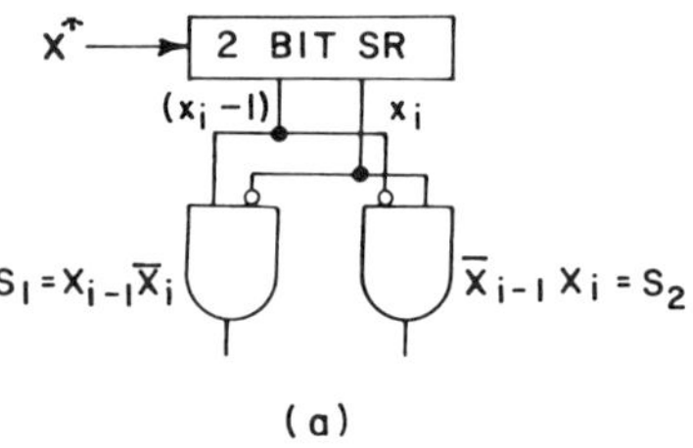

(a)

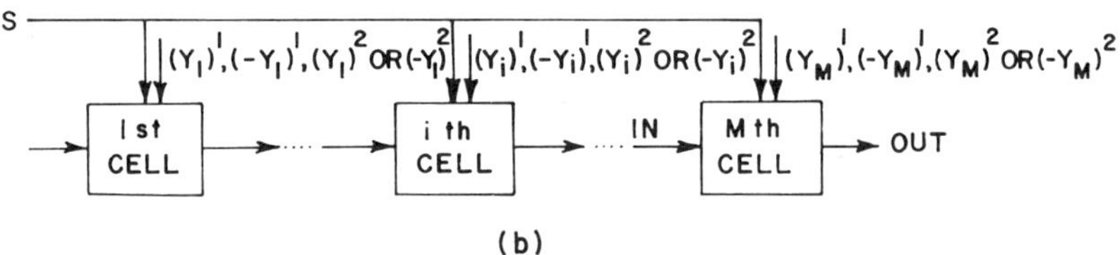

(b)

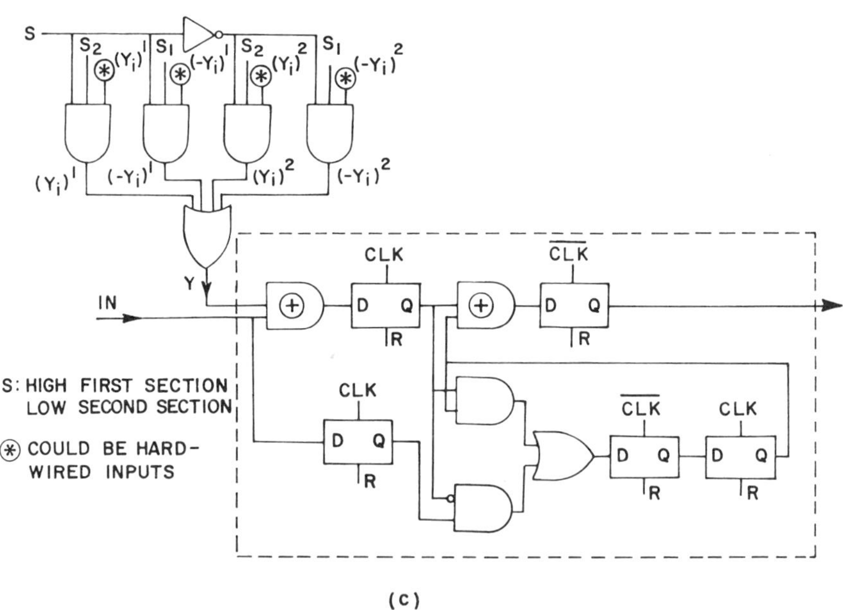

(c)

Figure 7.14 Pipeline multiplier using Booth's algorithm. (*a*) Obtaining the 2-bit word S_1S_2 from the serial input word x. (*b*) Multiplier module. (*c*) Multiplier cell and associated logic. The dotted block is a pipelined full-adder.

Multiplier Algorithm to Obtain YX

1. Add a reference "0" to precede the LSB of X (the data word) and repeat the sign bit of X, $M - 1$ places, resulting in X^*.
2. According to the bits $X_{i-1}X_i$ of X^*, perform the following operations in the multiplier cells:

	X_{i-1}	X_i		
(a)	1	0	Add $-Y$	to contents of multiplier cells
(b)	0	1	Add Y	and shift, repeat last multiplier
(c)	0	0		bit.
	1	1	Add 0	

3. At the end of the data word, reset the storage flip-flops of the multipliers.

Example:

$$\begin{array}{rl} & S\ \text{MSB}\ \ \text{LSB} \\ X = & 0\ .1\ 0\ 1\ 0 = +\tfrac{10}{16} \\ & \text{S MSB LSB} \\ Y = & 1\ .1\ \quad 1\ = \tfrac{1}{4} \\ -Y = & 0\ .0\ \quad 1 \\ X^* = & 000.1010(0) \text{ reference zero} \end{array}$$

X_{i-1}	X_i	Operation	Contents of Multiplier Cells	Output
0	0	(a)	0.00 before shift (BS)	
			0.00 after shift (AS)	0
1	0	(a)	0.01 BS	
			0.00 AS	1
0	1	(b)	1.11 BS	
			1.11 AS	1
1	0	(a)	1.10 BS	
			1.11 AS	0
0	1	(b)	0.01 BS	
			0.00 AS	1
0	0	(c)	1.11 BS	
			1.11 AS	1
0	0	(c)	1.11 BS	
			1.11 AS	1

S MSB LSB

Thus the output is 1. 110110 $= -\frac{10}{64}$

In Fig. 7.14 the maximum logic performed is at the input of the multiplier cell. It consists of detecting X_i and X_{i-1} and calculating S_1 and S_2 and then gating-in the multiplier coefficient. The total time allowed is 194 ns.

7.4.5 LSI / VLSI Design Consideration

As mentioned in Chapter 1 and in this chapter, in realizing subsystems in LSI/VLSI three factors have to be considered:

1. Speed of operation.
2. Chip area.
3. Power dissipation.

To increase the speed of operation, node capacitances have to be reduced (less area) and the driving capability of the active devices have to be increased (larger area). Thus, there is a maximum frequency of operation for a given LSI technology. The total chip area is a function of the technology and the required speed. Maximum chip area is closely related to the yield and the technology. For example, in general, the practical size of a CMOS LSI chip (which gives a reasonable yield) is larger than that of I^2L bipolar. The power dissipation in bipolar technology is mainly the dc power dissipation, while that of CMOS is mainly the ac power dissipation: cv^2f [39]. The maximum total power dissipation per chip per package should not exceed a predetermined value for conventional cooling of a package.

Based on the above consideration, it is possible to partition a given subsystem into different chips. For our example, it is necessary to study the different system design options in detail and then estimate that total power dissipation and silicon area before determining a chip partition. This exercise, because it is a function of the LSI technology under consideration and its layout rules, is left to the reader. However, Fig. 7.15 shows two proposed chip partitions for the digital processors DP and DP* of Fig. 7.6. Partition *A* is suitable for a standard-cell LSI approach, with two chips to be designed. Partition *B* is suitable for a high-density costum LSI approach where two chips are used and only one LSI design is required. In both partitions the integrated designs of the blocks P_1, P_2, and DF are identical to that of the blocks P_1^*, P_2^*, and DF* except for the metalization masks.

7.5 MICROPROCESSORS AND MICROCOMPUTERS

The dream of having "a computer-on-a-chip" came true with LSI in the early 1970s. A microcomputer is a stored-program computer that uses few LSI chips and utilizes a "microprocessor" as the central processing unit. A block diagram of a microcomputer, or a general-purpose computer for that matter, is shown in Fig. 7.16. It consists of four basic blocks: an I/O unit, a memory, an arithmetic processing unit, and a control unit. The I/O unit consists of input and output devices. Input devices are used to supply the data (needed for calculation) and instructions (to specify the type of calculation). Input devices include teletypewriter, keyboards, punched-paper-tape reader, card reader, and magnetic tape readers. Output devices supply output data and include teletypewriters, paper-tape punchers, card punchers, line printers, plotters, and magnetic tapes. The memory unit stores data and the program instructions. The

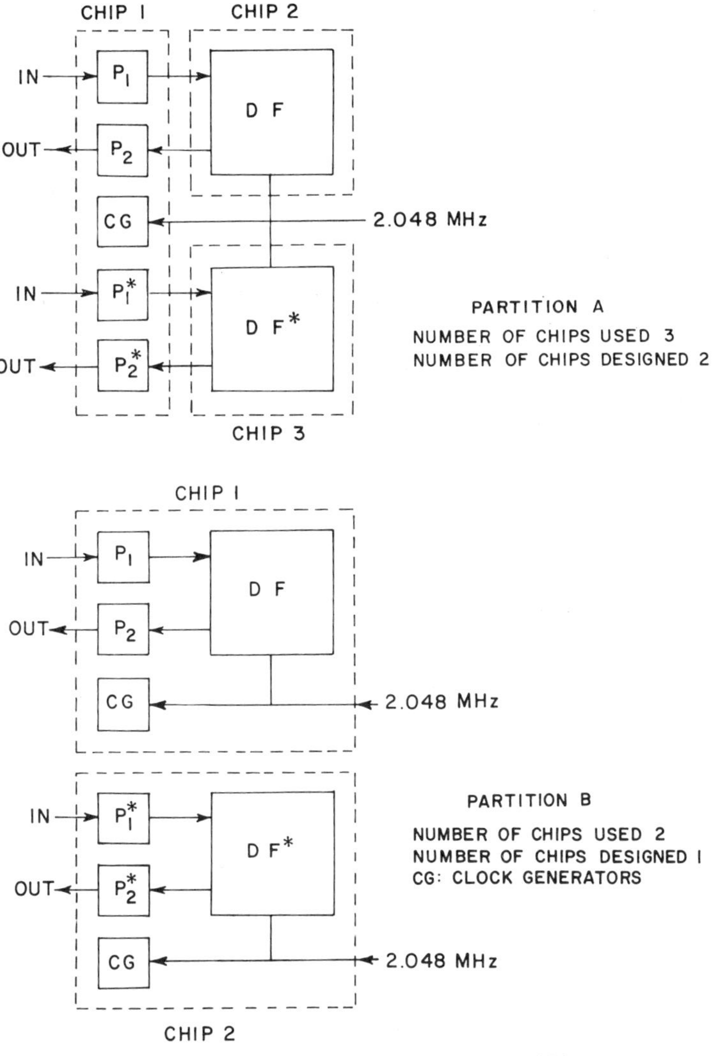

Figure 7.15 Chip partitioning for DP and DP* of Fig. 7.6

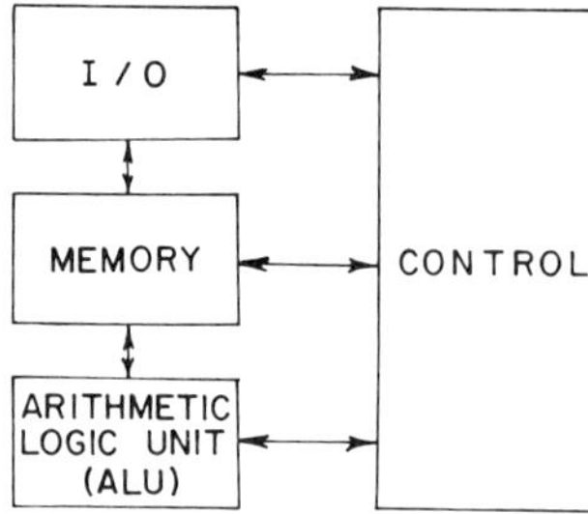

Figure 7.16 Microcomputer block diagram.

data are stored in random-access memories (RAMs) and the program instructions are usually stored in read-only memories (ROMs). Some of the instructions could be stored in user-programmable ROMs (PROMS). The ALU performs arithmetic and logical operations, for example, addition, subtraction, multiplication, division, AND, OR. The control unit receives instructions from the memory and controls the sequence of events during the computations.

As we have seen in Section 7.4 in dealing with LSI implementation of signal processors, subsystem architecture is an important consideration. Microcomputer architecture, unlike that of a signal processors where software considerations are minimal, encompasses both hardware and software. The interaction between hardware and software design, in the early stages of defining the architecture, should be emphasized. The hardware of a microcomputer is the "static" part of the system while the software is the "dynamic" part. Software, or machine instructions, are replacements of hardware components. For example, instead of using two particular hardware components, an instruction can specify that the operation is to be performed twice on a single hardware component. Trade-offs exist in choosing between realizing a function in hardware or software. For example, multiplication can be realized with a hardware multiplier or using few machine instructions to manipulate data as ADD/shift. The choice is affected by the maximum speed of operation of the LSI ships, the required throughput, power dissipation, area, and cost. The following section deals with some of these questions as related to microprocessor architecture.

7.5.1 Microprocessor Architectures

As a central processing unit, a microprocessor has the same basic architecture as a large-scale computer [13]. The basic blocks (Fig. 7.17) are: an ALU, an instruction decoder, and control and synchronization units. The instruction decoder is a ROM, which translates machine instruction code into microinstructions that are executed by the processors. The control and synchronization unit interprets the microinstructions and controls the executions of these microinstructions. Microprocessors use busses as means for transferring data, addresses, and control signals between system components. A microprocessor

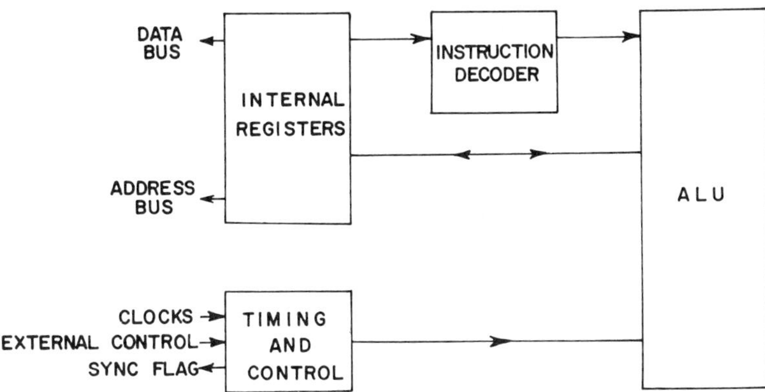

Figure 7.17 Microprocessor architecture.

uses some architectural features to enhance its data processing power, for example, resource sharing, direct memory access, and transfer and interrupts, and these are explained next.

In a microprocessor a number of resources must be shared by different jobs or users at different times. These include busses, registers, I/O pins, memory, or control programs. The resources are shared by means of time-division multiplexing. The control and synchronization unit defines specific time periods or subcycles during which certain operations are allowed to take place. This information is presented by a timing diagram of the system.

In some applications there is a considerable data transfer between memory and peripheral devices. It is efficient to transfer these data directly without process control. This is referred to as direct memory access (DMA). DMA is implemented by means of an external bus that connects the RAMs, the peripherals, and the microprocessor. When the microprocessor is not using that bus, an external hardware is activated to allow data transfer along the bus between the peripheral devices and the memory.

Certain applications of microcomputers require immediate response of the processor to an external condition. In such cases the processor must interrupt the program presently being executed and begin a new program to handle the external condition or "interrupt."

7.5.2 Microprocessor Software

The use of microprocessor software as a replacement for hardware components is one of the key advantages of microprocessor systems. Hardware functions can be simulated by software. Cost considerations that make a hardware design impractical for a given application are not applicable to software design. Hardware and software are not directly interchangeable, and each has its own characteristic. One advantage of software over hardware is that it is a nonrecurring cost item. Once a software has been designed and tested, it can be duplicated and stored in a memory of any number of systems. Thus the total development cost for the software is spread over the total number of systems.

Microcomputer/microprocessor software is a sequence of instructions that control the operation of the system. Unlike in large-scale computers, it is impractical to load these programs from external storage every time the machine is turned on. Microcomputers store these frequently used control programs permanently in ROMs (or a combination of ROMs and PROMs and in this case the system is microprogrammable). There are a number of different levels of microcomputer software: system executive, programs, subroutines, and microprograms.

The system executive is concerned with overall job, task, and data management for the system. It corresponds to the monitor, supervisor, and operating system in large computers. A program is a set of instructions for performing a specific operation. It consists usually of a number of individual routines or subroutines that perform a specific calculation or an algorithm. If the subroutine is used frequently, it is worthwhile to encode it as microinstructions in a microprogram.

7.5.3 Microcomputer Systems Design

Microcomputer systems design is the task of designing and developing a microcomputer system for specific applications using some basic elements; microprocessors, RAMs, ROMs, I/O devices, and interface components. The design includes choosing among the available off-the-shelf components, or designing custom LSI ships for performance, availability, security, and/or economical reasons. The design involves both hardware and software aspects and the different tasks can be summarized as follows:

1. Specifying the range of applications for which the system will be used. This establishes system parameters, for example, word size and throughput.
2. Determining if a microcomputer system would meet the specified system parameters. If so, selecting suitable off-the-shelf components (e.g., microprocessor) or deciding to design custom LSI ships to meet the system requirements.
3. Hardware design and software development.
4. Dubugging and testing.

Many factors influence the above design decisions, for example, cost, flexibility, compatibility, reliability, and speed.

Cost is one of the principal considerations that involves engineering and nonengineering factors. For example, development cost is a one-time expense, and the greater the number of systems produced, the lower the per-unit development cost.

Flexibility is a key advantage of microcomputer systems. The stored program can be changed easily by replacing ROMs or by reprogramming PROMs. The system can be designed to accommodate field tests and changes on location by having fusible-link ROMs.

Compatibility within the system itself and with the "outside world" is another consideration. A microcomputer-system designer is in fact a computer-system designer and the compatibility of his or her system with, for example, analog channels, users I/O devices, other microcomputers, and other pieces of hardware, has to be considered.

Reliability is a factor related to the working environment. A microcomputer reliability for space applications is obviously has to be higher than that used in educational aids.

Speed requirements are related to applications. A microcomputer may be required to work in real time. The LSI technology used to realize the different components of a microcomputer (e.g., microprocessors, RAMs, ROMs) affects the overall speed of processing data. Moreover, the type of architecture used (e.g., bit slice ALUs [38]) also affects the speed.

7.5.4 Example: A Bipolar Microprocessor

The example discussed here is TI SBP0400. It is a 4-bit slice microprocessor using I^2L (SBP stands for semiconductor bipolar processor). The architecture of the processor is organized around a multiple internal-bus structure and an internal PLA (programmable logic arrays). A bit-slice system architecture has the processor broken down into individual units, each performing the various computer functions on a "slice" of the data word. In our example this slice is 4-bit. Thus the architecture allows four microprocessors to process 16-bit data words.

The basic architecture of the SPB 0400 chip is shown in Fig. 7.18. The key distinction of the SBP 0400 over other microprocessors is the replacement of the instruction register by the internal programmable logic array. The user may define an extensive instruction set that is unique to his or her particular requirements and write programs using such instructions. The user's instruction is represented as a 9-bit "operation-select" work, which is placed on a 9-pin input to the internal PLA. The PLA, operating under clock control, decodes this 9-bit input into the specific operations required by the user's instruction, which is represented by a 20-bit internal-control word. This internal-control word provides the appropriate commands to the ALU, the registers, and the bus lines for executing the intended maching operation. A single 4-bit slice unit offers the user the possibility of some 512 discrete machine operations that may be executed in a single clock cycle, which is considerably more than the instruction sets of microprocessors.

The advantage of the programmed logic array in the SBP 0400 is twofold: first, it offers greater system integrity and security. Since the PLA is programmed at the factory by the designer, it would be extremely difficult to

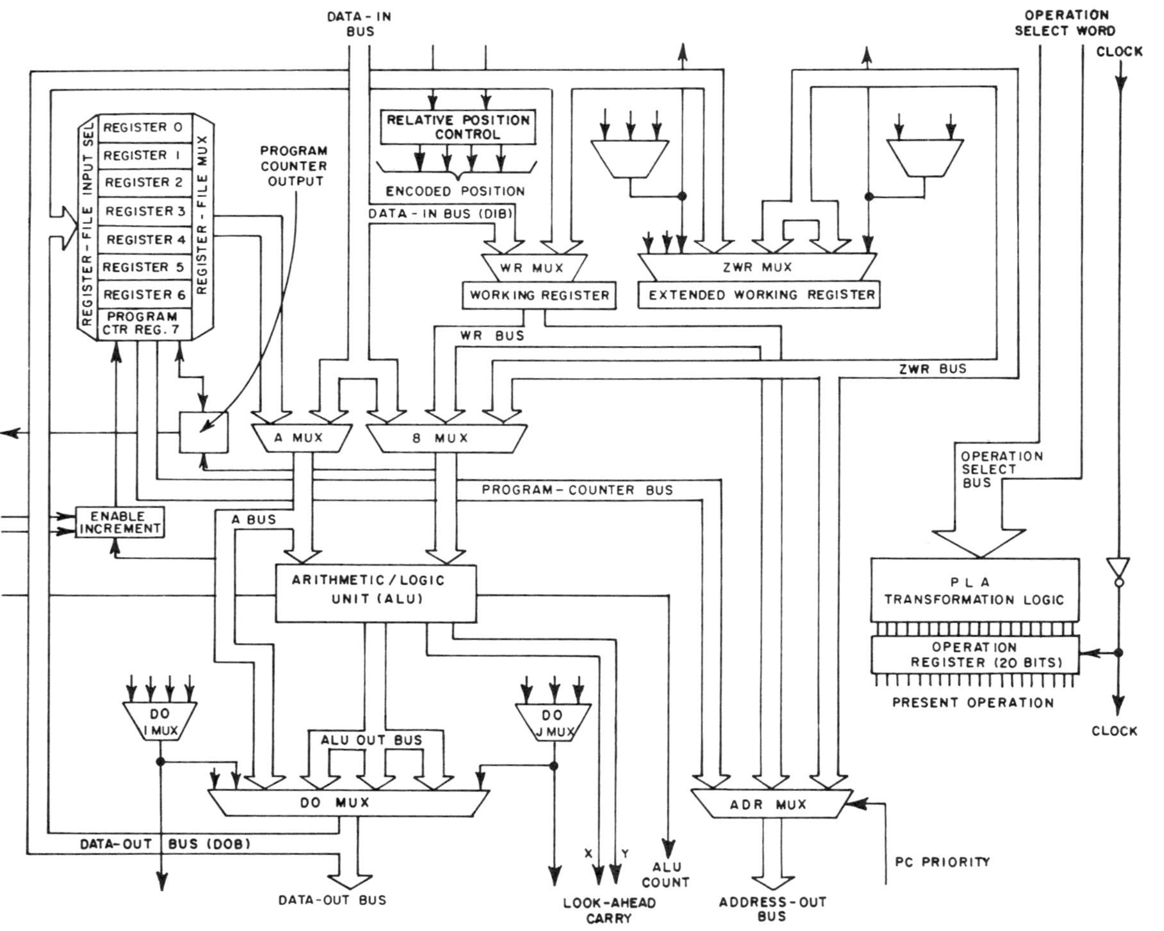

Figure 7.18 Bipolar microprocessor block diagram.

"reverse-engineer" or attempt to copy the system. Designers who have invested considerable sums in software development may now say that their efforts remain proprietary by incorporating such programs in terms of "macroinstructions" utilizing the PLA. Since the operation register of the PLA is internal to the SBP 0400 chip, it would be utterly impractical for a user to attempt to decode the PLA structure by examining register contents and ALU operations for predetermined operation-select-word inputs. The second advantage of the PLA in the SBP 0400 is to permit emulation of larger-scale computers having extended instruction sets. By appropriate coding the PLA such larger-scale computers may be emulated without significant overhead or operation degradation.

Turning to the block diagram in Fig. 7.18 we note the two basic inputs to the ALU unit: a multiplexed A input port, and a multiplexed B input port. The data-in bus has access to either the A port, the B port, or the working register. The eight general-purpose registers have access to the A port.

The output from the ALU is applied to a data output multiplexer (DO MUX), which transfer the data either externally of the chip or internally along the data-out bus (DOB) back to the general-purpose registers, working register, or extended working register.

REFERENCES

1. T. R. Blakeslee, *Digital Design with Standard MSI and LSI*, Wiley-Interscience, New York, 1975.
2. J. Eimbinder, *Semiconductor Memories*, Wiley-Interscience, New York, 1971.
3. D. A Hodges, *Semiconductor Memories*, IEEE Press, New York, 1972.
4. G. K. Kostopoulos, *Digital Engineering*, Wiley-Interscience, New York, 1975.
5. A. Barna and D. I. Porat, *Integrated Circuits in Digital Electronics*, Wiley-Interscience, New York, 1973.
6. L. R. Rabiner and C. M. Rader, *Digital Signal Processing*, IEEE Press, New York, 1972.
7. H. J. Blinchikoff and A. I. Zvereu, *Filtering in the Time and Frequency Domains*, Wiley-Interscience, New York, 1976.
8. R. E. Bogner and A. G. Constantinides, *Introduction to Digital Filtering*, Wiley, New York, 1975.
9. B. Gold and C. M. Rader, *Digital Processing of Signals*, McGraw-Hill, New York, 1969.
10. A. Antonion, *Digital Filters Analysis and Design*, McGraw-Hill, New York, 1979.
11. A. Barna and D. I. Porat, *Introduction to Microprocessors*, Wiley-Interscience, New York, 1976.
12. B. Soncek, *Microprocessors and Microcomputers*, Wiley-Interscience, New York, 1976.
13. D. R. McGlynn, *Microprocessors: Technology, Architecture and Applications*, Wiley-Interscience, New York, 1976.
14. F. Faggin, How VLSI Impacts Computer Architecture, *IEEE Spectrum* **15**, 28–31 (May 1978).
15. R. C. Minnick and R. A. Short, *Cellular Linear-Input Logic*, Stanford Research Institute, SRI Project 4122, February 1964.
16. R. C. Minnick *et al*., *Cellular Arrays for Logic and Storage*, Stanford Research Institute, SRI Project 5087, April 1966.

17. R. C. Minnick, *Cutpoint Cellular Logic*, *IEEE Trans. EC* **EC–13**, 685–698 (1964).

18. R. C. Minnick, Survey of Microcellular Research, *J. ACM* **14** 203–241 (1967).

19. W. H. Kautz *et al.*, Cellular Interconnection Arrays, *IEEE Trans. Computers* **C–17** 443–445 (1968).

20. A. Waksman, A Permutation Network, *J. ACM* **15** 159–163 (1968).

21. W. H. Kautz, A Cellular Threshold Array, *IEEE Trans. ED* **ED–16**, 680–682 (1967).

22. H. S. Stone, A Logic-in-Memory Computer, *IEEE Trans. Computers* **C-19** 73–78 (1970).

23. S. E. Wahlstrom, Programmable Arrays and Networks, *Electronics* **40**, 91–95 (December 11, 1967).

24. W. H. Kautz, Fault Testing and Diagnosis in Combinational Digital Circuits, *Proceedings of the First Annual IEEE Computer Conference*, Chicago, IL, September 1967, IEEE Press, New York, pp. 2–5.

25. W. H. Kautz, Testing for Faults in Combinational Cellular Logic Arrays, *Proceedings of the 8th Annual Symposium on Switching and Automata Theory*, Austin, Texas, October 1967, IEEE Press, New York, pp. 161–174.

26. J. Goldberg *et al.*, *Technique for the Realization of Ultra-Reliable Space-borne Computers*, Interim Scientific Report III, SRI Project 5580, June 1968.

27. W. H. Kautz, Cellular Logic-in-Memory Arrays, *IEEE Trans. Computers*, **C–18** (3), 719–727 (1969).

28. R. R. Seeber, Associative Self-Sorting Memory, *Proceedings of the 1960 Eastern Joint Computer Conf.*, Vol. 18, Western Periodicals, New York, pp. 179–188.

29. B. Elspas and R. A. Short, A Bound on the Run Measure of Switching Functions, *IEEE Trans. EC*, **ED–13**, 1–4 (Feb. 1964).

30. D. L. Shell, A High-Speed Sorting Procedure, *Commun. ACM*, **2** (7), 30–32 (1959).

31. R. C. Bose and R. J. Nelson, A Sorting Problem, *J. ACM*, **9** (2), 282–296 (1962).

32. L. R. Rabiner and C. M. Rader, *Digited Signal Processing*, IEEE Press, New York, 1972.

33. *Selected Papers in Digital Signal Processing II*, IEEE Press, New York, 1975.

34. H. Kaneko, A Unified Formulation of Segment Companding Laws and Synthesis of Codes and Digital Companders, *Bell Syst. Tech. J.*, **47** (7), 1555–1588 (1970).

35. M. I. Elmasry and R. C. Madter, Pipeline Digital Filtering at 250 MHz Bit-Rate, *The 1975 Midwest Symposium on Circuits and Systems*, Western Periodicals, New York, pp. 461–465.

36. A. Peled and B. Lin, A New Hardware Realization of Digital Filters, *IEEE Trans. Acoust., Speech and Signal Processing* **ASSP–22** 456–462 (1974).

37. L. B. Jakson, J. F. Kaiser, and H. S. McDonald, An Approach to the Implementation of Digital Filters, *IEEE Trans. on Audio and Electroacoustics*, **AU–16** (3) 413–421 (1968).

38. D. R. Appelt, Making It Compatible and Better, *Electronics*, October 11, 1979, pp. 131–136.

39. M. I. Elmasry, *Digital MOS Integrated Circuits*, IEEE Press, New York 1981.

Appendix A

Design and Processing of Integrated Circuits

A designer of an IC starts by studying the *requirements* to produce an IC for a given function. The study involves IC engineering considerations in addition to cost, competing products, market applications, manpower, and other resources. This general study is followed by specifying a *set of system/subsystem* constraints, for example, speed of operation, maximum power dissipation, value of power supply, and technology used. This phase also includes choosing a system architecture (see Chapter 7) and subsystem/logic algorithms for IC implementation.

This phase is followed by a *circuit design* phase where IC circuit structures for a given IC technology and a given set of performance specifications are designed. Computer-aided-design (CAD) circuit-analysis programs are used throughout the circuit-design phase.

Once the circuit design has been verified, *layout* of the chip(s) is to be started. CAD programs can be used in layout optimization since silicon area consumed affects yield and hence cost. In many cases another round of circuit simulation follows the layout in order to take into account the actual capacitive and resistive effects of the interconnections.

Following the IC design phase as indicated by (1) in Fig. A.1, another phase would start to *generate masks* to be used in the silicon-wafer processing. In fact, the layout data could be used to directly "write" on the wafer the pattern for each pattern level. This process is referred to as *artwork*, *IC photolithography*, or in general *IC lithography*.

The density of an IC is directly related the resolution of the IC lithography, and efforts are always being made to improve it. Figure A.2 shows how the average area of an IC and the average minimum line width have changed during the years. Each minimum line width is related to a technique that must be used to achieve such line width. This is shown in the box marked (2) in Fig. A.1 and explained as follows.

For 3–5-μm line widths conventional contact and *proximity printing* (where the mask is put into contact, or near contact, with the silicon wafer) can be used. This is indicated by block numbers 22, 23, 24, and 25 in Fig. A.1. In this system each mask level of an IC is digitized and used through a step-and-repeat process to generate a mask replica to be used in wafer processing.

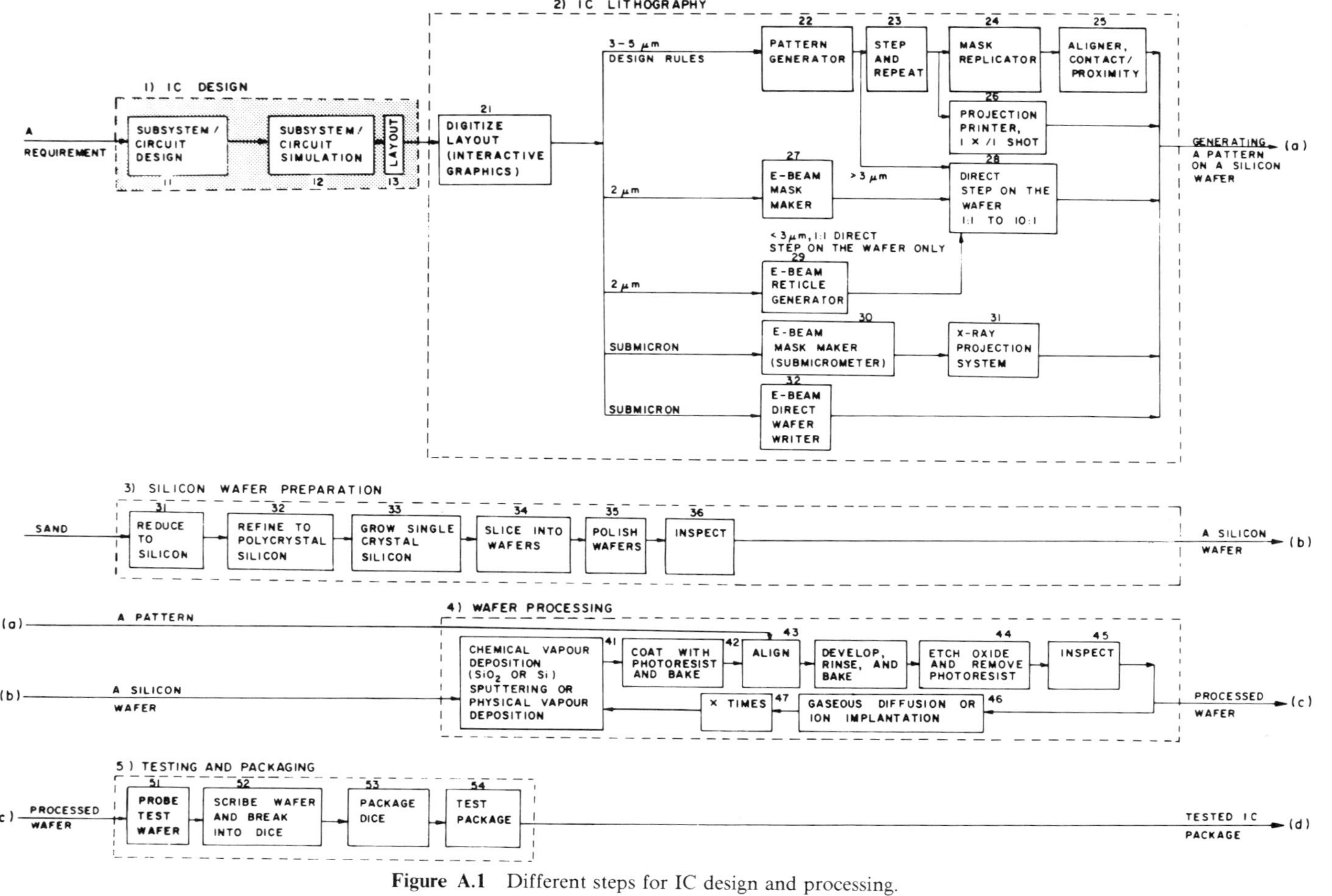

Figure A.1 Different steps for IC design and processing.

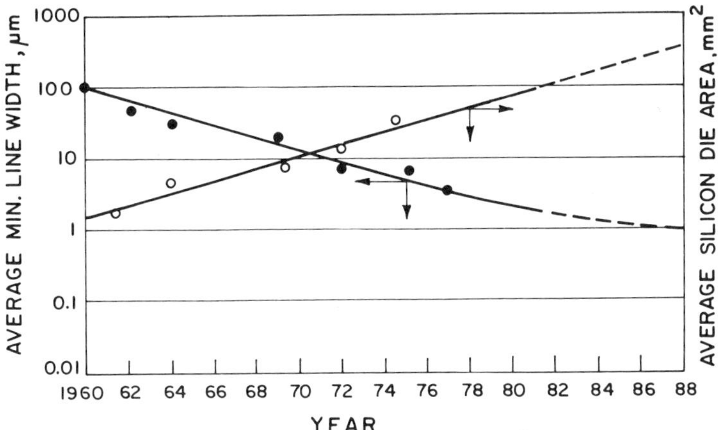

Figure A.2 Average silicon die area and average minimum width versus year.

For the same line widths, *optical-projection* lithograph systems (blocks 22, 23, and 26), where step-and-repeat optical reduction techniques are used, have recently begun to take over from conventional contact printing. The reason for using projection printing is the substantial increase in device yield. This is because the procedure uses chrome master masks ordinarily used to contact-print production masks. Thus imperfections are minimized. Projection printing also eliminates yield losses from such factors as stripping of mask emulsion onto the wafer, sticking of the photoresist to the emulsion, and pressing onto the emulsion of epitaxial spikes, silicon pieces, and dust particles (which then print onto succeeding wafers).

Optical-projection methods reach their resolution limit around 1μm. Thus for $\leqslant$ 2-μm design rules, *electron-beam mask makers* (blocks 27 or 30) or *electron-beam reticle generators* (block 29) are used. They are used in association with *a direct wafer stepper* (block 28), where an electron beam, under computer control, exposes patterns onto a wafer coated with electron-sensitive resist. The computer control uses the mask data generated by the E-beam mask maker (1:1) or uses the reticle made by the E-beam generator (10:1). Because the beam of electrons is infinitely and minutely adjustable, exposure can be tailored to the shape of the pattern and to the shape of the wafer. Thus nonflat surfaces in large wafers can be tolerated resulting in low mask-defect rate.

Another lithography method for submicron design rules is the use of *X-ray exposure* of a wafer (block 31) with a mask produced by an E-beam mask maker. The method is simple and inexpensive but the masks themselves are expensive.

In the future, the functions of the direct wafer stepper and the E-beam generator can be combined into one: *an E-beam direct wafer writer* (block 32). The main advantages of using direct E-beam wafer writer systems are better line resolution (down to 0.25 μm), faster turn around time, fewer defects, and less contamination.

During silicon-wafer processing, selective regions of a chip, according to a given pattern, are identified for processing. This is done by covering the wafer by silicon oxide or silicon nitride (block 41), coating it with photoresist (block 42), exposing it to a pattern (block 43), and finally etching oxide and removing photoresist from selected areas. Removing the photoresist and etching off the protective layers of silicon oxide or silicon nitride is another critical step in IC processing. The standard method of stripping and etching is to use acids, which have given the procedure its name of *wet etching*. Ideally the etching step should be performed to selectively etch the SiO_2 layer without affecting the underlying silicon layer. However, most wet etchants are good for line dimensions of the order of 2–5 μm, as they are isotropic (they etch in all directions at the same rate) and they tend to undercut masked areas by anywhere from 0.5 to 1.5 μm. A further problem, which gets more critical with finer dimensions, is that wet etchants may leave a certain amount of residue on the wafer that must be cleaned off.

Recently wet etching has been replaced with a *dry* procedure that is called *plasma etching*. Plasma etching involves the use of gas plasma produced by a high-voltage RF discharge. It offers several advantages over wet etching including finer line resolutions (well below 1 μm) and minimization of the resist undercutting problem. In some plasma-etching systems, etching can be anisotropic so that steep well-defined narrow line geometries can be etched with little, or no, horizontal etching of undercut masked areas.

Not only have etching techniques recently received attention but special *resist materials* have also been optimized for direct-exposure E-beam systems. These materials are polymeric chemicals that are developed in a solution, after being exposed to electron beams. In the case of negative resists, the unexposed portion of the resist is dissolved in the solution. With positive resists, the exposed portion is dissolved in the solution. The quality of a resist is determined mainly by its *sensitivity*, *resolution*, and *compatibility* with variant wafer fabrication processes. Most negative E-beam resists have good sensitivity but suffer poor resolution (1-μm limit) and poor dry-etch resistance. Positive resists, on the other hand, have the highest resolution (better than 0.1 μm) but suffer from poor sensitivities. Efforts continue to develop optimum resist material with both good resolution and high sensitivity.

REFERENCES

1. J. Lyman, Demands of LSI are Turning Chip Makers Towards Automation, Production Innovations, *Electronics*, **50** 81–92 (July 21, 1977).
2. R. Allan, Semiconductors: Toeing the (Microfine) Line", *IEEE Spectrum*, **14**, 34–40 (Dec. 1977).
3. J. Lymann, Scaling the Barriers to VLSI's Fine Lines, *Electronics*, **53**, 115–126 (June 19, 1980).

Appendix B

Ion Implantation*

B.1 INTRODUCTION

To begin, let us define what we mean by "ion implantation." As employed in moden semiconductor technology, ion implantation is the introduction of energetic charged atomic particles into a substrate for the purpose of changing the electrical, metallurgical, or chemical properties of the target. The typical ion energies E considered are 3 keV $\gtrsim E \gtrsim$ 400 keV. Doses ϕ vary from 10^{11} ions/cm^2 $\lesssim \phi \lesssim 10^{16}$ ions/cm^2. The practical use of ion implantation so far is based chiefly on the greater control of impurity density and profile afforded by the substitution of a high-energy "physical" process for a low-energy "chemical" process such as the more commonly employed diffused predeposition technique. These and other advantages (and disadvantages) are summarized in Table B1. The high energies associated with ion implantation allow any impinging impurity ion to enter and penetrate any substrate despite chemical-barrier effects that might tend to preclude this motion. As in all ion-beam techniques, parameters such as energy, time-integrated current, and target and projectile atomic mass and number, become more important in determining the initial depth profile than temperature, solubility, chemical diffusion constant, and reaction rate. Because the ion beam may be electrically steered or the substrates mechanically moved, and because the beam current and energy are easily measured with great precision, the new technique readily lends itself to computer control, reducing much of the handling, guess work and "black art" necessary with earlier technologies.

B.2 RANGE AND STRAGGLE

A brief summary of the important principles associated with ion implantation is given here according to the LSS theory of Lindhard, Scharff, and Schiøtt [1, 2]. The reader is referred to References 3–10 for details.

If we ignore the effects of the crystal lattice and consider the substrate to be amorphous (providing the ion beam is incident in a direction that deviates more than a critical angle θ about 5° from one of the major crystal axes), there are two basic stopping mechanisms to the implanted ions: the first is collisions

*This section with permission is based on a review article by K. A. Pickar.

with target nuclei which cause deflections of the projectile ions (nuclear stopping); the second is the interaction with electrons in the target which do not cause substantial deviations in the projectile path (electronic stopping).

Nuclear stopping is more important for heavier ions and lower energies and electronic stopping for lighter ions and higher energies. For example, electronic stopping exceeds nuclear stopping for energies $\gtrsim 17$ keV in the case of boron and $\gtrsim 800$ keV in the case of arsenic [11]. Assuming a Gaussian spatial distribution for the implanted ions and assuming only nuclear stopping, a universal curve using reduced parameters is obtained for the total *range* (total length of the ion track including zigs and zags) and the total *straggle* (the statistical fluctuation in this length) as a function of incident energy. Corrections are made to account for the electronic stopping contribution and to convert from total range and straggle to perpendicular range R_p, perpendicular straggle, ΔR_p, and transverse straggle, ΔR_T. The relationship of these quantities is shown in Fig. B.1. By R_p we mean the median perpendicular distance of the implanted ions, beneath the surface, or approximately, the depth of the peak of the concentration profile. By ΔR_p we refer to the width of the (assumed) Gaussian profile curve in the Z direction. For a series of ions incident along the negative Z axis, at the rate s ions/s, the scattering will cause the ions to deviate from this axis. The width of the Gaussian in this transverse direction X is ΔR_T. This last parameter is important in determining the distribution near a mask edge or for nonorthogonal implantations. Then the doping density $n(Z, X)$ for this infinitesimally small beam diameter is given as

$$n(Z, X) = \frac{s}{(2\pi)^{3/2}\Delta R_p \Delta R_T} \exp\left[-\left(\frac{Z - R_p}{\sqrt{2}\,\Delta R_p}\right)^2\right] \exp\left[-\left(\frac{x}{\sqrt{2}\,\Delta R_T}\right)^2\right]$$

For an implantation from an infinitesimal beam which is scanned uniformly across the target surface, the X dependence of the doping density drops out in regions several ΔR_T from the edge of the scanned area.

Table B1 Ion Implantation in Device Technology

Advantages	Disadvantages
1. Precise control over total dose, depth profile, and area uniformity, giving good reproducibility.	1. Expensive, relatively complicated machinery.
2. Wide choice of mask materials.	2. Junctions not automatically passivated.
3. Low-temperature processing.	
4. Implanted junctions can be made self-aligned to edge of mask.	
5. Vacuum cleanliness except for polymerized hydrocarbons from oil pumps.	

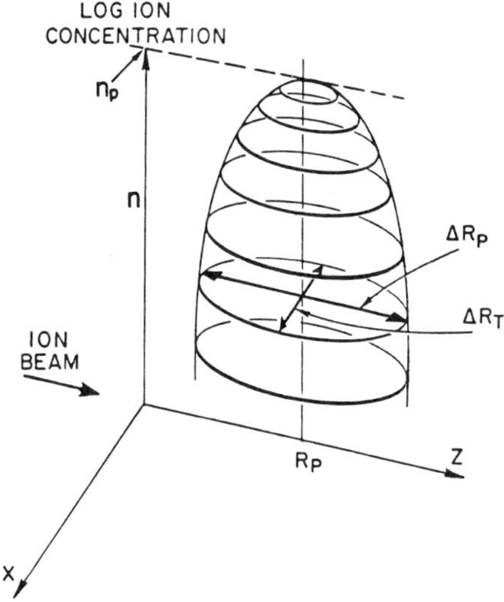

Figure B.1 Definitions of ion implantation parameters.

The doping density may then be expressed as

$$n(Z) = \frac{\phi}{(2\pi)^{1/2}\Delta R_p} \exp\left[-\left(\frac{Z - R_p}{\sqrt{2}\,\Delta R_p}\right)^2\right]$$

which describes the Gaussian distribution in doping density of the implanted ions. The quantity ϕ is the *dose* (i.e., the total number of ions per unit area impinging on the surface). The quantity $\phi/(2\pi)^{1/2}\Delta R_p$ is the *peak doping concentration* n_p.

B.3 DEPARTURES FROM LSS THEORY

We consider now some higher-order effects in the ion implantation process not covered by LSS.

1. Exponentially Decaying Tails

In the case of silicon substrates, the Gaussian distribution has been shown to be inaccurate when the doping density falls to about an order of magnitude below the peak concentration. Although this is not important for many applications, for others, such as the prediction of accurate junction depths, the use of a Gaussian approximation can lead to serious errors. For example Fig. B.2 shows experimental distributions for implanted and annealed boron [12]. A "tail" in the distribution is observed for the deeply penetrating ions. This effect has been attributed to enhanced diffusion due to the large amount of interstitial impurities present in the radiation damaged material and/or

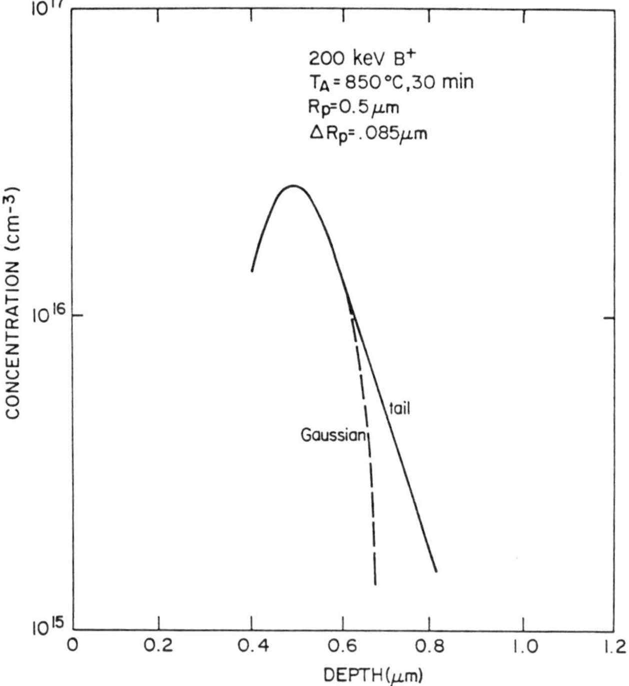

Figure B.2 Experimental distributions for implanted and annealed boron.

scattering of the impinging impurity ion into preferred crystallographic directions ("channeling").

2. Empirical Electronic Stopping Correction

Experimentally [13] it is found that the magnitude of the electronic stopping oscillates with atomic number. For boron into silicon this effect can cause a large deviation from the predicted LSS range. For heavy ions, $M \gtrsim M(\mathrm{Si})$ the oscillation is still pronounced, but the electronic stopping contribution to the total stopping power becomes less important and so the correction is negligible.

3. Discrepancies Between Impurity and Carrier Distributions

It must also be noted that the LSS distributions are calculated for the As-implanted ions and may not coincide with the distribution of *electrically active* impurities. For example, Fig. B.3 [14] shows the total phosphorus concentration profile (open circles) as determined by neutron activation and the electrically active phosphorus concentration (closed circles) as determined from Hall measurements. In this experiment, the phosphorus was implanted with the substrate at 600°C. The incomplete activation seen is characteristic of high-temperature implants.

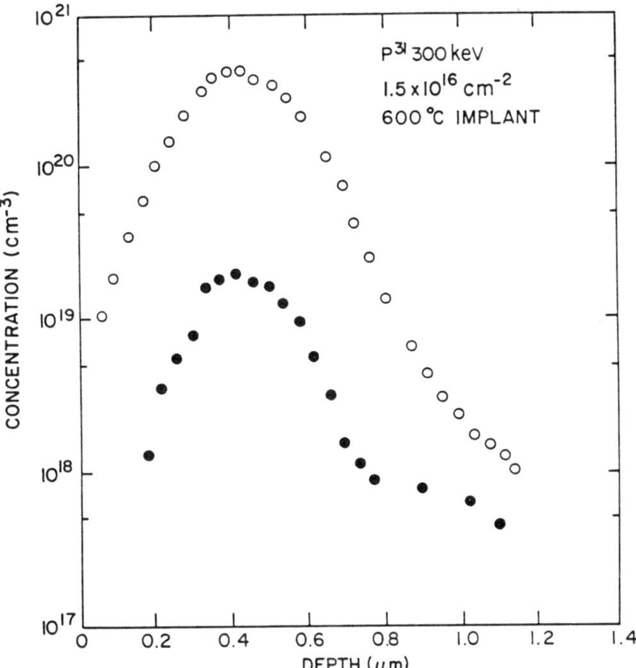

Figure B.3 Phosphorus concentration profile.

B.4 VALUES OF RANGE PARAMETERS

Figures B.4*a*,–B.4*c* give values of R_p, ΔR_p, ΔR_T for B, P, As in Si. The results for R_p, ΔR_p are those computed by Johnson and Gibbons [15].

B.5 EFFECT OF DIFFUSION

In many applications there are several high-temperature steps which follow the implantation. The final distribution may then be at least partially determined by diffusion effects. These can be combined with the straggle as

$$L = \left(2\sum_i D_i t_i + \left(\Delta R_p\right)^2\right)^{1/2}$$

where L is the effective diffusion length and D_i, t_i are the ith diffusion time. It is assumed that D is not dependent on radiation damage or doping concentration.

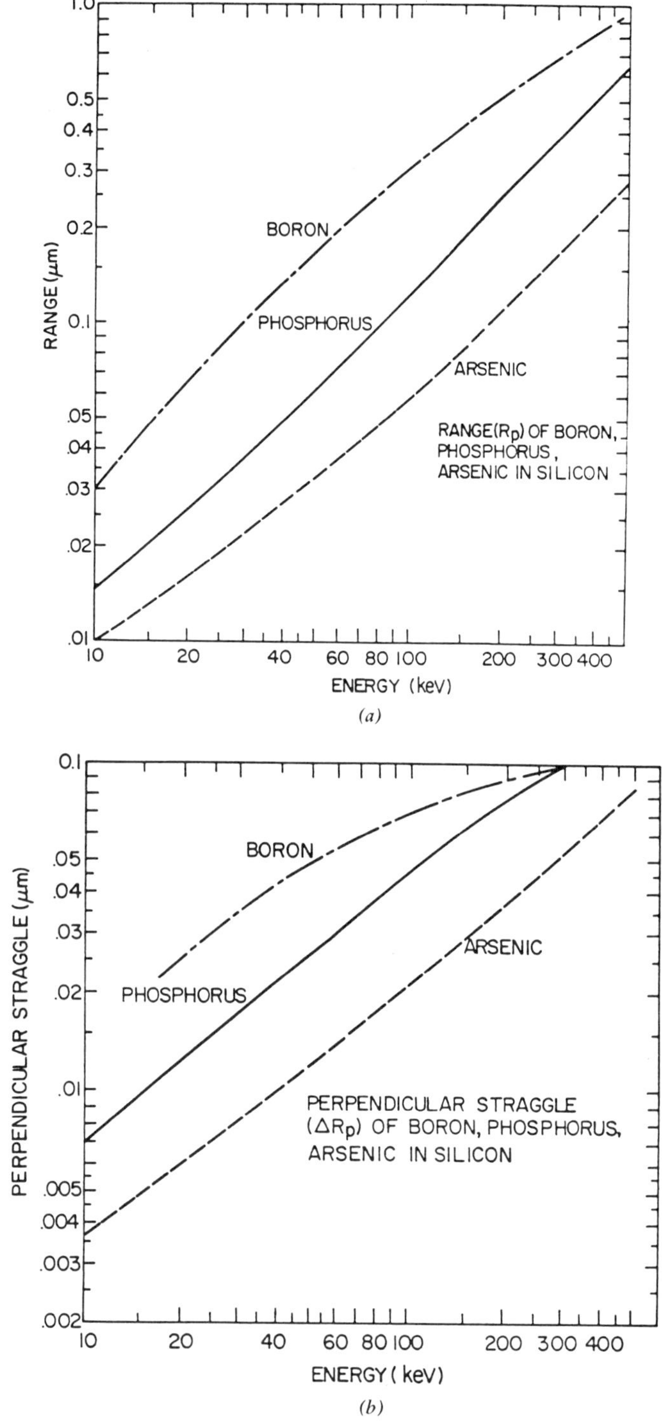

Figure B.4 (*a*) Range R_p of boron, phosphorus, and arsenic in silicon. (*b*) Perpendicular straggle ΔR_p of boron, phosphorus, and arsenic in silicon. (*c*) Transverse straggle ΔR_T of boron, phosphorus, and arsenic in silicon.

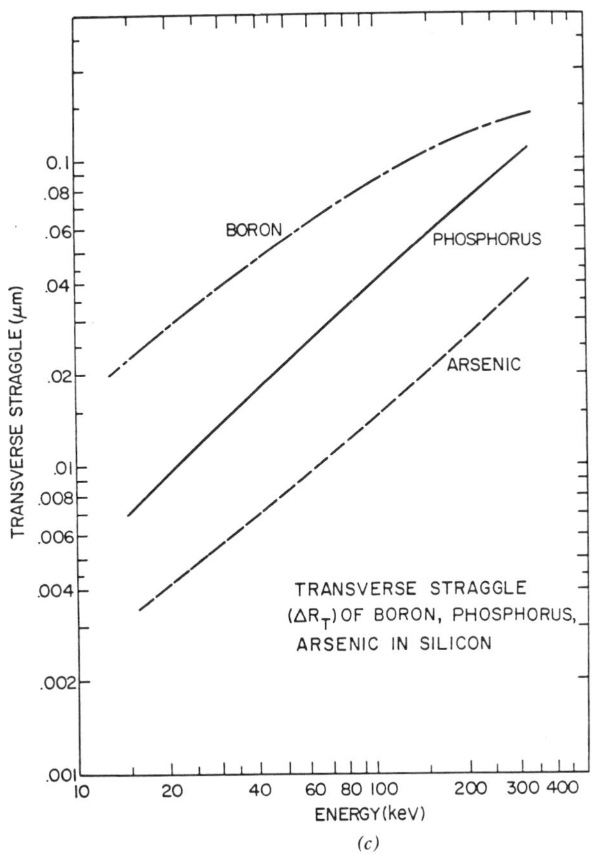

(*c*)

Figure B.4 (*Continued*)

B.6 MASKING FILMS

For films used as ion implantation masks, the minimum mask thickness necessary to stop the impinging ions is given by $d_{\text{min}} \simeq R + 4\Delta R_p$. Using this and the LSS results for B^+, P^+, As^+ in photoresist, SiO_2, Si_3N_4, curves of minimum masking thickness versus energy may be constructed. Figures B.5*a*–B.5*c* give these results.

B.7 CHANNELING

To obtain other profiles, especially deeper ones, the channeling phenomenon may be exploited. In channeling, incident ions impinging along a major crystallographic axis are guided between walls of atoms in a crystal row. Because a channeled ion does not approach close to the rows of atoms but is gently steered away by the smoothed repulsive potential of many wall atoms, the only stopping mechanisms that need be considered is electronic stopping.

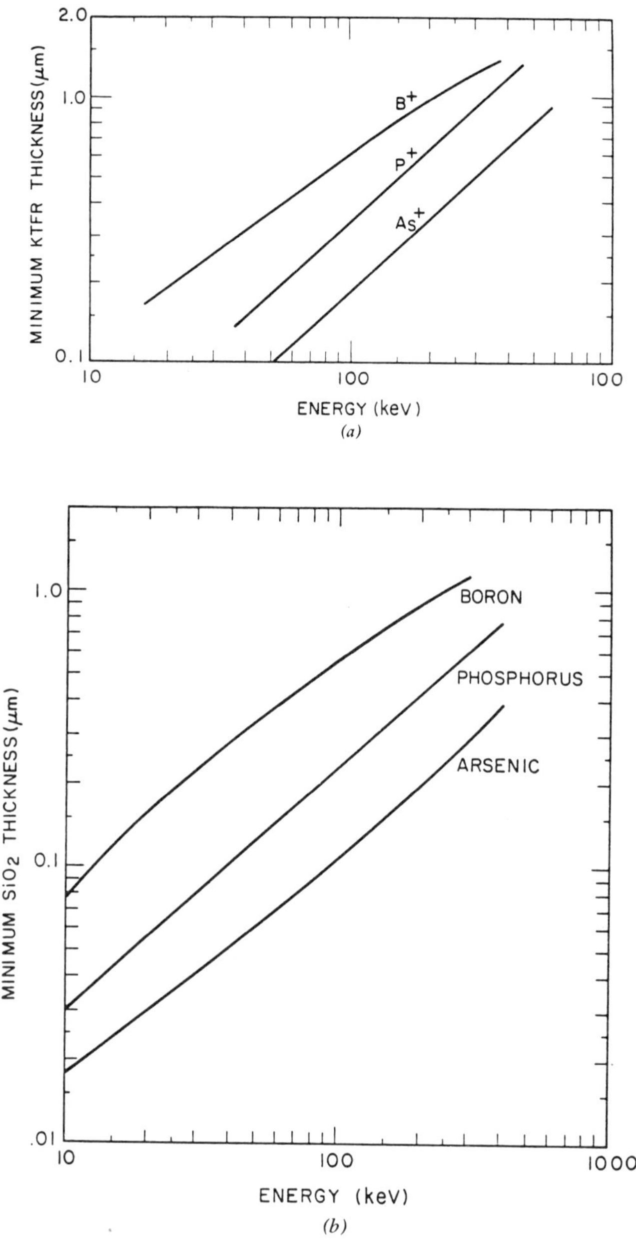

Figure B.5 (*a*) Minimum KTFR masking thickness versus energy. (*b*) Minimum SiO_2 masking thickness versus energy. (*c*) Minimum Si_3N_4 masking thickness versus energy.

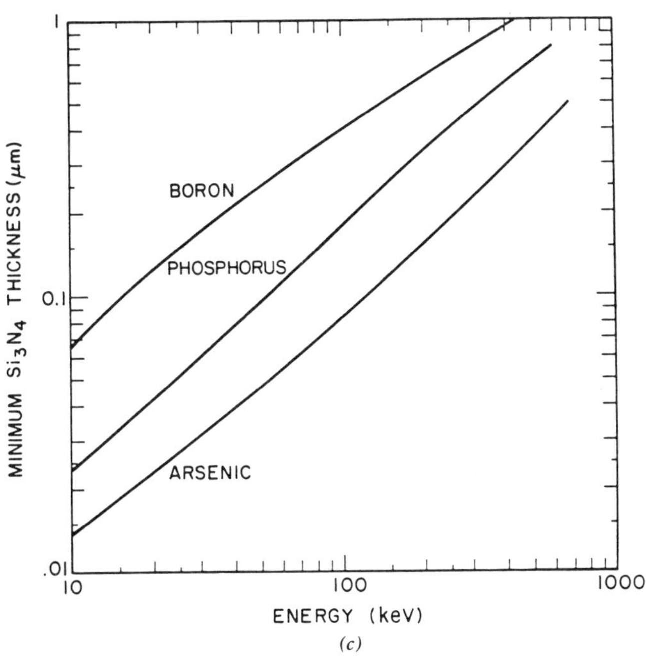

(c)

Figure B.5 *(Continued)*

Inasmuch as electronic stopping decreases with increasing ionic mass, the reverse situation exists for the penetration of these heavier ions that for the "normal" nonchanneling stopping, that is, heavier ions, penetrate deeper. It is difficult to predict the "maximum" range $R_{\max}$ for channeling because the electronic stopping power, which determines the range, is not accurately predicted by LSS, a theory which assumes an amorphous substrate. In addition the actual details of the distribution depend on the number of atoms that are dechanneled from cumulative transverse effects due to electronic interactions, striking the surface atoms (only about 2–5% for a properly aligned crystal and low dose), defects or impurities in the bulk (i.e., atoms blocking the channels), surface contamination, thermal vibrations, etc. Defects can be present in the grown material or arise from radiation damage produced by previous implantations. Figure B.6 shows the normalized depth distributions of phosphorus ions incident in a channeling direction as a function of dose. The fraction that is channeled decreases markedly with dose, because of the accumulation of damage. The peak close to the surface is located at the amorphous range R_p.

Because of the complexities involved in the channeling process and perhaps because of the difficulty in aligning the crystal in a production environment, channeling implantation has not yet been commercially exploited for device manufacture.

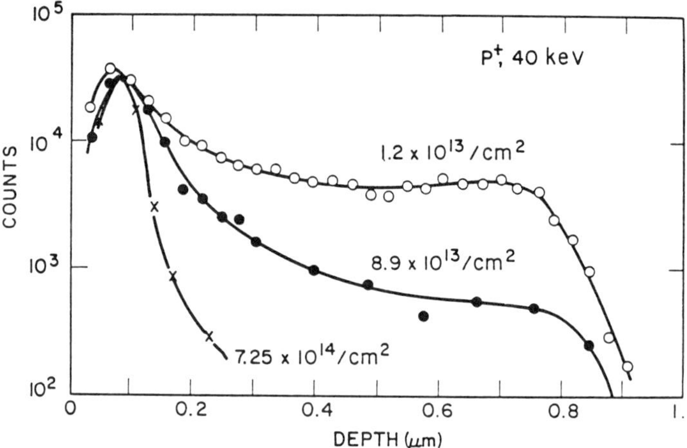

Figure B.6 Normalized depth distributions of phosphorus ions incident in a channeling direction as function of dose.

B.8 DAMAGE

There are two important questions that one naturally asks when approaching the subject of ion implantation. The first is where do the ions themselves end up? This question was addressed in the previous sections. Of equal importance is the problem of effects on the host lattice itself. Ion implantation is an inherently violent process. For nuclear stopping, the energy of the implanting ion goes into displacing target atoms from their lattice positions. These recoiling atoms may possess appreciable fractions of the incident ion energy (depending on the relative masses and whether the target atom is unfortunate enough to be the victim of a relatively rare "head-on" collision). These in turn cause cascades of secondary recoils to form a tree of displaced substrate atoms. This damage effect, in general, is more important than any chemical effects of the implanted impurity ion in the preannealed sample as hundreds of thousands of target atoms may be displaced by one incident ion. At the same time, as the implantation is typically carried out at a finite temperature, some annealing processes are proceeding *in situ*. For example, in the case of silicon, single vacancies and interstitials are highly mobile even at room temperature. Thus the amount of unannealed damage depends on the interaction of the recoil atoms with each other to form divacancies and higher, order clusters that are stable at room temperature. The situation is complicated by the fact that the various means of experimental observations of damage give results that often cannot be correlated, as the actual defect structure is quite complex and the same defect may not be visible by more than one technique. This is especially true of the defect structures that cause electrical effects such as a decrease in minority-carrier lifetime, carrier removal, and mobility degradation. A review by Gibbons [16] treats these matters in great detail.

B.9 ION IMPLANTED *pn* JUNCTIONS

To illustrate how ion implantation may be successfully meshed with conventional planar technology to fabricate *pn* junctions diodes, the limitations of both are examined. One such limitation, when ion implantation is employed, is the absence of automatic junction passivation. Because diffusion occurs roughly to the same distance in a transverse direction to the surface as in the perpendicular direction, the junction is "tucked under" the oxide and only the highly doped diffused region is exposed to the bare surface. This is true even when the oxide edge profile is irregular as shown in Fig. B.7*a*. However, ion implanted junctions are only shallowly passivated, if at all, and the junction quality may depend on such factors as the shape of the mask edge (see Fig. B.7*b*).

Therefore, to ensure the passivation of ion implanted junctions some methods must be used other than a one-step implantation into a bare substrate with a conventional thick oxide mask. Several choices are available.

Method 1: Predeposition—Drive-in

Implant, then follow with a high-temperature-diffusion step to cause the junction edge to diffuse under the oxide (Fig. B.8*a*). This of course precludes low-temperature processing.

Method 2: Shaped Oxide Edge

Implant at a sufficiently high energy to give penetration under gradually sloped oxide edge. Follow with a low-energy implant to ensure all the surface is converted (Fig. B.8*b*). The edge may be obtained by using a fast-etching

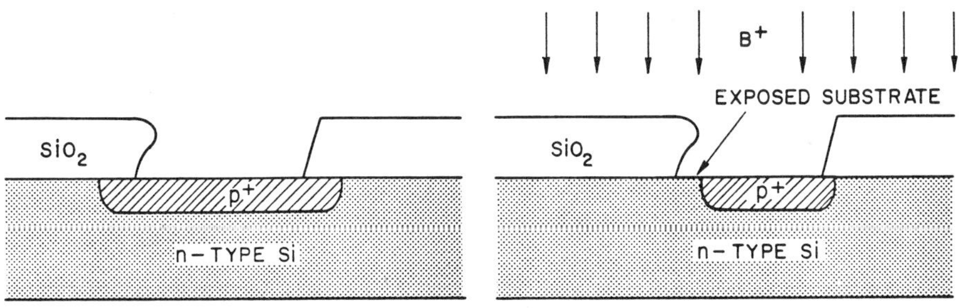

Figure B.7 Diffused and implanted *pn* junction.

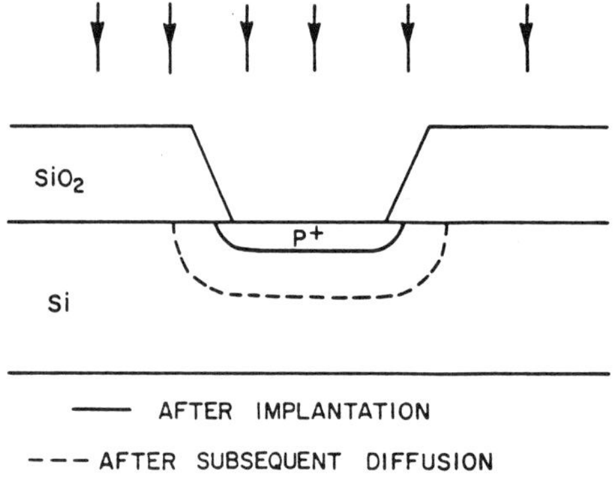

POST-IMPLANTATION DIFFUSION

(a)

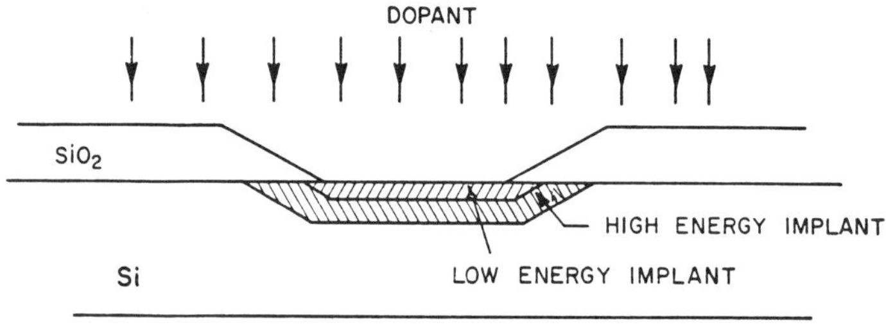

SLOPED OXIDE

(b)

p+

SiO_2 SiO_2

Si

LOCAL OXIDATION REGION
IMPLANTION MASK

(c)

Figure B.8 (*a*) Post-implantation diffusion. (*b*) Sloped oxide. (*c*) Local oxidation. (*d*) Implant through oxide. (*e*) Prediffused guard rings.

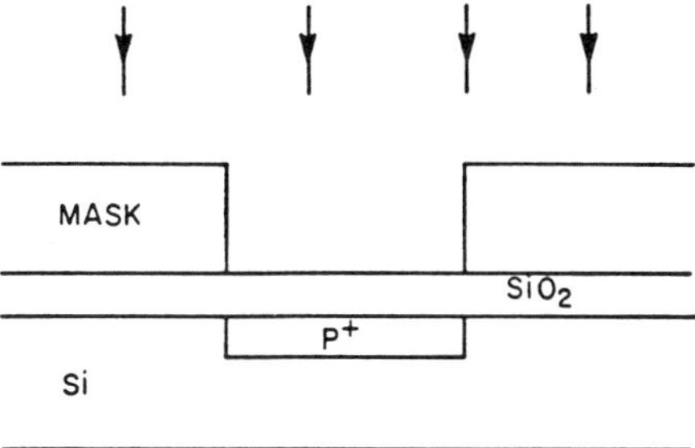

IMPLANT THROUGH OXIDE

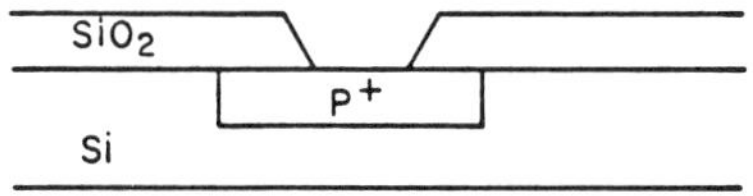

REMOVE MASK, ETCH WINDOW

TWO STEP PROCESS

(d)

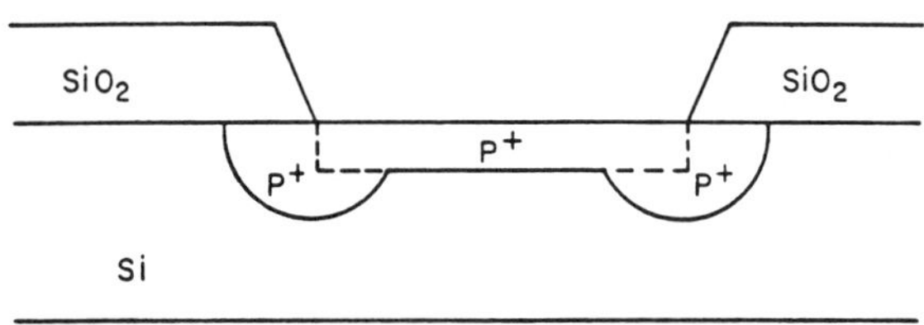

PRE-DIFFUSED GUARD RINGS

(e)

Figure B.8 (*Continued*)

surface layer. Alternatively, the local oxidation technique [17, 18] gives a "bird-beak" edge and the thick oxide fabricated by this method may serve as a suitable mask (Fig. B.8*c*).

Method 3: Double Lithography

Implant through the passivating oxide and then open windows within the boundaries of the implantation (Fig. B.8*d*). The method requires an extra photolithographic step.

Method 4: Guard Ring

Implant into a structure with previously diffused or implanted-diffused guard rings (Fig. B.8*e*).

B.10 ION IMPLANTED RESISTORS

Ion implanted resistors [19–26] are generally made by implanting through a thin passivating oxide layer (Method 2). Where contacts are desired, heavily diffused or implanted regions are formed (Method 4). These resistors have several advantages over conventional diffused resistors.

1. The control afforded allows low doses to be employed resulting in higher sheet resistances and hence smaller area consumption. The resistance

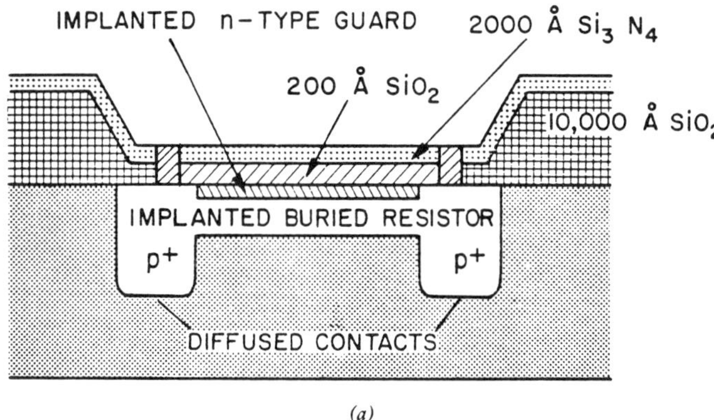

(*a*)

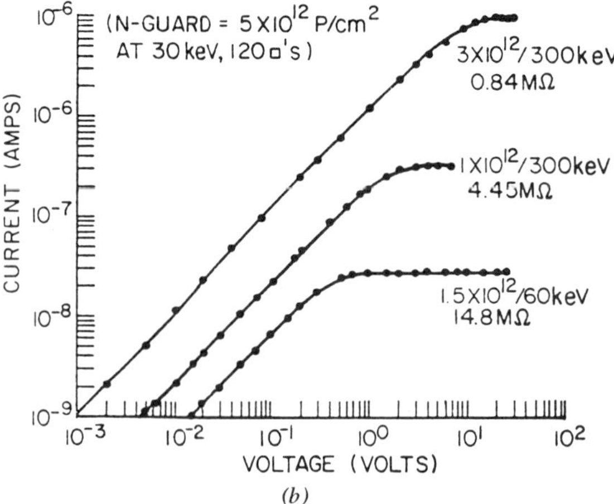

(*b*)

Figure B.9 (*a*) Implanted resistor. (*b*) *I-V* characteristic of implanted resistors.

limitation due to voltage nonlinearity is set by depletion into the implanted region. This occurs from the *pn* junction which the implanted layer forms with the substrate. For 1 Ω-cm substrate material the limit is about 20 kΩ/□ for an implanted resistor at $V \lesssim 1.0$ V.

2. A limit in reproducibility is set by variations in the positive interface charge causing fluctuations in the surface depletion of the B^+ implanted layer. This occurs for a dose $\phi \sim Q_{ss}/q \sim 10^{11}$ cm^{-2}, corresponding to sheet resistances of $1/e\mu \sim 10$ kΩ/□. For diffused resistors, the limit is typically $\lesssim 900$ Ω/□ set by the reproducibility of the diffusion.

The surface-depletion-effect limit may be relaxed by the implantation of a shallow n^+ shielding layer [20] which allows a constant depletion of the buried implanted resistor (Fig. B.9*a*) independent of variations in interfacial charge. This permits the reproducible fabrication of resistors up to 125 kΩ/□. Another advantage of the shielding layer is that metal lines may pass over the resistors with no parasitic channels formed. The *I-V* characteristics of these doubly implanted resistors is shown in Fig. B.9*b*. The saturation effect is exactly equivalent to pinch-off in a junction-field-effect transistor.

3. By altering the doping under a Schottky barrier [27] implantation can be used to fabricate nonohmic resistors where the reverse leakage current is controlled by the doping. This gives the largest saving in area, as shown in Fig. B.10. These resistors were used as the load elements of a bipolar memory cell. The disadvantages of using Schottky barrier resistors for other nondigital applications are their highly nonlinear nature and the temperature-sensitive behavior of the current–voltage characteristics. The temperature coefficient of resistivity is 3%/°C.

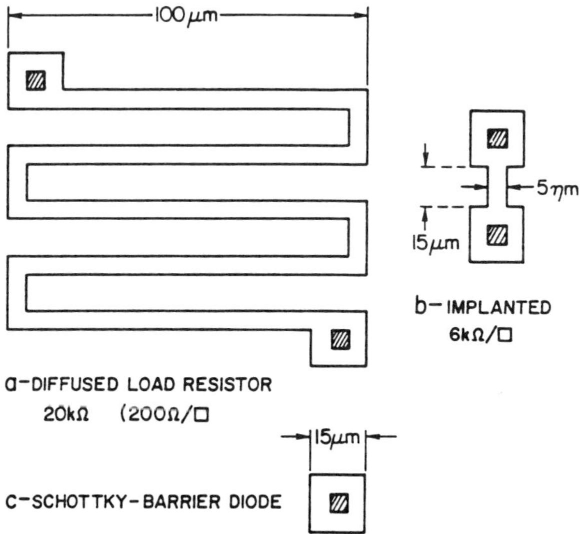

Figure B.10 (*a*) Diffused load resistor. (*b*) Implanted resistor. (*c*) Schottky-barrier diode.

B.11 ION IMPLANTED BIPOLAR TRANSISTORS

Early ion implanted bipolar transistors [28, 29] used implanted phosphorus or boron emitters. Because of the pronounced tail in the phosphorus distribution, unannealable (amorphous) damage was created in the base leading to excess recombination. In addition, the slow fall off in doping adversely affected the emitter efficiency. For these reasons, either the current gain h_{FE} or frequency response was low on these devices.

Although arsenic, as implanted, also has a tail in its distribution, the tail is not so deeply penetrating as phosphorus. In addition, arsenic possesses (as does phosphorus) a concentration-dependent diffusion constant. As a result, if the postimplant drive-in is sufficient, the tail is consumed by the fast-diffusing high-concentration front. Thus, arsenic has a sharper concentration fall-off after drive-in than before.

The procedure followed by Payne *et al.* [30] included separate boron implants for active and inactive base regions, followed by an emitter implant ($\sim 10^{16}$ As^+). The As concentration gradient is estimated as $dC_{As}/dx \gtrsim 10^{23}$ cm^{-4} at a concentration of $C_{As} = 4 \times 10^{17}$ cm^{-3}. The effect of a drive-in is shown to decrease the back-diffusion of holes into the emitter. Figure B.11 shows a plot of the product of current gain and base width ($h_{FE} \times W_B$) as a function of annealing temperature for the implanted emitter structure. For uniform lifetime and uniform base doping this product should be constant.

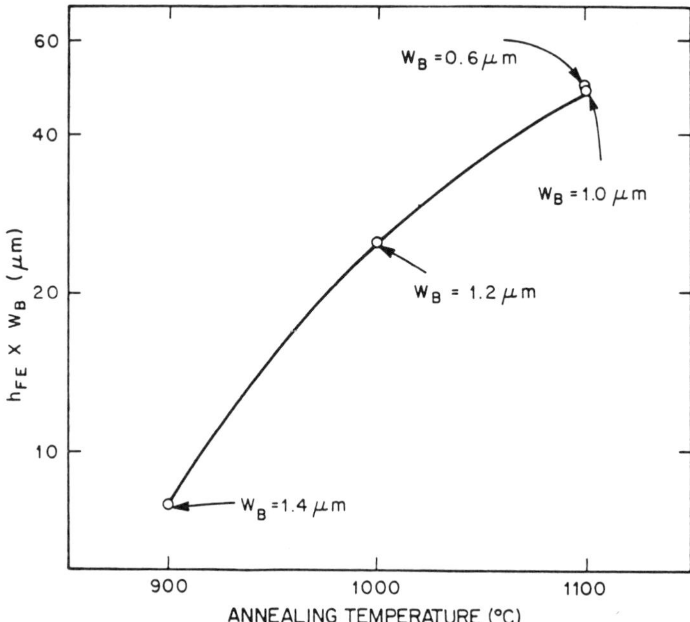

Figure B.11 Gain × base width versus annealing temperature.

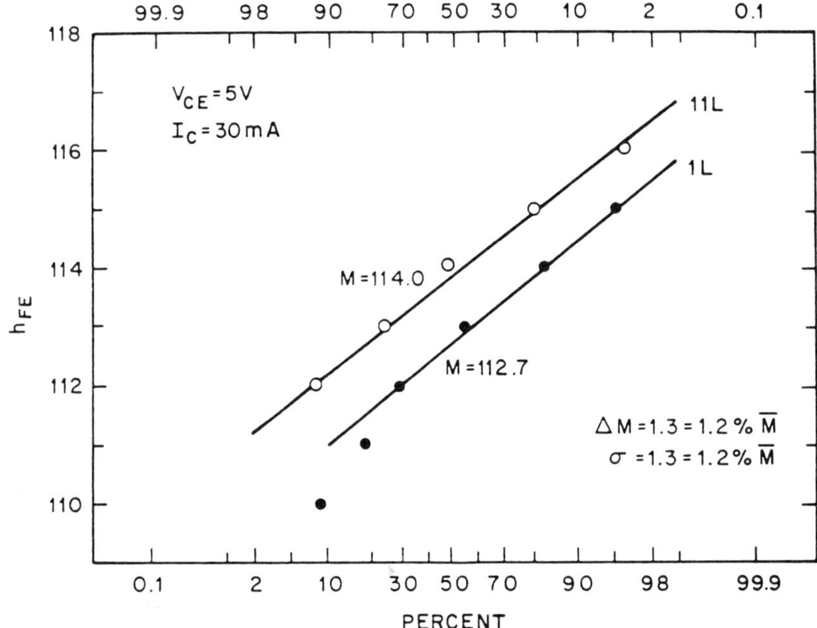

Figure B.12 Gain versus percentage change.

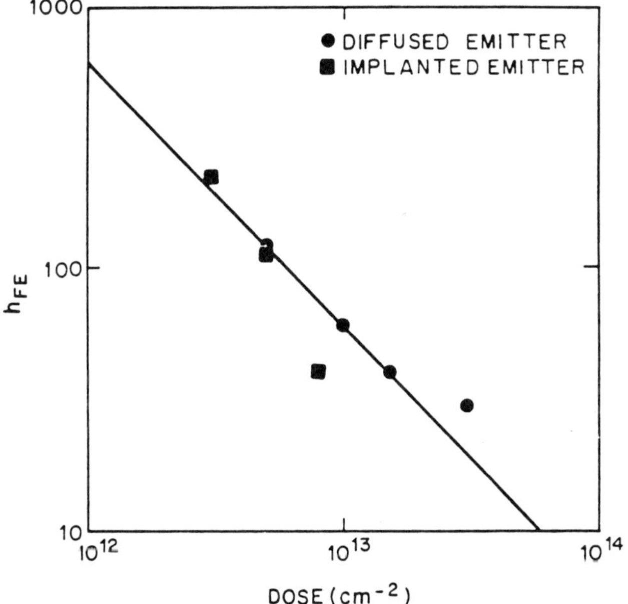

Figure B.13 Gain versus dose.

The increase seen in Fig. B.11 can be explained in the following manner. After a 900°C anneal, little As diffusion occurs, and there arises an excess back-diffusion of holes from the damaged junction region which depresses the current gain. For higher temperatures, as the junction moves below the damaged region, this excess back-diffusion disappears, and the current gain rises.

An additional advantage of the abrupt emitter profile is to allow the gain to be selected by varying the total doping in the base according to

$$h_{FE} = k\phi^{-1}$$

that is, the current gain may be chosen by varying the dose of only the active base implant.

Under these conditions, excellent control in h_{FE} can also be maintained over a slice and from slice to slice. Figure B.12 illustrates this control on two wafers of 2-GHz transistors implanted to the same dose. Both the standard deviations and the differences between the mean values of each wafer are 1.2%.

Ion implantation allows active and inactive base implants to be separately optimized [31, 32], thus maintaining excellent control over current gain by varying the active base implant. A further improvement can be derived by also implanting the emitter (1–5×10^{15} As/cm^2, 15 keV) followed by a drive-in to give an emitter–base junction depth of 1600 Å. Varying the base implantation dose gives the $1/\phi$ dependence for h_{FE} expected (Fig. B.13).

Implanted-base *pnp* transistors, for use in complementary circuits, have also been successfully fabricated. The p^+ implantation was performed at 50 keV to a dose of 5×10^{13}/cm^2. The boron emitter was diffused at 1000°C to a depth of ~ 0.11 μm. The cut-off frequency for these devices was ~ 8 GHz.

REFERENCES

1. J. Lindhard, M. Scharff, and H. E. Schiøtt, *Mat. Fys. Medd. Dan. Vid. Selsk.* **33** (14) 1–39 (1963).
2. J. Lindhard and M. Scharff, *Phys. Rev* **124**, 128 (1961).
3. J. W. Mayer, L. Eriksson, and J. A. Davies, *Ion Implantation in Semiconductors*, Academic Press, New York, 1970.
4. G. Dearnaley, J. H. Freeman, R. S. Nelson, and J. Stephen, *Ion Implantation*, North Holland, Amsterdam, 1973.
5. G. Dearnaley, *Rep. Prog. Phys.* **32**, 405 (1969).
6. J. F. Gibbons, *Proc. IEEE*, **56**, 295 (1968).
7. J. W. Mayer and O. J. Marsh, in *Applied Solid State Science Advances in Materials and Device Research*, R. Wolfe, ed., Academic Press, New York, 1969, Vol. I, p. 239.
8. F. H. Eisen and L. T. Chadderton, eds., *Ion Implantation*, Gordon and Breach, London, 1971. These also appeared in *Radiation Effects* **7** (1971) and constitute the Proceedings of the First International Conference.
9. I. Ruge and J. Graul, eds., *Ion Implantation in Semiconductors*, Springer-Verlag, Berlin, 1971, the Proceedings of the Second International Conference.
10. Billy L. Crowder, ed., *Ion Implantation in Semiconductors and Other Materials*, Plenum Press, New York-London, 1973, the Proceedings of the Third International Conference.

11. J. W. Mayer, L. Eriksson, and J. A. Davies, *Ion Implantation in Semiconductors*, Academic Press, New York, 1970, p. 24.
12. T. E. Seidel, in *Ion Implantation in Semiconductors*, I. Ruge and J. Graul, eds., Springer-Verlag, Berlin, 1971, p. 47.
13. F. H. Eisen, *Can. J. Phys.* **46**, 561 (1968).
14. B. L. Crowder and J. M. Fairfield, *J. Electrochem. Soc.* **117**, 363 (1970).
15. W. S. Johnson and J. F. Gibbons, *Projected Range Statistics in Semiconductors*, distributed by Stanford University Bookstore, 1969.
16. J. F. Gibbons, *Proc. IEEE* **60**, 1062 (1972).
17. J. A. Appels, E. Kooi, M. M. Paffen, J. J. Schatorje, and W. H. C. G. Verkuijlen, *Philips Res. Reports* **25**, 118 (1970).
18. J. A. Appels and M. M. Paffen, *Philips Research Reports* **26**, 157 (1971).
19. J. D. MacDougall, K. E. Manchester, and P. E. Roughan, *Proc. IEEE* **57**, 1538 (1969).
20. T. E. Seidel and W. C. Gibson, *IEEE Trans. Electronic Devices* **ED-20**, 744 (1973).
21. K. Rosendal, in *Ion Implantation*, (F. H. Eisen and L. T. Chadderton, eds., Gordon and Breach, London, 1971, p. 399.
22. D. P. Oosthoek, J. A. Den Boer, and W. K. Hofker, in *European Conference on Ion Implantation*, Peter Pereginus, Stevenage, England, 1970, p. 88.
23. K. H. Nicholas, B. J. Goldsmith, J. H. Freeman, G. A. Gard, J. Stephen, and B. J. Smith, *J. Phys. E. (G.B.)* **5**, 309 (1972).
24. J. W. Hanson, R. J. Huber, and J. N. Fordemwalt, *J. Vac. Sci. Technol.* **10**, 944 (1973).
25. K. H. Nicholas and R. A. Ford, in *Ion Implantation in Semiconductors*, I. Ruge and J. Graul, eds., Springer-Verlag, Berlin, 1971, p. 357.
26. H. G. Dill, R. W. Bower, and T. N. Toombs, in *Ion Implantation*, F. H. Eisen and L. T. Chadderton, eds., Gordon and Breach, London, 1971, p. 349.
27. D. A. Hodges, M. P. Lepselter, D. J. Lynes, R. W. MacDonald, A. U. MacRae, and H. A. Waggener, *IEEE J. Solid-State Circuits* **SC-4**, 280 (1969).
28. J. L. Assemat in *Ion Implantation in Semiconductors*, I. Ruge and J. Graul, eds., Springer-Verlag, 1971, p. 351.
29. J. A. Kerr and L. N. Large, in *Internation Conference on Applications of Ion Beams to Semiconductor Technology*, Grenoble, Editions Ophrys, 1967, p. 601.
30. R. S. Payne, R. J. Scavuzzo, K. H. Olson, J. M. Nacci, and R. A. Moline, *IEEE Trans. Electron Devices*, **ED-21**, 273, 1974.
31. J. A. Archer, *Electron Lett.* **8**, 499 (1972).
32. M. I. Elmasry and D. J. Roulston, in *Solid State Electronics* **24**, 371 (1981).

Appendix C

Digital Bipolar Integrated Circuit Projects

In the following projects make any necessary but reasonable assumptions, for example, value of available power supply, layout design rules (see Appendix D), process parameters, operating speed, allowable power dissipation, operating temperatures, etc. Use any level of analysis: first-order analytical equations and/or computer simulations. If IC fabrication facilities are available, layout, fabricate, test, and evaluate your design. The course instructor can generate similar projects to highlight advances in bipolar technology, devices, structure, and/or circuits.

1. Design a ST^2L two-input logic gate to operate from a minimum value of power supply. Study the sensitivity of V_0, V_1, and delay times to variations in temperature.
2. Design T^2L and ECL interface circuits for T^2L/ECL and ECL/T^2L. Compare the designs regarding area, maximum speed of operation, and power dissipation.
3. Design a ST^2L cross-coupled flip-flop memory cell to operate from a minimum value of power supply and to have minimum power dissipation. Calculate area and speed of operation.
4. Design a T^2L driver circuit to drive a capacitive load of 20 pf with a logic swing of 5 V. Calculate power dissipation, area, and rise and fall times.
5. Design a 10-bit shift register using T^2L, I^2L, ECL, and EFL to operate at a 10-MHz clock rate. Compare regarding area, minimum value of power supply required, and power dissipation.
6. Design temperature-compensated ST^2L, I^2L, and ECL logic gates. Compare regarding ease of compensation.
7. Compare the effect of scaling layout dimensions on the performance of T^2L, I^2L, ST^2L, ECL, and EFL [1–4] and comment on which family benefit most from this scaling and why.
8. Compare the performance of Schottky-diode transistor logic (SDTL) [5] to ST^2L, I^2L, and EFL.
9. Design a minimum area full-adder in I^2L, SI^2L, and ISL. Compare regarding power dissipation, area, and maximum speed of operation.

10. Design a full-adder in I^2L using conventional logic gates and using threshold gates. Compare regarding area, power dissipation, and speed of operation.

11. Make a general comparison between LSI bipolar logic families, NMOS, and CMOS dynamic and static LSI [6] and nonsilicon-based logic circuits [7].

12. In EFL, the logic performed per unit area is higher than that for ECL. Compare ST^2L to SI^2L based on this criterion.

13. Calculate the value of the minimum power supply for T^2L, ST^2L, I^2L, SI^2L, ECL, one-level EFL, and two-level EFL. Compare the speed-power product at these minimum values of power supply for each of the above families.

15. Design a mask programmable logic array [8–12] which consists of uncommitted logic gates in ST^2L, I^2L and EFL. Compare the number of gates in a 100 mil × 100 mil chip using the same design rules. Allow reasonable area for random interconnections on the chip. Calculate total power dissipation and speed of operation.

16. For the logic array of project 15, calculate the ratio of the total active area A_A to the total isolation area A_{ISO} for the same number of gates for ST^2L, I^2L, and EFL.

17. It is required to evaluate T^2L-based, I^2L-based, and ECL-based bipolar technologies for digital/linear applications [13].

18. Ion implantating the base of bipolar transistors can improve the performance of the transistor in the upward and downward modes of operation [14]. Discuss the effect of these improvements on T^2L, I^2L, and ECL.

19. The use of polysilicon as emitter or base regions of bipolar transistors has been reported [15–16]. Design a logic gate using poly I^2L structures and compare to SDTL [5] regarding maximum speed of operation, area, and power dissipation.

20. Figure C.1 shows a block diagram of serial full-adder. Realize using I^2L and EFL and compare regarding area, speed, and speed-power product.

21. Repeat project 20 for the serial subtractor shown in Fig. C.2. Design a universal block to realize both functions: a serial full-adder and a serial substrator.

22. In a given digital bipolar family/technology the potential increase in the level of integration is function of such factors as yield, processing steps, and power dissipation [18, 19]. Discuss the effects of such factors on the following:

1. I^2L/dielectric isolation technology.
2. EFL/double metallization technology.

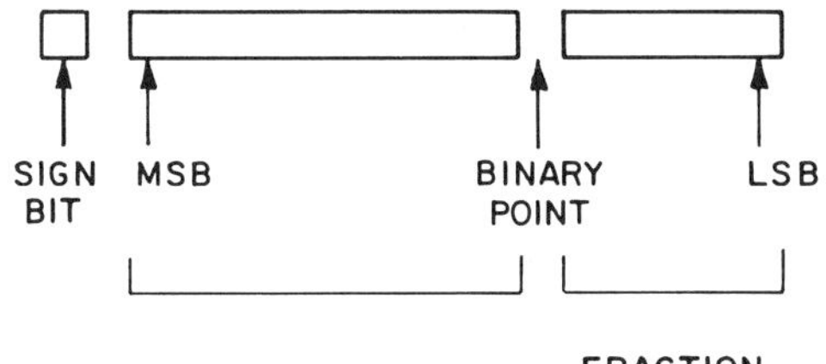

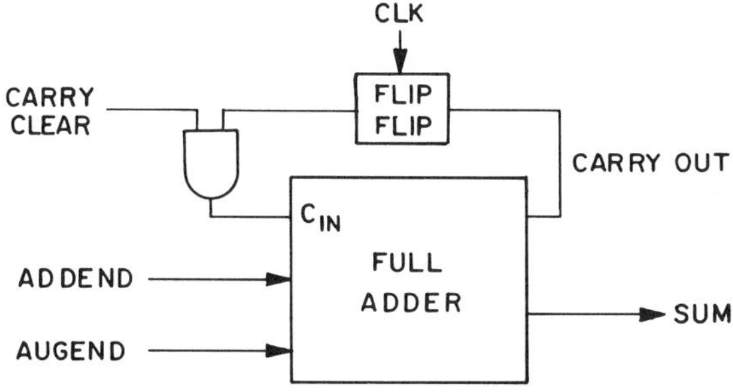

Figure C.1 Block diagram of serial full-adder.

23. Discuss how sensitive the bipolar transistor performance is to the overall doping profile of the device and to the method of contacting the different transistor regions in an IC environment [19].
24. Design an EFL pipeline multiplier to generate $a \times b$, where a and b are 8-bit words [20].
25. Select a suitable architecture for an I^2L multiplier for $a \times b$, where a and b are 8-bit words. Compare to that of project 24 regarding area, speed of operation, and speed-power product.
26. The speed of operating a bipolar logic family can be adjusted to compensate for processing variations [21]. Design such compensating circuits for ST^2L, I^2L, and EFL.

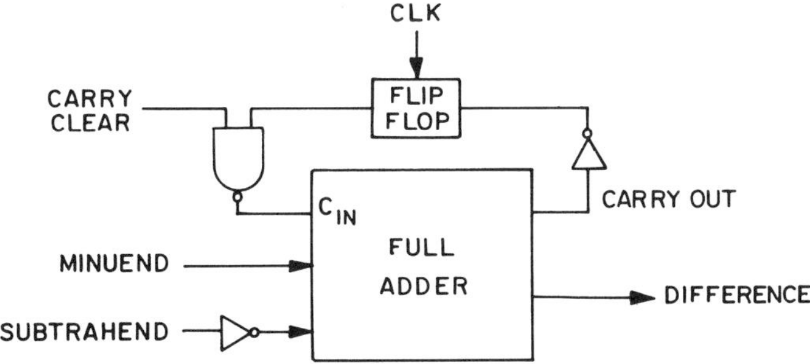

Figure C.2 Block diagram of serial subtractor.

27. Emitter-coupled injection logic (ECIL) [22] and static induction transistor logic (SITL) [23] have recently been introduced. Calculate the logic levels of these families, power-supply requirements, power dissipation, and speed-power product. Compare to I^2L and EFL. Discuss compatibility with I^2L and EFL.

28. Collector-coupled Structures (C^2S) are recently introduced as high-density bipolar structures suitable for mixed-bipolar-MOS (BIMOS) technologies [24]. Analyze the dc and the transient performance of C^2S and compare to EFL and I^2L. Can C^2S be adapted to I^2L-based bipolar technologies? Why?

29. Compare different circuit techniques to prevent a bipolar transistor from saturation. Use the following in the comparison: ease of integration, speed-power product, and minimum value of V_{DD}.

30. Show how the exclusive-OR function can be realized with conventional T^2L, I^2L, ECL, and EFL. Use circuit invoation based on these families to integrate the exclusive-OR function in a minimum area.

REFERENCES

1. P. A. H. Hart, T. Van't Hof, and F. M. Klaassen, Device Down Scaling and Expected Circuit Performance, *IEEE J. Solid State Circuits*, **SC–14** 343–351 (1979).
2. S. A. Evans, Scaling I^2L for VLSI, *IEEE J. Solid State Circuits* **SC–14** 318–327 (1979).
3. S. S. Rafail, M. I. Elmasry, and E. L. Heasell, Integrated Injection Logic for VLSI, *IEEE Trans. ED* **27**, 1301–1303 (1980).
4. P. M. Soloman and D. D. Tang, Bipolar Circuit Scaling, *IEEE ISSCC*, **27**, 1980, Lewis Winner, FL, pp. 86–87.
5. K. Okada, K. Aomura, T. Nakamura, and H. Shiba, A New Polysilicon Process for a Bipolar Device: PSA Technology, *IEEE J. Solid State Circuits* **SC–14**, 307–311 (1979).
6. M. I. Elmasry, *Digital MOS Integrated Circuits*, IEEE Press, New York, 1981.
7. B. G. Bosch, Gigabit Electronics—A Review, *IEEE Proceedings* **67**, 340–379 (1979).
8. M. Usami *et al.*, A 5000-Gate I^2L Masterslice LSI, *IEEE ISSCC*, **26**, 1979, Lewis Winner, FL, pp. 58–59.
9. R. J. Blumberg and S. Brenner, A 1500 Gate Random Logic Masterslice, *IEEE ISSCC*, **26**, 1979, Lewis Winner, FL, pp. 60–61.
10. W. Braeckelmann *et al*, A Subnanosecond Masterslice Array Offering Logic plus Memory, *IEEE ISSCC*, **26**, 1979, Lewis Winner, FL, pp. 64–65.
11. C. Davis *et al.* IBM System/370 Bipolar Gate Array Micro-Processor Chip, *IEEE ICCC*, 1980, IEEE Press, New York, pp. 669–673.
12. A. H. Dansky, Bipolar Circuit Design for VLSI Gate Arrays, *IEEE ICCC*, *1980*, IEEE Press, New York, pp. 674–677.
13. M. I. Elmasry *et al.*, Integrated Injection Logic for a Linear/Digital LSI Environment, *IEEE Trans. on ED* **25**, 351–357 (1978).
14. M. I. Elmasry and D. J. Roulston, Base Component of Gain and Delay Time in Base Implanted Bipolar Transistors, *Solid State Electronics*, **24**, 371–375 (1981).
15. K. Okada *et al.*, A Polysilicon Self-Aligned Process for Bipolar LSIs, *IEEE ICCC*, *1980*, IEEE Press, New York, pp. 689–692.

16. R. D. Davies and J. D. Meindl, Poly I^2L: A High Speed Linear Compatible Structure, *IEEE ISSCC, 1977*, Lewis Winner, New York, Vol. 20, pp. 218–219.

17. H. Murrmann, *Modern Bipolar Technology for High-Performance ICs*, Simens Forsch u. Entwickl, Ber. Bd. 5 (1976) Nr. 6, Springer-Verlag, 1976, pp. 353–359.

18. P. W. J. Verhofstadt, Evaluation of Technology Options for LSI Processing Elements, *IEEE Proc*, **64**, 842–851 (1976).

19. T. N. Ning and R. D. Isaac, Effect of Emitter Contact on Current Gain of Silicon Bipolar Devices, *IEEE IEDM, 1979*, IEEE Press, New York, pp. 473–476.

20. M. I. Elmasry and R. C. Madter, Pipeline Digital Filtering at 250 MHz Bit Rate, *Proceedings of the* 1975 *Midwest Symposium on Circuits and Systems*, Western Periodicals, New York, pp. 461–465.

21. E. Berndlmaier *et al.*, Delay Regulation: A Performance Concept, *IEEE ICCC, 1980*, IEEE Press, New York, pp. 701–704.

22. W. Kim and W. L. Engl, Emitter Coupled Injection Logic, *IEEE ISSCC, 1980*, IEEE Press, New York, pp. 62–63.

23. Y. Horiba *et al.*, Static Induction Transistor Logic Compatible with IGHz ECL Circuits, *IEEE ISSCC, 1980*, IEEE Press, New York, pp. 64–65.

24. E. Z. Hamdy and M. I. Elmasry, Bipolar Structures for BIMOS Technologies, *IEEE JSSC*, April 1980, Lewis Winner, FL, pp. 229–236.

Appendix D

Symbolic Layout of Bipolar Integrated Circuits

In Chapters 4, 5, and 6 some typical examples of bipolar cell layouts are shown. In dealing with the layout of complex LSI/VLSI bipolar chips, there are different levels of representing the layout information of individual cells. At times, detailed layout geometries are important, for example, to calculate resistances of an inactive base of a transistor. But, symbolic layout diagrams of the topology of the cells can convey important information regarding interconnection topologies and can be used in compacting the cell areas according to layout design rules. This high-level representation of cell layouts is especially useful in planning the floor plan of an IC chip. The technique has been applied to NMOS circuits using stick diagrams.* This appendix offers a bipolar stick diagram notation. Examples are followed. Moreover, simple design rules using a universal layout parameter λ are given for typical junction isolation and dielectric isolation bipolar technologies.

D.1 STICK DIAGRAM NOTATION

Figure D.1 shows the stick diagram notations using black and white coding and color coding.

1. Notation for the n^+ underlayer, which defines the active area of the device. The areas between the underlayer regions are occupied by isolation regions.
2. Notation for the location of the n^+ collector region of an *npn* transistor and its associated ohmic contact. The same notation is used for an n^+ base region of a *pnp* transistor and its associated ohmic contact. It is also used to show the location of an n^+ region (and its associated ohmic contact) connecting an n well, where p resistors are fabricated, to a

*C. Mead and L. Conway, Introduction to VLSI Systems, Addison-Wesley, Reading, MA, 1980.

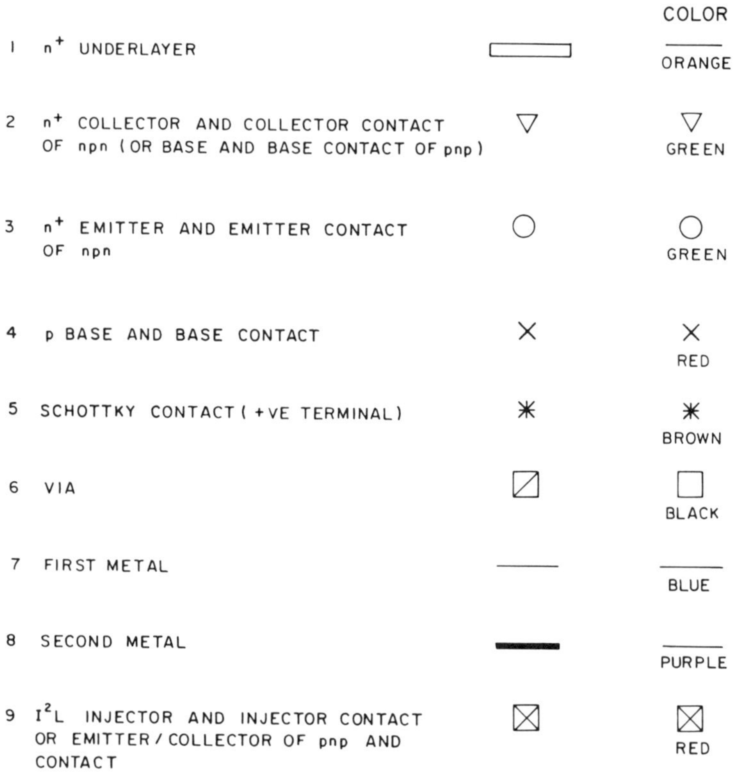

			COLOR
1	n^+ UNDERLAYER	▭	ORANGE
2	n^+ COLLECTOR AND COLLECTOR CONTACT OF npn (OR BASE AND BASE CONTACT OF pnp)	▽	▽ GREEN
3	n^+ EMITTER AND EMITTER CONTACT OF npn	○	○ GREEN
4	p BASE AND BASE CONTACT	×	× RED
5	SCHOTTKY CONTACT (+VE TERMINAL)	✳	✳ BROWN
6	VIA	◪	□ BLACK
7	FIRST METAL	———	BLUE
8	SECOND METAL	▬▬▬	PURPLE
9	I^2L INJECTOR AND INJECTOR CONTACT OR EMITTER/COLLECTOR OF pnp AND CONTACT	⊠	⊠ RED

Figure D.1 Stick diagram notation for bipolar transistors.

positive potential. Note that the collector contact of a downward *npn* transistor is used as an emitter contact of an upward *npn* I^2L transistor.

3. Notation for the location of the n^+ emitter region of an *npn* transistor and its associated ohmic contact. This is also used to indicate the collectors of an *npn* I^2L transistor.
4. Notation for the location of the *p* base and its associated ohmic contact. The base region extends to cover the areas under the emitters in the case of *npn* transistors; in the case of *p* resistors, the base region extends to the next resistor ohmic contact, as shown in Fig. D.2.
5. Notation for the location of a Schottky contact, which is the positive terminal of a Schottky diode formed in an *n*-epi region.
6. Notation for the location of vias, used to connect a first layer of metal to a second layer.
7. Notation for the location of the first metal used for interconnections.

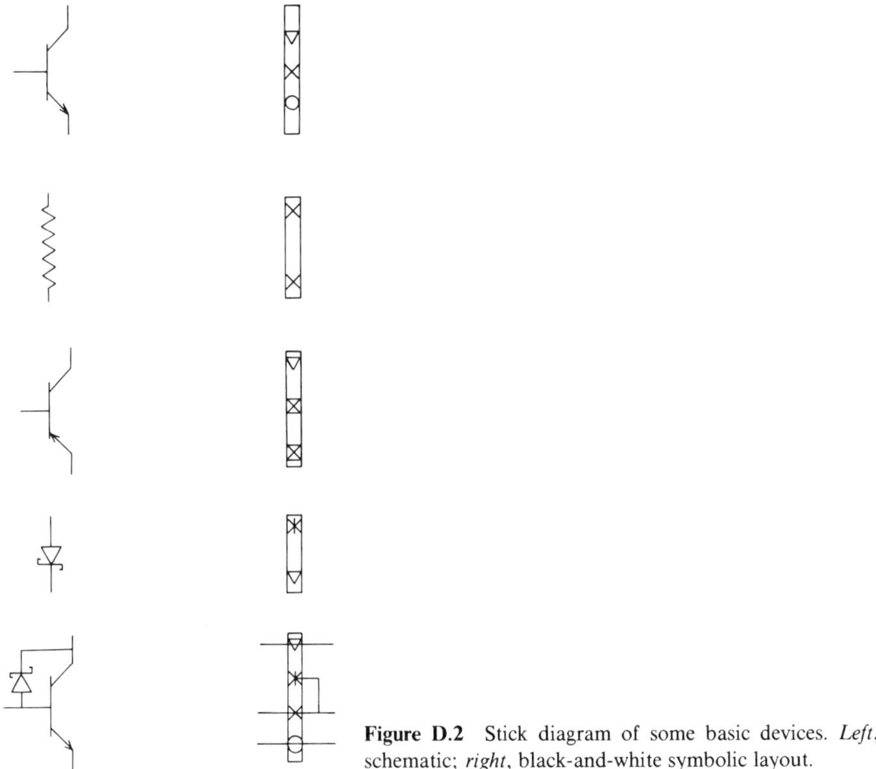

Figure D.2 Stick diagram of some basic devices. *Left*, schematic; *right*, black-and-white symbolic layout.

8. Notation for the location of the second metal used for interconnections.
9. Notation for the location of p^+ injectors and associated injector contacts, also used to indicate the emitters and collectors of *pnp* transistors.

Examples. Figure D.2 shows the stick diagrams of some basic elements using the black and white coding. Symbols are depicted for a downward *npn* transistor, a base resistor a lateral *pnp* transistor, a Schottky diode, and a Schottky transistor. In the case of a base resistor, the value of the resistor is a function of the spacing between the two base contacts. The Schottky diode is fabricated in an isolated *n*-epi region where the negative terminal of the diode is made by contacting the *n* region through an n^+ diffusion region and an ohmic contact. In the case of a Schottky transistor, the base terminal makes an ohmic contact to the base region and a Schottky rectifying contact to the *n* collector region. A useful exercise is to reproduce the stick diagrams of Figs. D.2 and the following figures using the color coding of Fig. D.1.

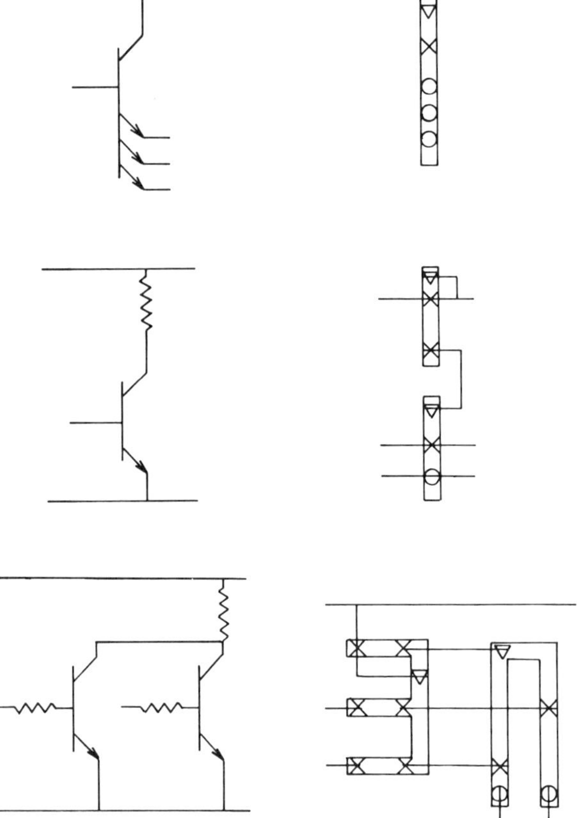

Figure D.3 Stick diagram of some basic structures. *Left*, schematic; *right*, black-and-white symbolic layout.

Figure D.3 shows the stick diagrams of more complex building blocks: a multiemetter *npn* transistor, a transistor/load resistor, and two-transistor inverter, where the two *npn* transistors share the same collector region hence the same underlayer region. All the resistors are fabricated within the same isolated n region, which is connected to the most positive potential of the circuit through an n^+ diffusion region and an ohmic contact.

Figure D.4 shows the stick diagrams of two simple gates: a Schottky integrated logic gate and a Schottky diode gate. A second layer of metallization is used for V_{CC} and ground.

Figure D.5 shows the stick diagram of a basic I^2L gate with two fan-out collectors. The injector resistor is assumed to be off-chip and is connected to the V_{CC} line. The stick diagram shows clearly the simplicity of the layout of an I^2L gate.

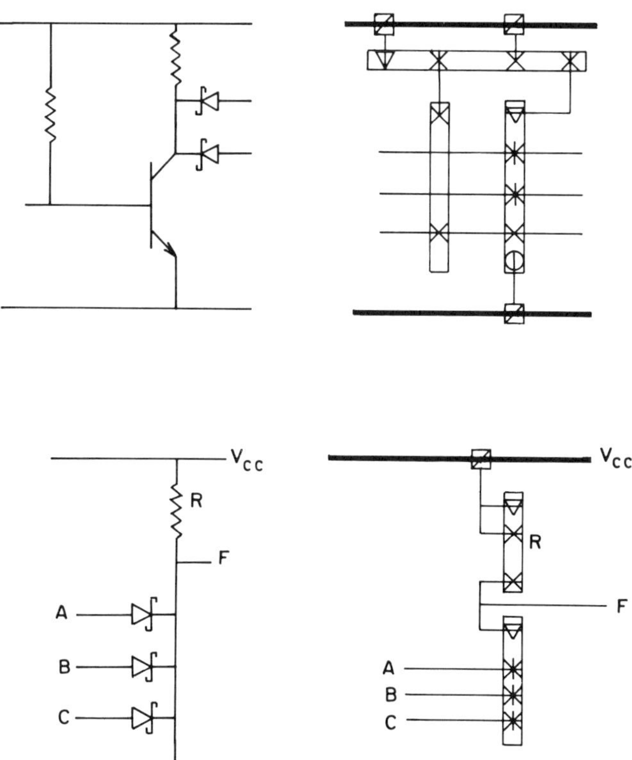

Figure D.4 Stick diagram of two bipolar circuits. *Left*, schematic; *right*, black-and-white symbolic layout.

Figure D.6 shows the stick diagram of a basic ST^2L gate, and Figure D.7 shows the stick diagram of a basic ECL gate. In the ECL gate, the resistors R_1, R_2, R_3, and R_5 are fabricated in the same *n*-epi region. Similarly, the resistors R_4 and R_6 are fabricated in the same *n*-epi region. The *n*-epi-region terminal of R_4 and R_6 is also the collector terminal of D_2. Thus in this case the *n*-well of R_4 and R_6 is at the same potential as the collector of D_2.

D.2 LAYOUT AREAS

Figure D.8 shows a key for layout areas using both black and white coding and color coding. This key is used to develop the design rules for two typical bipolar technologies, one with junction isolation and a single layer of metallization and the other with dielectric isolation, Schottky diodes, and two layer of metallization.

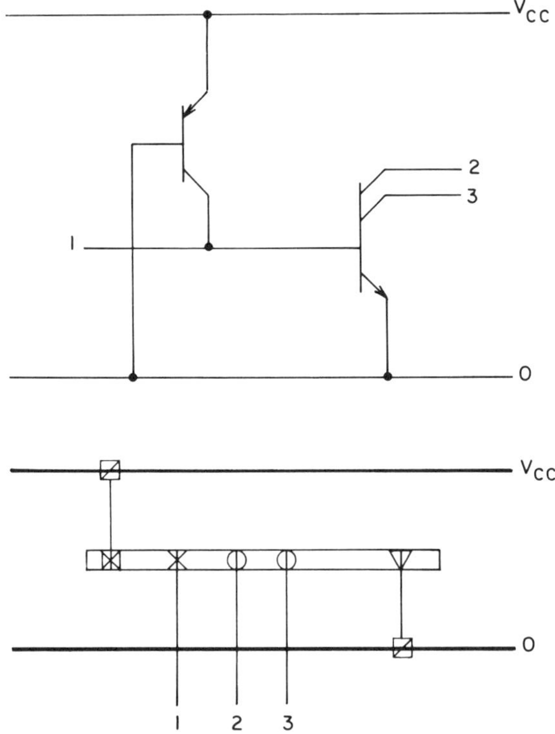

Figure D.5 Stick diagram of I^2L gate. *Top*, schematic; *bottom*, black-and-white symbolic layout.

D.3 DESIGN RULES FOR JUNCTION ISOLATION BIPOLAR TECHNOLOGY

Figure D.9 shows a simplified set of design rules for a typical bipolar technology with junction isolation and single metallization. The design rules are given in terms of a universal scaling parameter λ. The minimum size of a contact is $2\lambda \times 2\lambda$ and the minimum width of metal, (base, emitter, collector) diffusion and of p^+ isolation is 2λ. The minimum overlap of an emitter (collector of I^2L) with a base is 2λ. The minimum contact overlap with a diffusion is λ. The minimum n^+ underlayer overlap with a base is also λ. All the minimum spacings are 2λ except the minimum spacing of an n^+ underlayer and a p^+ isolation region, with is 4λ. This is because the p^+ isolation is deeper and more tolerance is needed for its side diffusion.

Processing Steps. It is useful at this point to examine the basic processing steps for a typical bipolar technology with junction isolation and relate these steps to the different mask levels and their associated design rules. Figure D.10 shows the basic processing steps.

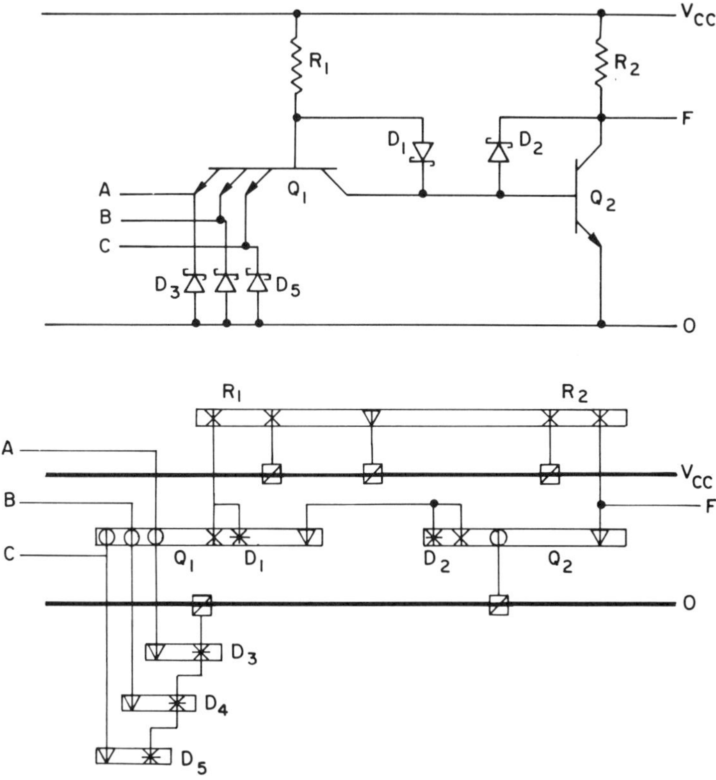

Figure D.6 Stick diagram of ST^2L gate. *Top*, schematic; *bottom*, black-and-white symbolic layout.

The starting material is a p substrate, and a mask is used to define n^+ underlayer regions. There regions are isolated later and, along with their n-epitaxial regions, are used to define the active areas. Following this step an n-epitaxial layer is grown over the whole wafer, thus no mask is needed. This is followed by patterning the silicon oxide with the isolation mask, which defines the areas where the p^+ diffusion occurs. There is a minimum spacing between the n^+ underlayer region and the isolation areas. Following this step the base mask is used to define the base areas. The n^+ underlayer areas have to overlap the base mask. Next an n^+ mask is used to define the areas of the emitters and collectors. The base area has to both overlap the emitter area and be spaced from the collector region. Following this step a mask is used to define contact hole areas. The diffusion areas have to overlap their associated contact holes. Finally a mask is used to pattern the metallization layer. The metal areas must also overlap the contact holes. The final diagram of Fig. D.10 is a composite top view of an *npn* transistor fabricated using the simplified processing steps. Figure D.11 shows (to scale) the different mask levels; the crosses at the tops of the diagrams are mask alignment marks.

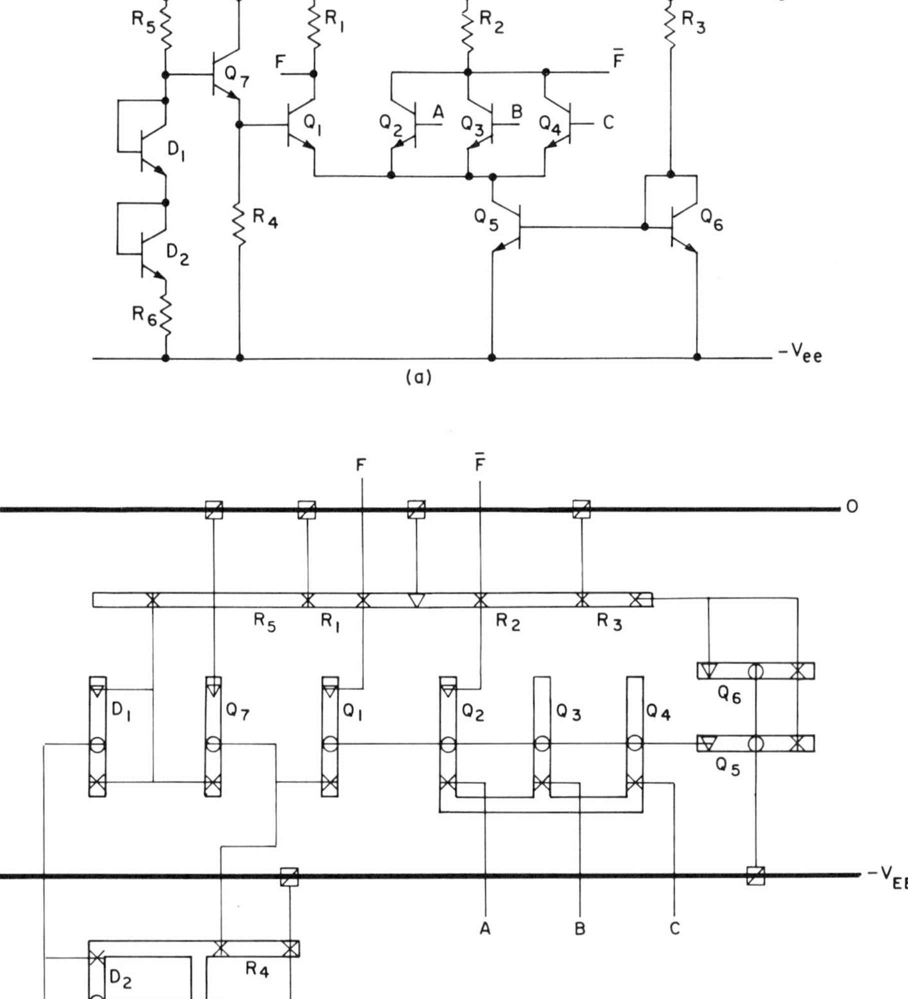

Figure D.7 Stick diagram of ECL gate. *Top*, schematic; *bottom*, black-and-white symbolic layout.

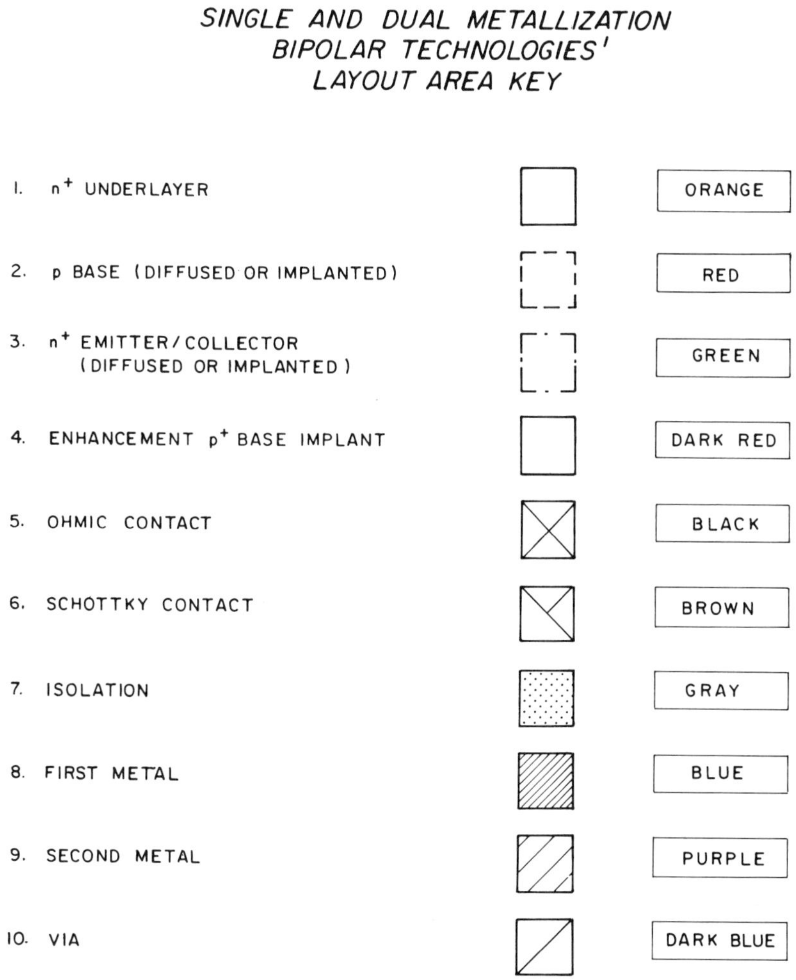

SINGLE AND DUAL METALLIZATION BIPOLAR TECHNOLOGIES' LAYOUT AREA KEY

1.	n^+ UNDERLAYER	ORANGE
2.	p BASE (DIFFUSED OR IMPLANTED)	RED
3.	n^+ EMITTER/COLLECTOR (DIFFUSED OR IMPLANTED)	GREEN
4.	ENHANCEMENT p^+ BASE IMPLANT	DARK RED
5.	OHMIC CONTACT	BLACK
6.	SCHOTTKY CONTACT	BROWN
7.	ISOLATION	GRAY
8.	FIRST METAL	BLUE
9.	SECOND METAL	PURPLE
10.	VIA	DARK BLUE

Figure D.8 Key for layout areas for single and dual metallization bipolar technologies. *Left column*, black-and-white symbols; *right column*, color codes.

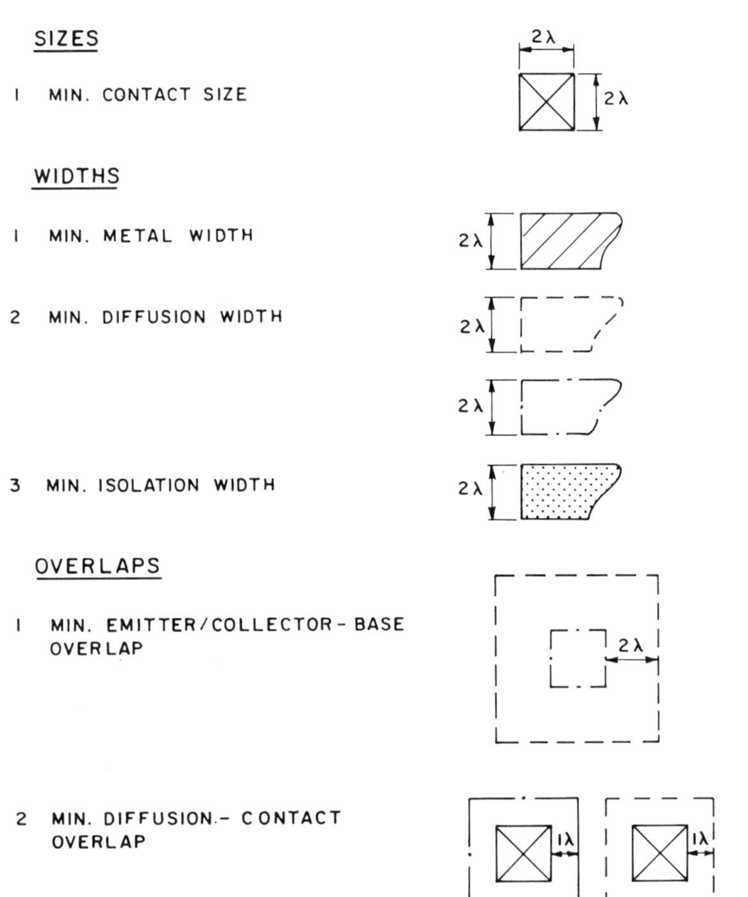

Figure D.9 Design rules for a bipolar technology with junction isolation and single metallization.

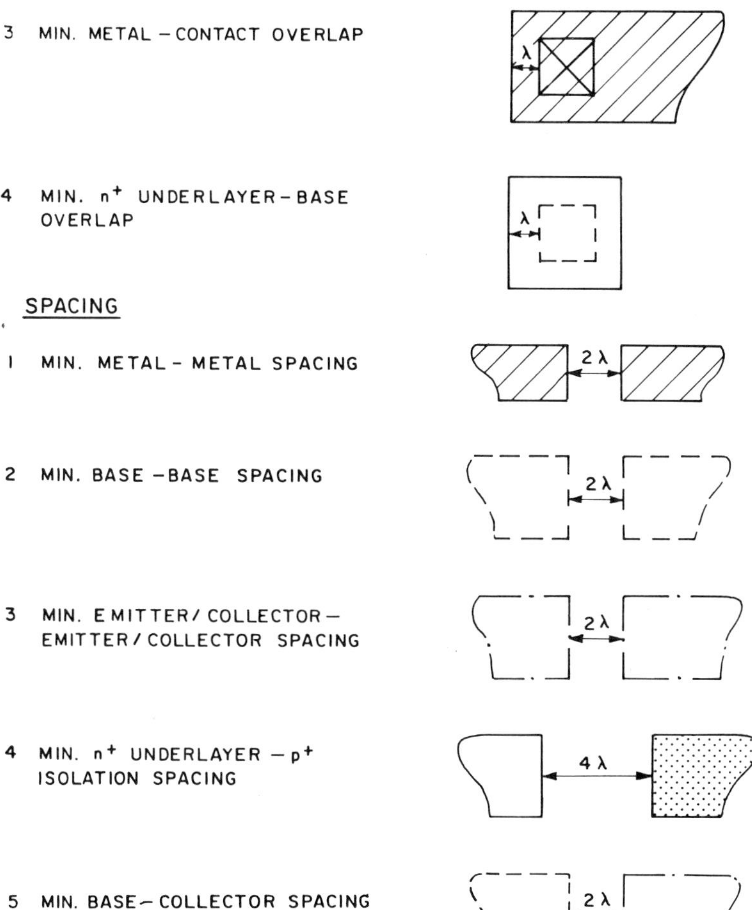

Figure D.9 (*Continued*)

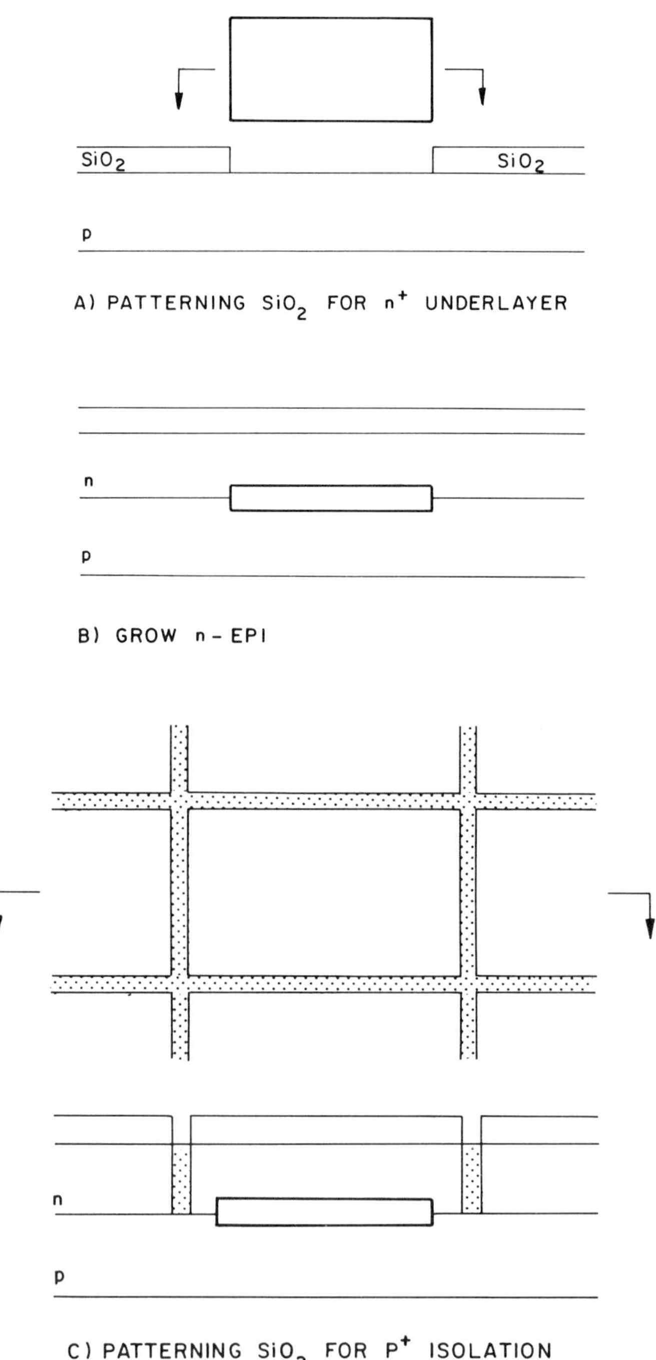

Figure D.10 Simplified processing steps for a bipolar technology with junction isolation and single metallization.

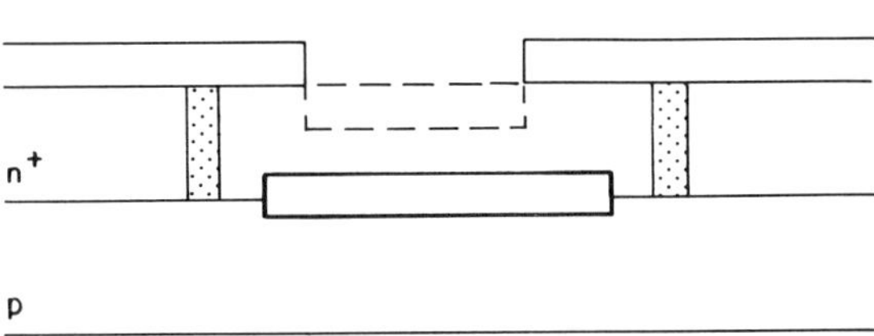

D) PATTERNING SiO_2 FOR P^+ BASE

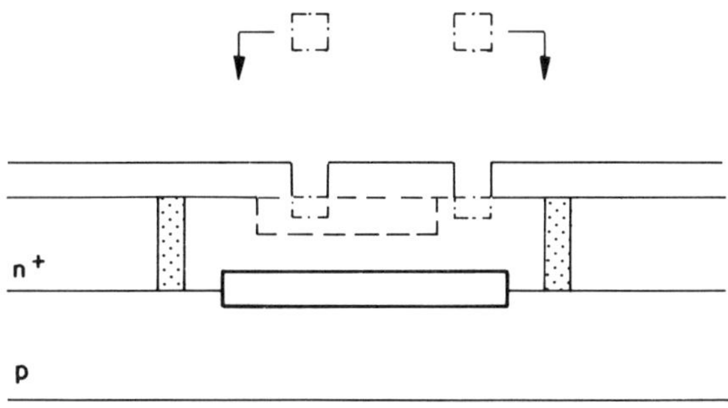

E) PATTERNING SiO_2 FOR n^+ EMITTER AND COLLECTOR

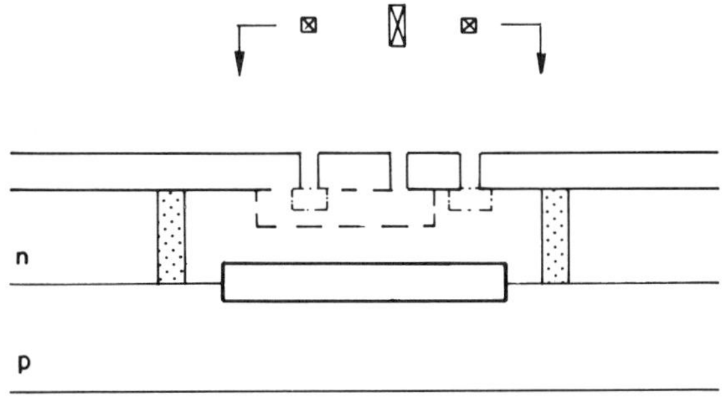

F) PATTERNING SiO_2 FOR CONTACT HOLES

Figure D.10 *(Continued)*

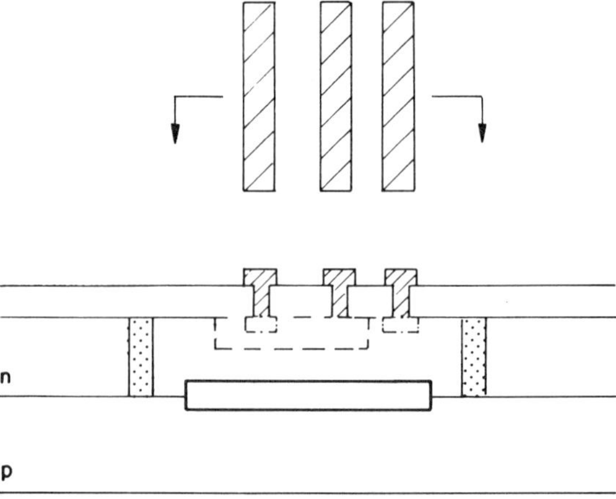

G) PATTERNING METAL LAYER

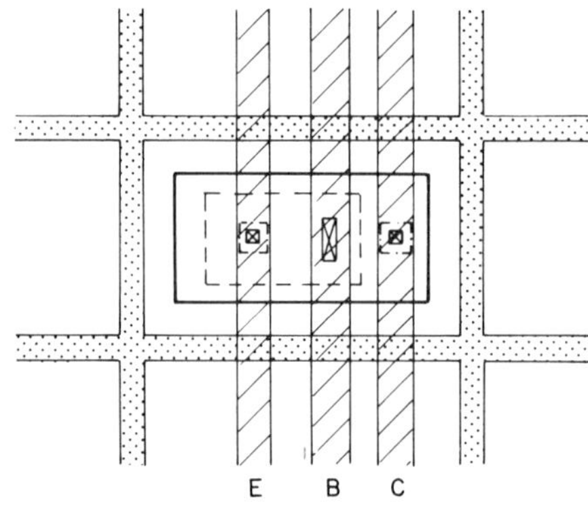

(H) COMPOSITE TOP VIEW

Figure D.10 (*Continued*)

D.5 DESIGN RULES FOR DIELECTRIC ISOLATION BIPOLAR TECHNOLOGY

Figure D.12 depicts a simplified set of design rules for a typical bipolar technology with dielectric isolation, dual metallization, and Schottky diodes. In the minimum sizes, the sizes of a minimum via and a Schottky contact are larger than the sizes of an ohmic contact or the minimum diffusion areas. The minimum width of a second metal is larger than that of the first metal. The

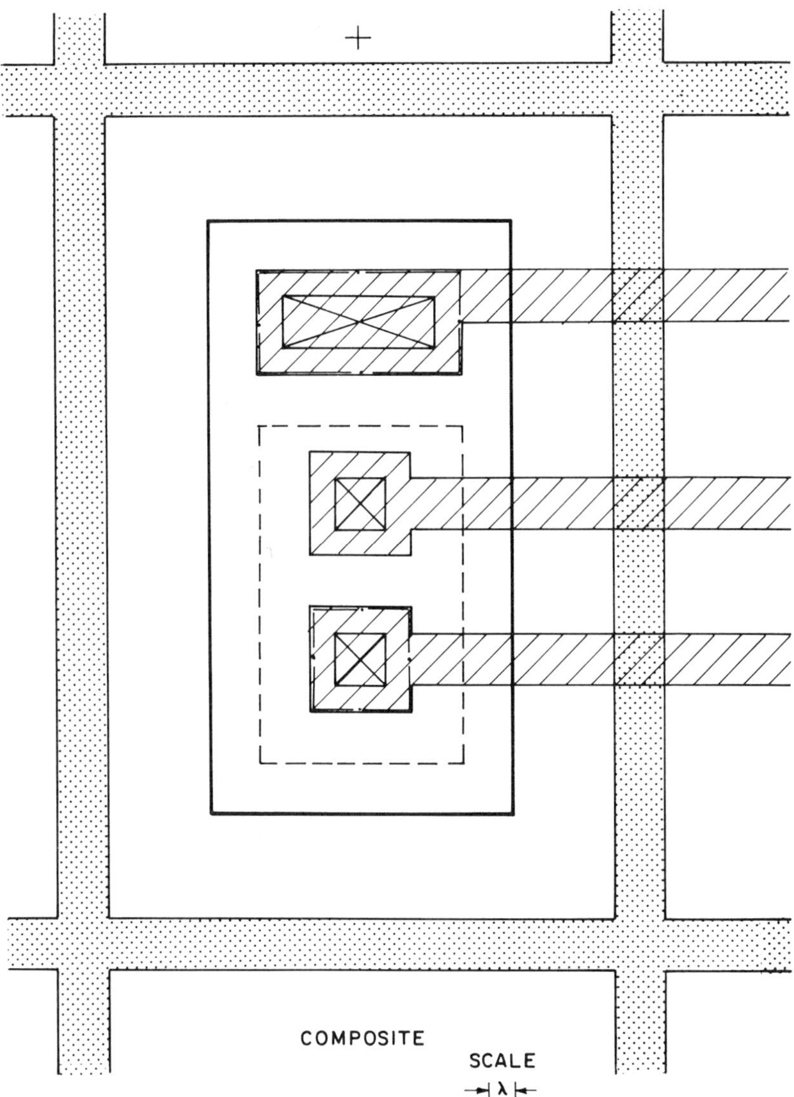

Figure D.11 Composite layout to scale, using λ, of an *npn* transistor and its associated mask levels. The Simplified Processing Steps of Fig. D.10 are used.

Figure D.11 (*Continued*)

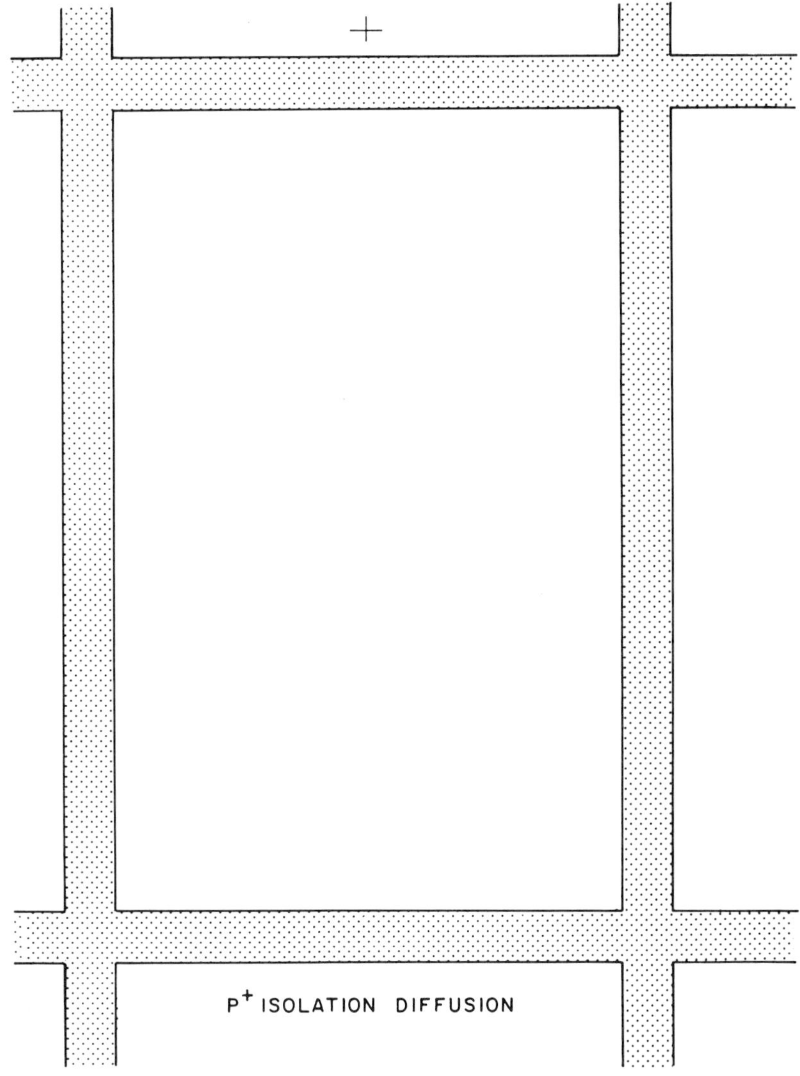

Figure D.11 (*Continued*)

BASE DIFFUSION

Figure D.11 (*Continued*)

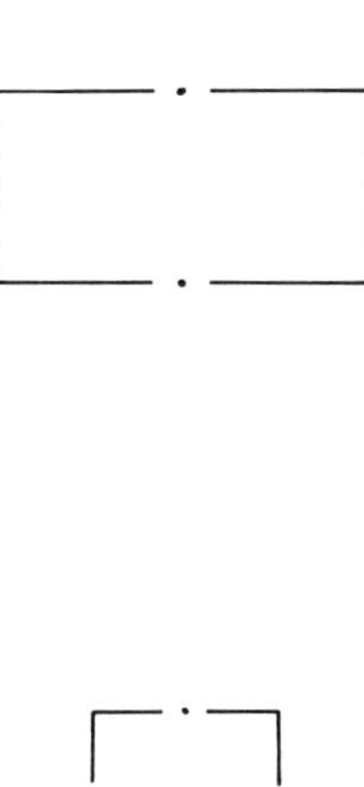

Figure D.11 (*Continued*)

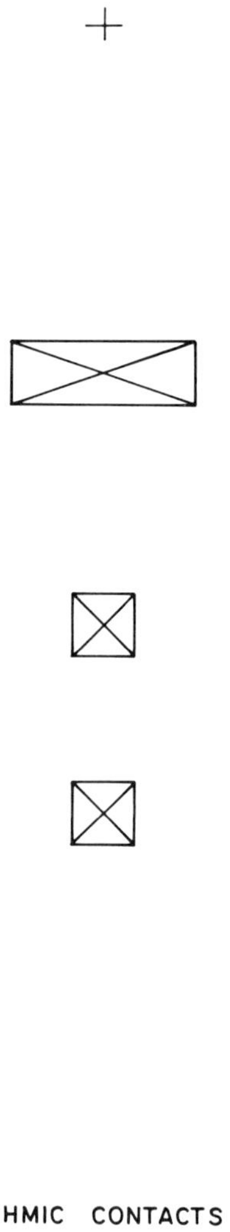

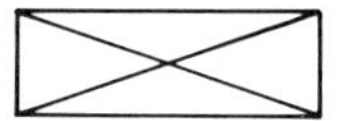

OHMIC CONTACTS

Figure D.11 (*Continued*)

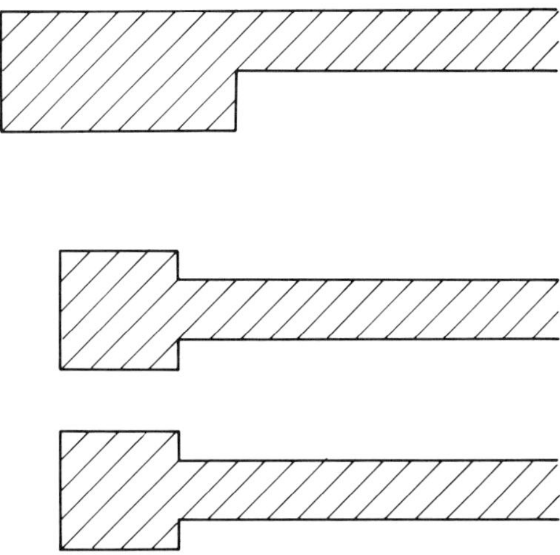

METALLIZATION

Figure D.11 (*Continued*)

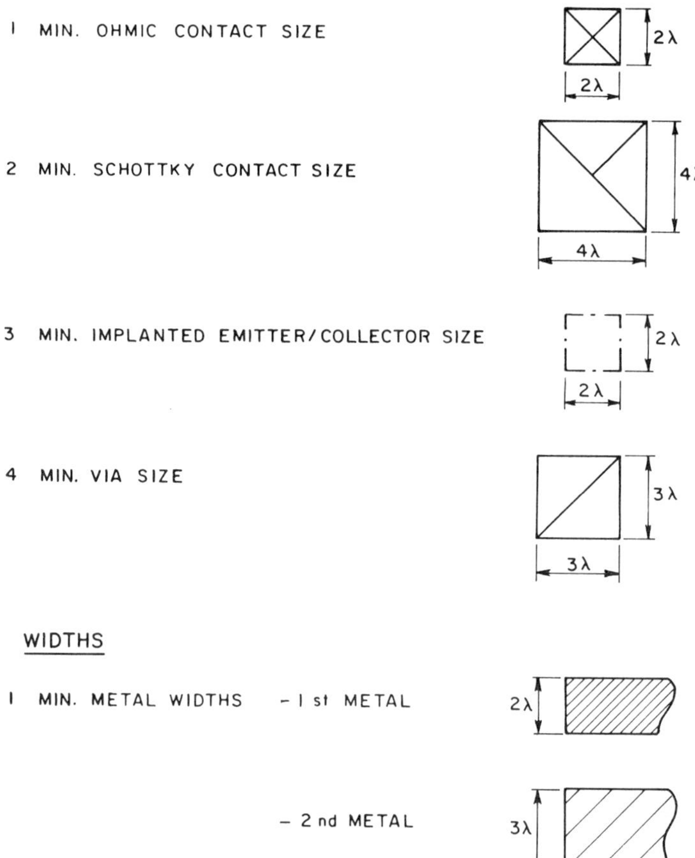

Figure D.12 Design rules for a bipolar technology with dielectric isolation and dual metallization.

minimum isolation width is 2λ if it is within an active device and 4λ if it is between active devices (this is shown in the Fig. D.14 composite diagram). The minimum overlap of a metal with a Schottky contact is 2λ. The base and collector/emitter masks have to overlap the isolation area with a minimum of 1λ. The implantation mask for base, collector, or emitter has to overlap the contact area with a minimum of 1λ. The second and first layers of metallization have to overlap a via with a minimum of 3λ and 1λ, respectively. The larger overlap for the second metal layer is due to the larger tolerance on the alignment of the second metal, since it is a later processing step. The base region has to overlap the emitter/collector areas with a minimum of 1λ, whereas the emitter/collector can coincide with the underlayer area. A

2 MIN. IMPLANTED BASE WIDTH

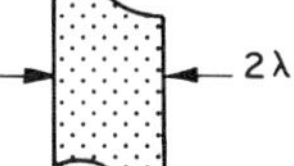

3 MIN. ISOLATION WIDTH WHEN ISOLATED AREAS HAVE SAME n^+ UNDERLAYER

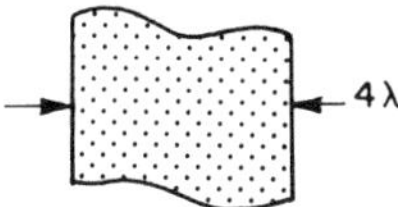

4 MIN. ISOLATION WIDTH WHEN ISOLATED AREAS HAVE ISOLATED n^+ UNDERLAYERS

4λ

OVERLAPS

1 MIN. METAL-SCHOTTKY CONTACT OVERLAY

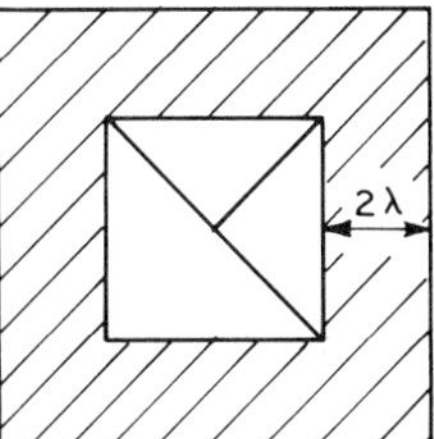

2 MIN. BASE IMPLANT-ISOLATION OVERLAP

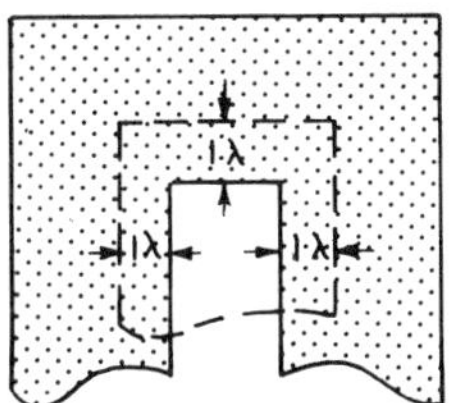

3 MIN. COLLECTOR/EMITTER-ISOLATION OVERLAP

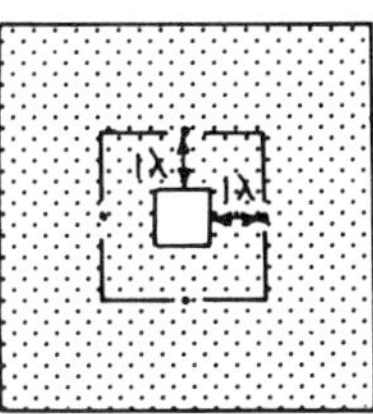

Figure D.12 (*Continued*)

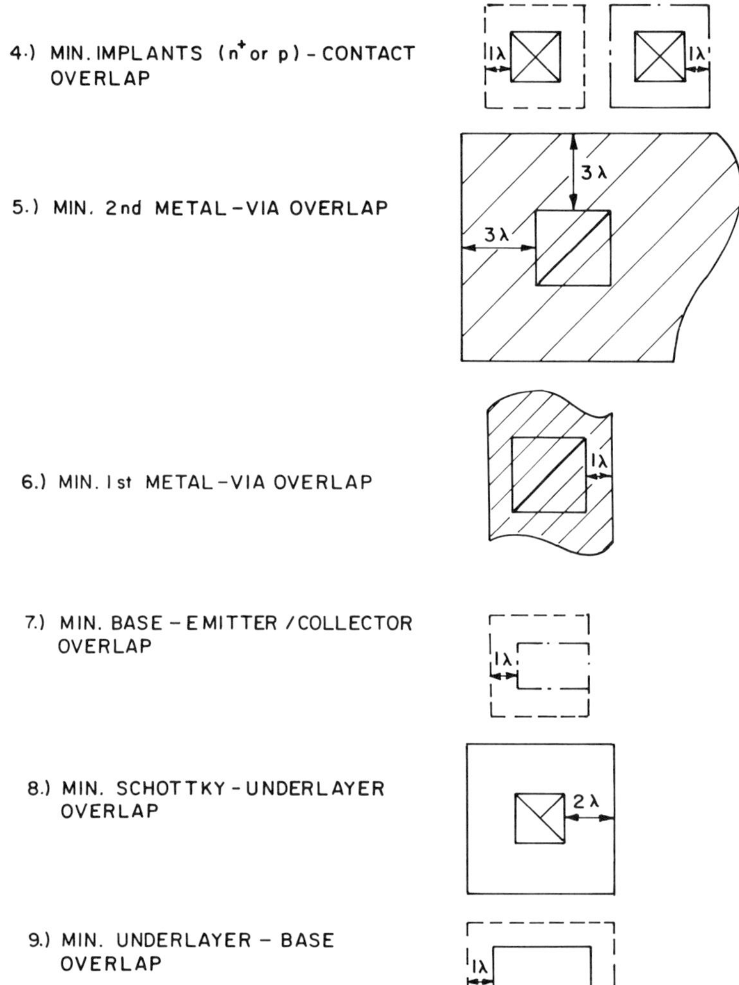

Figure D.12 (*Continued*)

10.) MIN. ISOLATION-UNDERLAYER OVERLAP

SPACING

1.) MIN. METAL-METAL SPACING

- 1st METAL

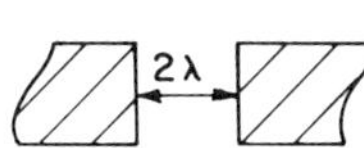

- 2nd METAL

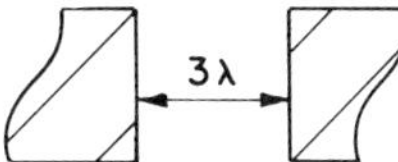

2.) MIN. EMITTER IMPLANT-BASE SPACING

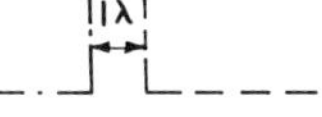

3.) MIN. BASE-BASE SPACING

4.) MIN. VIA-BASE, ENHANCEMENT, EMITTER/COLLECTOR SPACING

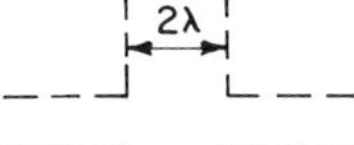

5.) MIN. EMITTER SPACING

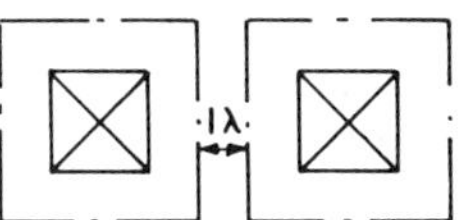

6.) MIN. BASE ENHANCEMENT-EMITTER/COLLECTOR SPACING

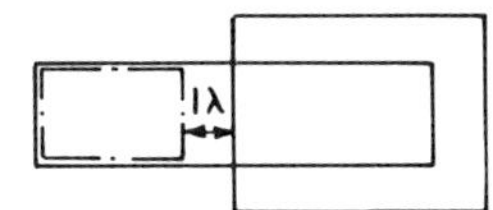

Figure D.12 (*Continued*)

Schottky contact has to overlap with an underlayer area by a minimum of 2λ, and the base and the enhancement implant of the inactive base have to overlap the underlayer area with a minimum of 1λ. The minimum spacings are 1λ to 2λ except for the second metal spacing, which is 3λ.

Processing Steps. Figure. D.13 shows a simplified set of processing steps for a bipolar technology with dielectric isolation. A starting material is a *p* substrate, and a mask is used to define n^+ underlayer diffusion areas. Then an epitaxial *n* region is grown, and a mask is used for defining the dielectric silicon oxide isolation areas. Note that the collector of an *npn* transistor is connected through the n^+ underlayer area to the rest of the transistor, as shown in part C. The minimum width of the isolation region between the collector and the rest of the transistor is 2λ, and the minimum width of the isolation region between transistors is 4λ. The isolation mask overlaps the underlayer areas by 1λ. Next a mask is used to define the base region of the transistor; this base mask has to overlap the dielectric isolation region with a minimum of 1λ. Then a mask is used to define the n^+ collector and emitter ion implantation areas; the mask has to overlap the isolation region with a minimum of 1λ. Then a mask is used to define a p^+ implantation area in the inactive base to reduce its resistance. This is followed with masks to define contact holes, first layer of metallization, vias, and second layer of metallization. Figure D.13(K) is a composite top view. Figure D.14 shows (to scale) a composite top view and its associated masks. The crosses at the tops of the diagrams are mask alignment marks.

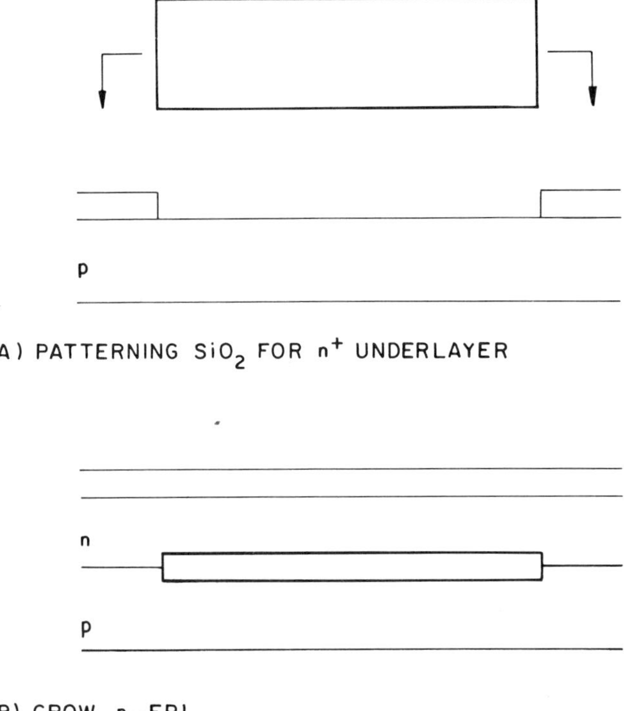

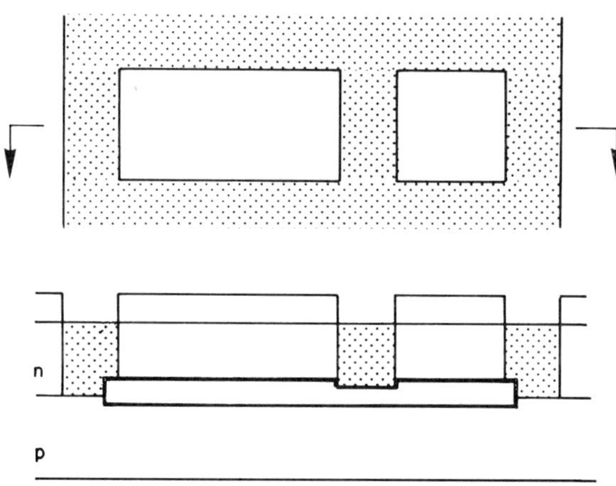

Figure D.13 Simplified processing steps for a bipolar technology with dielectric isolation and dual metallization.

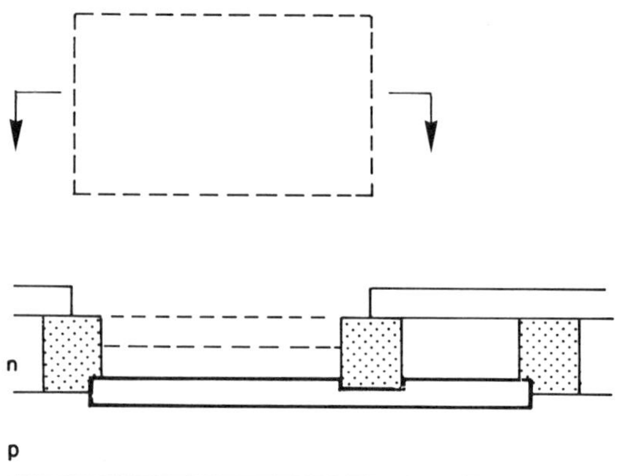

D) PATTERNING SiO_2 FOR P BASE

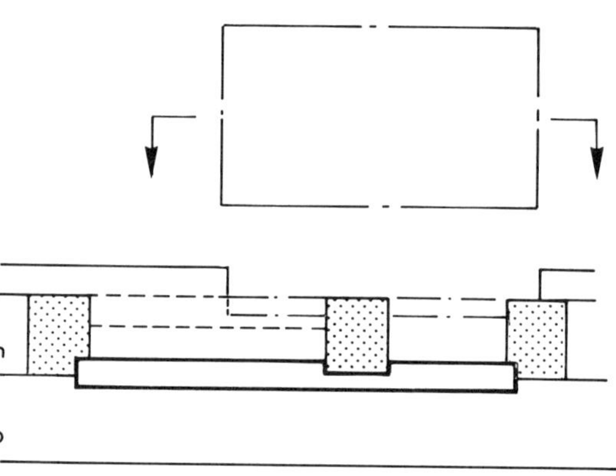

E) PATTERNING SiO_2 FOR n^+ EMITTER AND COLLECTOR

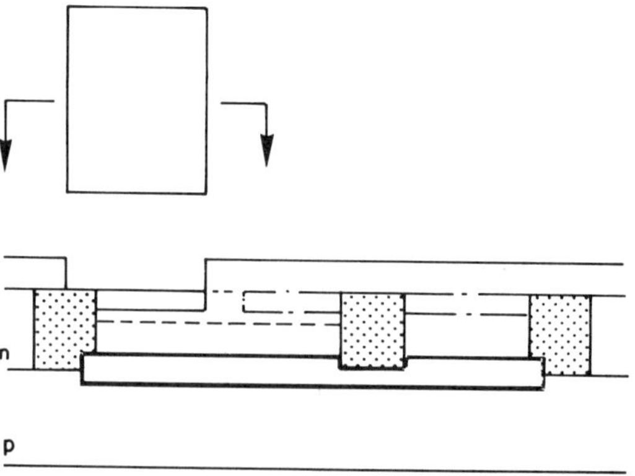

F) PATTERNING SiO_2 FOR ENHANCEMENT P^+ BASE

Figure D.13 (*Continued*)

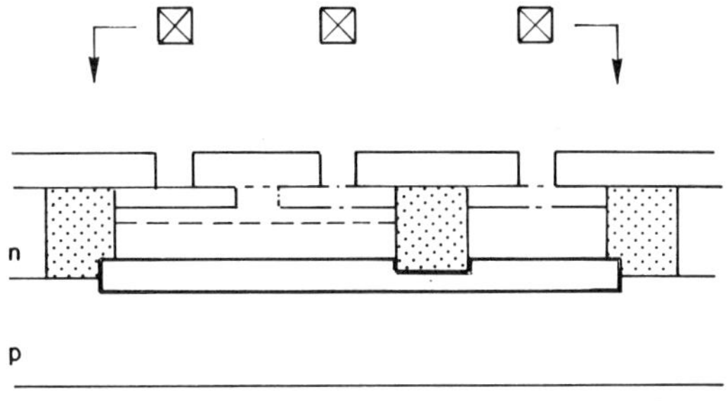

G) PATTERNING SiO_2 FOR CONTACT HOLES

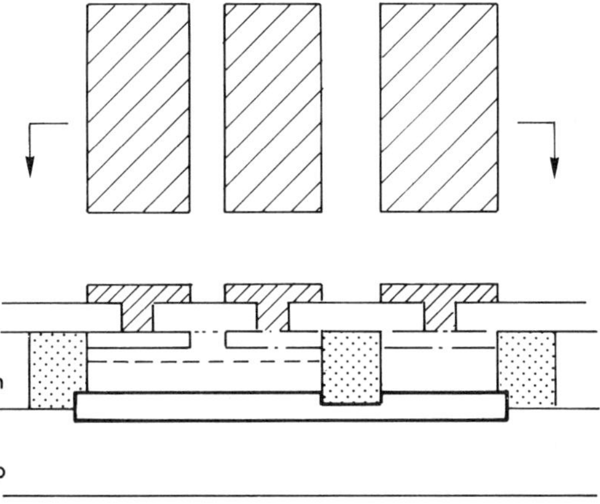

H) PATTERNING FIRST METAL LAYER

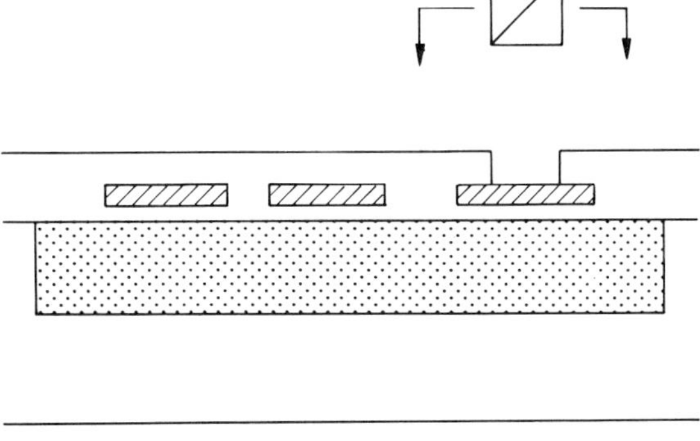

I) PATTERNING SiO_2 FOR VIAS

Figure D.13 (*Continued*)

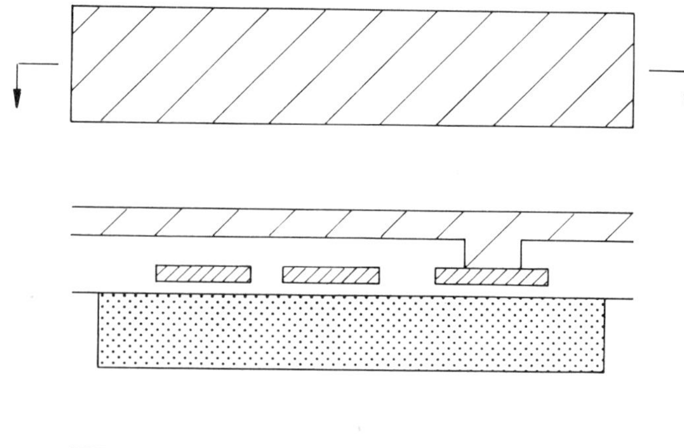

J) PATTERNING SECOND METAL LAYER

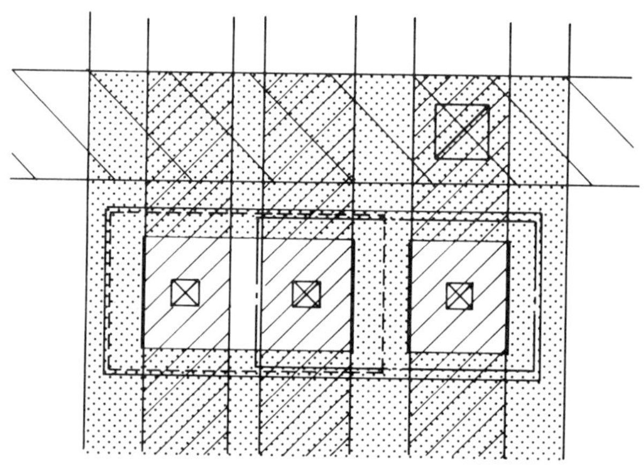

(K) COMPOSITE TOP VIEW

Figure D.13 (*Continued*)

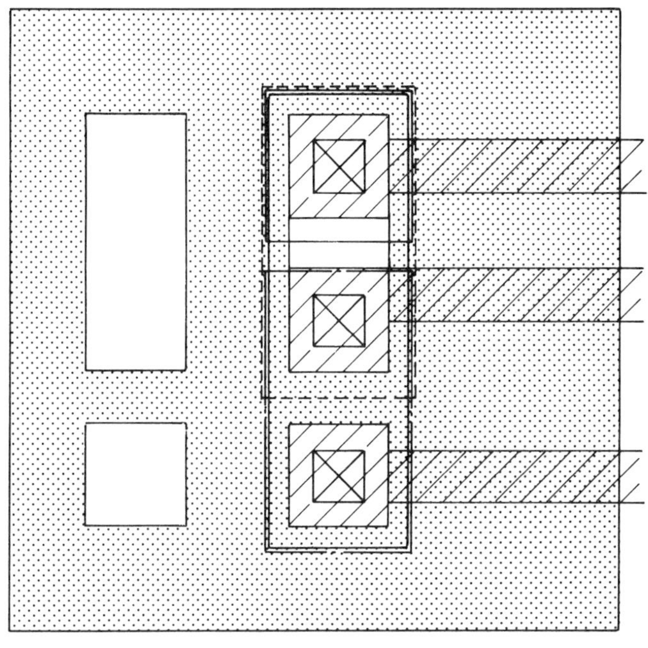

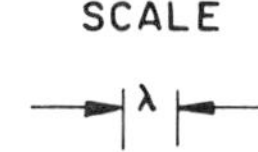

Figure D.14 Composite layout to scale, using λ, of an *npn* transistor and its associated mask levels. The Simplified Processing Steps of Fig. D.13 are used. Only one layer of metallization is used.

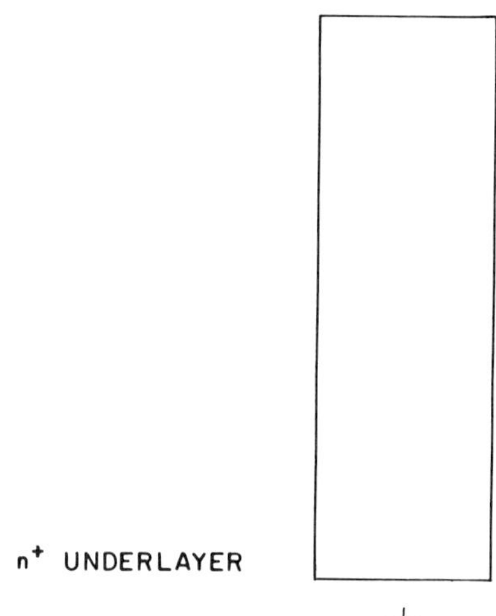

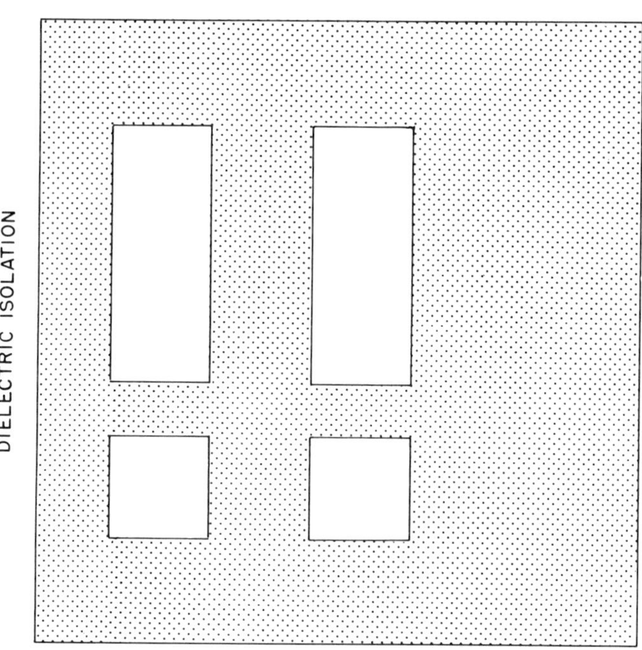

Figure D.14 *(Continued)*

BASE IMPLANT

EMITTER AND COLLECTOR IMPLANT

Figure D.14 (*Continued*)

Figure D.14 (*Continued*)

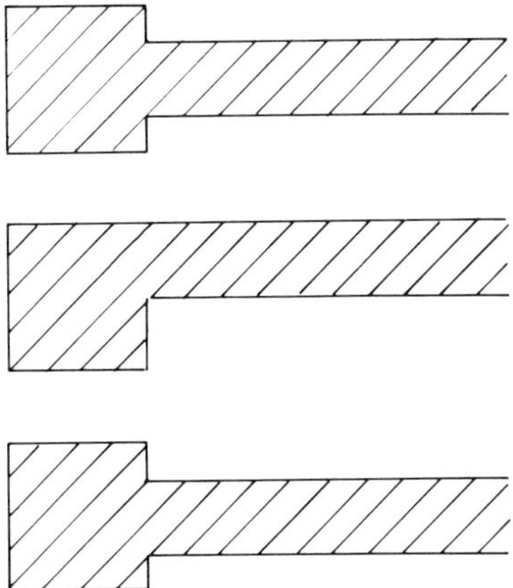

FIRST METALLIZATION

Figure D.14 (*Continued*)

Index